ELECTRIC POWER SYSTEMS

ELECTRIC POWER SYSTEMS

Syed A. Nasar
and
F.C. Trutt

Department of Electrical Engineering
University of Kentucky
Lexington, Kentucky

CRC Press
Boca Raton London New York Washingt[illegible].

Library of Congress Cataloging-in-Publication Data

Nasar, S. A.
Electric power systems / S.A. Nasar and F.C. Trutt.
p. cm.
Includes bibliographical references and index.
ISBN 0-8493-1666-9 (alk. paper)
1. Electric power systems. I. Trutt, Frederick C. II. Title.
TK1001.N373 1998
621.31—dc21 98-38265
CIP

Direct all inquiries to CRC Press LLC, 2000 Corporate Blvd., N.W., Boca Raton, Florida 33431.

International Standard Book Number 0-8493-1666-9
Library of Congress Card Number 98-38265
Printed in the United States of America 1 2 3 4 5 6 7 8 9 0
Printed on acid-free paper

PREFACE

During the last forty years, the field of Electrical Engineering has become very diversified and is much broader in scope now than ever before. For instance, compared to publishing very few technical journals in the sixties, the Institute of Electrical and Electronics Engineers (IEEE) now issues over fifty different transactions, each almost totally dedicated to one special area in Electrical Engineering. With emerging new topic areas, ranging from microelectro-mechanics to light-wave technology, the number of Electrical Engineering courses available to university students has considerably increased. In order to have sufficient breadth, the luxury of a two-semester elective sequence for each topical area can hardly be afforded within the current Electrical Engineering undergraduate curriculum. In order to keep pace with the progress in technology, we must adapt to provide the student with fundamental knowledge in several areas. Although a two-semester course sequence is still desirable to achieve some depth, the student may have time to take only one elective or required course in a given area. In accordance with these constraints, this book is intended for a one-semester senior-level, introductory course in electric power systems. As listed in the bibliography at the end of the book, several excellent texts are available for a two-semester sequence. It is assumed that the student has completed basic courses covering steady-state and transient networks, machinery, transformers and transmission lines, although these topics are reviewed and expanded at various points in the text. These topics may be included or omitted in accordance with student backgrounds and instructor preferences.

In order to achieve our objective of presenting an introduction to electric power systems in one semester, we focus on four major topics in the text. These are power flow, symmetrical components, fault calculations, and power system stability. Other important topics such as a review and the in-depth modeling of transformers, synchronous machines, and transmission lines are delegated to appendixes. Using commercially available software packages, illustrative computer solutions for simple utility and industrial systems are included. Copies of these programs for educational use are either complimentary or are available at greatly reduced cost to institutions wishing to incorporate computer analysis and design techniques into this course work. To reiterate, the emphasis in this text is on the fundamentals. Computer techniques are used as a facility to solve complex problems not amenable to longhand calculations.

To briefly review the contents of this text, Chapter 1 provides an introduction to basic concepts relating to power and energy. It is intended to give an overview and it may be covered in-depth or as a simple reading assignment.

Chapter 2 presents a brief summary of ac circuit analysis. Emphasis is on three-phase circuits, with special reference to complex power in three-phase circuits. This chapter could have been delegated to an appendix, and may be skipped by those having a sound knowledge of three-phase ac circuits. However, experience has shown that the majority of students are not so well prepared. Often a review of this topic is necessary.

Various components of a power system and their simplified models are discussed in Chapter 3. Single-line and reactance diagrams representing a power system with the interconnecting components are also presented in this chapter, which ends with a discussion of per-unit quantities and per-unit representation of a system.

It must be mentioned again that, in addition to the control, protection and loading elements, the three basic components of a power system are transmission lines, transformers, and synchronous machines. Because in most curricula, the course in power systems is offered after a course in electric machines, students often have a background in transformers and synchronous machines. Therefore, these two topics are considered in Appendixes II and III, respectively. Appendix IV presents a discussion of power transmission lines operating under steady-state. Various representations of transmission lines are given in this Appendix. These topics are therefore available for reference or review without disturbing the continuity of the text.

Power flow is presented in Chapter 4. Most power flow studies are based on numerical methods. Some of these methods are introduced in this chapter, with examples illustrating the concepts and the use of standard computer programs. Section 4.7 contains an in-depth discussion of decoupled power flow which may be skipped by instructors wishing to emphasize other topics.

Chapter 5 deals with balanced and unbalanced fault calculations. This is followed by a brief survey of power system protection. The chapter ends with a discussion of bus impedance matrix formation by algorithm. It is available for reference and may be skipped by those instructors who wish to emphasize applications.

Analytical and numerical solutions to power system stability problems are presented in Chapter 6. The text ends with a brief overview of economic power dispatch and control of power systems which may be omitted by those wishing to emphasize the major topics.

Throughout the text, many worked-out examples are given to illustrate the various concepts. As mentioned earlier, examples are also presented to illustrate the use of computer techniques. A wide range of problems is included at the end of each chapter. These problems may be adapted to hand calculation, machine computation, or design applications as appropriate.

Commercial software available from several vendors has been used to generate computer solutions at various points in the text. The use of such programs for homework or design projects has the significant advantage that students will be applying software that may be identical to the packages that they will use when they become practicing engineers. A facility for using these tools is readily developed without sacrificing the need for understanding the basic concepts. Such low-cost educational software is currently available from (listed in alphabetical order):

EDSA Micro Corporation
11440 West Bernardo Court
Suite 370
San Diego, CA 92127 USA
(800) 362-0603
E-mail: adib@edsa.com
Home Page: www.edsa.com

Operation Technology, Inc.
23705 Birtcher Drive
Lake Forest, CA 92630 USA
(714) 462-0100
E-mail: oti@etap.com
Home Page: www.etap.com

Complimentary packages for educational use are currently available from:

SKM Systems Analysis, Inc.
225 S. Sepulveda Blvd.
Suite 350
P.O. Box 3376
Manhattan Beach, CA 90266-1376
(310) 372-0088
E-mail: pwrtools@skm.com

THE AUTHORS

Syed A. Nasar is Professor, Department of Electrical Engineering, University of Kentucky. He earned his Ph.D. degree in Electrical Engineering from the University of California at Berkeley. He is the author or co-author of 33 books and over 100 journal papers in Electrical Engineering. He is a Fellow of IEEE and the Institution of Electrical Engineering, United Kingdom. Professor Nasar has been involved in teaching, research, and consulting in the area of electric machines for over 30 years.

Frederick C. Trutt received his undergraduate and graduate educations in Electrical Engineering at the University of Delaware, Newark. He served as an Instructor of Electrical Engineering while completing his graduate studies and earned the Ph.D. degree in 1964. Upon graduation, Dr. Trutt was employed by the E.I. DuPont de Nemours Corporation as a Research Engineer. He then served as a Captain on active duty in the U.S. Army Artillery Corps and subsequently assumed a position with the U.S. Army Mobility Equipment Research and Development Center, Fort Belvoir, VA. He later joined the U.S. Army Advanced Material Concepts Agency as a Project Engineer and Team Chief. In 1972, he began his postgraduate academic career as a Faculty Member at The Pennsylvania State University, University Park, PA where he became a Professor of Electrical Engineering engaged in teaching and research pertaining to power systems and rotating machinery. He is presently a Professor of Engineering at the University of Kentucky where he has also served as the Associate Dean for Academic Affairs in the College of Engineering, Chair of the Department of Electrical Engineering, and Acting Chair of the Department of Chemical Engineering. Dr. Trutt is the author or co-author of more than 70 journal papers in Electrical Engineering and has been involved in teaching, research, and consulting in the areas of power systems and electric machines for more than 30 years. He is a Fellow of the IEEE and a past President of the IEEE Industry Applications Society.

CONTENTS

CHAPTER 1

INTRODUCTION

In this book, our goal is to develop a qualitative and quantitative understanding of electric power systems. The study of electric power systems is concerned with the generation, transmission, distribution, and utilization of electric power. Thus, the scope of study is schematically represented by Figure 1.1, which shows a coal-fired electric power plant providing electricity to a customer. We will study the major electrical components of this system, as well as the overall system itself. The main electrical components of this system are generators, transformers and transmission lines. Study of their roles in power system operation will be the subject matter of this book. Whereas Figure 1.1 illustrates the theme of the book, we will also study interconnected systems whereby several power generating systems are interconnected with each other. Such an interconnection between individual systems is necessary for an economic and reliable delivery of electric energy to the consumer. Referring to Figure 1.1, we notice that thermal energy is being converted to electric energy at the generating station. The electric energy is then transmitted to a substation which may be at a location near the consumer. In general, the generating station is located away from the consumer. Thus, one of the major advantages of utilizing electric power is that it may be conveniently transmitted from the point of generation to the point of utilization. This aspect leads us to the study of power transmission

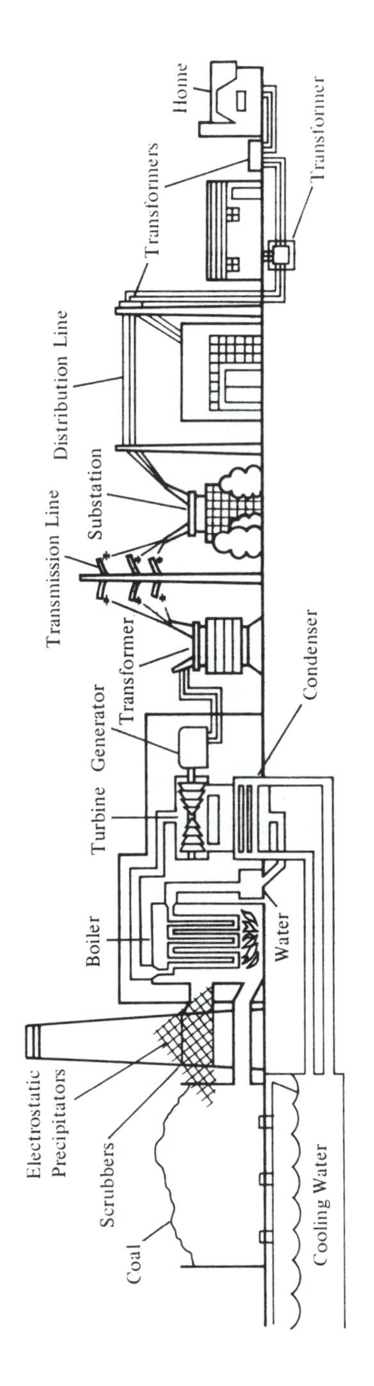

FIGURE 1.1
A coal-fired steam generating station supplying a home.

lines. In our study of a power system we will include its normal as well as abnormal behavior. This latter mode of behavior determines the choice of protective equipment in a power system. In summary, we will present a discussion of the operation of transmission lines, power system protection equipment and the overall behavior of a power system. Before we embark on this study, in the present chapter, we review certain basic concepts relevant to energy and power. Furthermore, in order to assess the energy perspective quantitatively, we first introduce certain units and basic definitions. Next, we will briefly discuss the concept of growth rates. This concept is useful in planning for the future. We will end this chapter with a discussion of major energy sources and limitations to electric power generation.

1.1 ENERGY AND POWER

Let a force F be applied to a mass so as to move the mass through a linear displacement l in the direction of F. Then the work Q done by the force is defined as the product Fl; that is,

$$Q = Fl \tag{1.1}$$

If the displacement is not in the direction of F, then the work done is the product of the displacement and the component of the force along the displacement, that is,

$$Q = Fl \cos \alpha \tag{1.2}$$

where α is the angle that F makes with l. Work is measured in joules (J). From (1.1), one joule is the work done by a force of one newton in moving a body through a distance of one meter in the direction of the force: 1J = 1 Nm.

The *energy* of a body is its capacity to do work. Energy has the same unit as work, although several other units are used for different forms of energy. For electric energy, the fundamental unit is the watt-second (Ws), where

$$1 \text{ Ws} = 1 \text{ J} \tag{1.3}$$

More commonly, however, electric energy is measured in kilowatthours (kWh). From (1.3) we have

$$1 \text{ kWh} = 3.6 \times 10^6 \text{ J} \tag{1.4}$$

The two most important forms of mechanical energy are kinetic energy and potential energy. A body possesses kinetic energy (KE) by virtue of its motion, such that an object of mass M (in kilograms), moving with a velocity u (in meters per second), has the kinetic energy

$$\mathrm{KE} = \frac{1}{2} M u^2 \ \mathrm{J} \tag{1.5}$$

A body possesses potential energy (PE) by virtue of its position. Gravitational potential energy, for instance, results from an object's position in a gravitational field. A body of mass M (in kilograms) at a height h (in meters) above the earth's surface has a gravitational PE given by

$$\mathrm{PE} = Mgh \ \mathrm{J} \tag{1.6}$$

where g is the acceleration due to gravity, in meters [per second per second].

Thermal energy is usually measured in calories (cal). By definition, one calorie is the amount of heat required to raise the temperature of one gram of water at 15°C through one Celsius degree. A more common unit is the kilocalorie (kcal). Experimentally, it has been found that

$$1 \ \mathrm{cal} = 4.186 \ \mathrm{J} \tag{1.7}$$

Yet another unit of thermal energy is the British thermal unit (Btu), which is related to the joule and the calorie as follows:

$$1 \ \mathrm{Btu} = 1.055 \times 10^3 \ \mathrm{J} = 0.252 \times 10^3 \ \mathrm{cal} \tag{1.8}$$

Because the joule and the calorie are relatively small units, thermal energy and electric energy are generally expressed in terms of the British thermal unit and kilowatthour (or even megawatthour), respectively. A still larger unit of energy is the quad, which stands for "quadrillion British thermal units." The mutual relationships among these various units are

$$1 \ \mathrm{quad} = 10^{15} \ \mathrm{Btu} = 1.055 \times 10^{18} \ \mathrm{J} \tag{1.9}$$

(Some authors define 1 quad as 10^{18} Btu.)

Power is defined as the time rate at which work is done. Alternatively, power is the time rate of change of energy. Thus the instantaneous power p may be computed as

$$p = \frac{dQ}{dt} \tag{1.10}$$

where Q represents either work or energy. The SI unit of power is the watt (W); one watt is equivalent to one joule per second:

$$1 \text{ W} = 1 \text{ J}/s \tag{1.11}$$

Multiples of the watt commonly used in power engineering are the kilowatt and the megawatt. The power ratings (or outputs) of electric motors are expressed in horsepower (hp), where

$$1 \text{ hp} = 745.7 \text{ W} \tag{1.12}$$

Example 1.1 The net energy requirement for the United States in 1990 was approximately 2.81×10^6 GWh. What is the equivalent of this energy in British thermal units and in quads?

Solution

Since

$$1 \text{ GWh} = 10^9 \text{ Wh} = 10^6 \text{ kWh}$$

we have

$$2.81 \times 10^6 \text{ GWh} = 2.81 \times 10^{12} \text{ kWh}$$

Then, from (1.4)

$$2.81 \times 10^{12} \text{ kWh} = 3.6 \times 10^6 \times 2.81 \times 10^{12} \text{ J} = 10.116 \times 10^{18} \text{ J}$$

From (1.8) and (1.9) we finally obtain

$$10.116 \times 10^{18} \text{ J} = \frac{10.116}{1.055} \times 10^{15} \text{ Btu} = 9.589 \times 10^{15} \text{ Btu}$$
$$= 9.589 \text{ quad}$$

Example 1.2 Coal has an average energy content of 940Wyears/ton, and natural gas has an energy content of 0.036Wyear/ft^3. If 56 percent of the net energy requirement of Example 1.1 were to be met with coal, 10 percent with gas, and 34 percent with other sources (such as hydro, nuclear, etc.), what amounts of coal and gas would be required?

Solution

From Example 1.1,

$$2.81 \times 10^6 \text{ GWh} = \frac{2.81 \times 10^{15}}{365 \times 24} \text{ Wyears} = 3.208 \times 10^{11} \text{ Wyears}$$

Hence, we have

$$\text{Energy to be supplied by coal} = 0.56 \times 3.208 \times 10^{11} = 1.796 \times 10^{11} \text{ Wyears}$$

$$\text{Energy to be supplied by gas} = 0.1 \times 3.208 \times 10^{11} = 3.208 \times 10^{10} \text{ Wyears}$$

Consequently,

$$\text{Amount of coal required} = \frac{1.796 \times 10^{11}}{940} = 1.91 \times 10^{8} \text{ tons}$$

$$\text{Amount of gas required} = \frac{3.208 \times 10^{10}}{0.036} = 8.91 \times 10^{11} \text{ ft}^3$$

Example 1.3 A certain amount of fuel can be converted into 6×10^{-3} quads of energy in a power station. If the average load on the station over a 24-h period is 60 MW, determine how long (in days) the fuel will last. Assume a 15 percent overall efficiency for the power station, where efficiency is defined as the ratio of output power to input power.

Solution

From (1.9) and (1.11), the energy available from the fuel is

$$6 \times 10^{-3} \text{ quad} = 6 \times 10^{-3} \times 1.055 \times 10^{18} \text{ Ws}$$

$$= \frac{6 \times 10^{-3} \times 1.055 \times 10^{18}}{60 \times 60 \times 10^{6}} \text{ MWh} = 1.76 \times 10^{6} \text{ MWh}$$

In 24 h, the station produces 60×24 =1440 MWh of energy. At 15 percent efficiency, this requires a daily input (from the fuel) of 1440/0.15 = 9600 MWh. Hence, the fuel will be consumed in $1.76 \times 10^6/9600 = 183.3$ days.

Example 1.4 In 1990, the total electric energy consumed, in the United States, from various sources was 2.81×10^6 GWhr. If the power plant average efficiency is 0.12, calculate the total (equivalent) energy input to the plant, in quad, required during 1990.

Solution

$$\text{Energy produced} = 2.81 \times 10^{6} \text{ GWhr}$$

$$= 2.81 \times 10^{12} \text{ kWhr}$$

From (1.4) and (1.9), 1 quad = 2.93×10^{11} kWhr

Hence,

$$2.81 \times 10^{12} \text{ kWhr} = \frac{2.81 \times 10^{12}}{2.93 \times 10^{11}} = 9.59 \text{ quad}$$

$$= \text{output energy}$$

or

$$\text{input energy} = \frac{\text{output energy}}{\text{efficiency}} = \frac{9.59}{0.12} = 79.91 \text{ quad.}$$

Example 1.5 The following is the projection for electric energy generation in the year 2000 in the United States from various sources: coal, 57 percent; natural gas, 10 percent; hydro, 7.5 percent; oil, 3.5 percent; other sources, 22 percent. The total generation in the year 2000 is estimated at 3.56×10^6 GWhr. Calculate the amounts of coal, natural gas, and oil required to generate the estimated energy (in the year 2000) if the energy contents for coal, gas, and oil in watt-years are, respectively, 937/ton, $0.036/\text{ft}^3$, and 168/barrel. Assume a 15 percent overall efficiency of the power plant.

Solution

$$\text{Total input energy} = \frac{3.56 \times 10^6}{0.15} = 23.73 \times 10^6 \text{ GWhr}$$

$$\text{Energy supplied by coal} = 23.73 \times 10^6 \times 0.57 = 13.528 \times 10^6 \text{ GWhr}$$

$$\begin{aligned}\text{Energy content of coal (per ton)} &= 937 \text{ W yr} \\ &= 937 \times 365 \times 24 \times 10^{-9} \\ &= 8.208 \times 10^{-3} \text{ GWhr}\end{aligned}$$

$$\text{Amount of coal required} = \frac{13.528 \times 10^6}{8.208 \times 10^{-3}} = 1.648 \times 10^9 \text{ tons}$$

$$\text{Energy supplied by gas} = 23.73 \times 10^6 \times 0.1 = 2.373 \times 10^6 \text{ GWhr}$$

$$\begin{aligned}\text{Energy content of gas (per ft}^3) &= 0.036 \text{ W yr} \\ &= 0.036 \times 365 \times 24 \times 10^{-9} = 315.36 \times 10^{-9} \text{ GWhr}\end{aligned}$$

$$\text{Energy supplied by gas} = \frac{2.373 \times 10^6}{315.36 \times 10^{-9}} = 7.525 \times 10^{12} \text{ ft}^3$$

$$\begin{aligned}\text{Energy supplied by oil} &= 23.73 \times 10^6 \times 0.035 \\ &= 0.83055 \times 10^6 \text{ GWhr}\end{aligned}$$

$$\text{Energy content of oil (per barrel)} = 168 \text{ W yr}$$

$$\text{Amount of oil} = 168 \times 365 \times 24 \times 10^{-9} = 1.472 \times 10^{-3} \text{ GWhr}$$

$$\text{Amount of oil required} = \frac{0.83055 \times 10^6}{1.472 \times 10^{-3}} = 0.564 \times 10^9 \text{ barrels.}$$

1.2 GROWTH RATES

In planning to accommodate future electric energy needs, it is necessary that we have an estimate of the rate at which those needs will grow. Fig 1.2 shows the history and a typical energy-requirement projection for the world.

To understand the concept of growth rates and its significance we proceed as follows. Suppose a certain quantity M grows at a rate that is proportional to the amount of M present. Mathematically, this statement may be expressed as

$$\frac{dM}{dt} = aM \tag{1.13}$$

where a is the constant of proportionality, known as the per-unit growth rate. It is estimated that electricity consumption throughout the world has grown by 3.4 percent per year over the last two decades. The solution to (1.13) may be written as

$$M = M_0 e^{at} \tag{1.14}$$

where M_0 is the value of M at $t = 0$. At any two values of time, t_1 and t_2, the inverse ratio of the corresponding M_1 and M_2 is

$$\frac{M_2}{M_1} = e^{a(t_2 - t_1)} \tag{1.15}$$

From (1.15) we may obtain the *doubling time* t_d such that $M_2 = 2M_1$ and $t_2 - t_1 = t_d$. It is

$$t_d = \frac{\ln 2}{a} = \frac{0.693}{a} \tag{1.16}$$

Power system planners also need to know how much power will be demanded in the future. The peak power demand for the United States over several years is shown in Figure 1.3. We can approximate this curve by the equation:

$$P = P_0 e^{bt} \tag{1.17}$$

where P_0 is the peak power at time $t = 0$, and b is the per-unit growth rate for peak power. The area under this curve over a given period is a measure of the energy Q consumed during that period.

From (1.16) and (1.17) it follows that if the per-unit growth rate has not changed then the energy consumed in one doubling period equals the energy consumed for the entire time period to that doubling period. In particular, evaluating the energy Q_1 consumed up to time t_1 and the energy Q_2 consumed during the doubling time $t_d = t_2 - t_1$, we obtain, from (1.17)

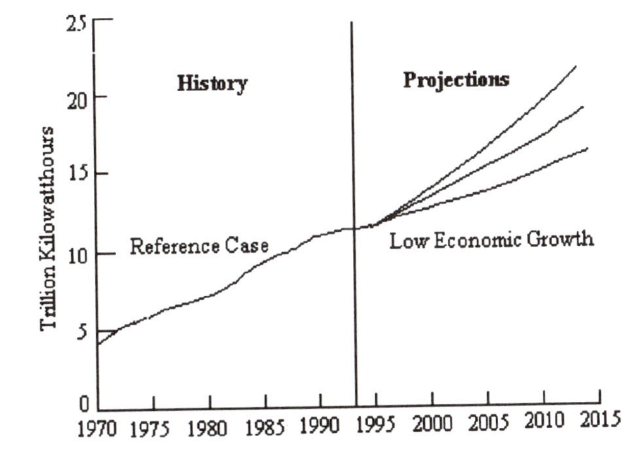

FIGURE 1.2
Annual energy requirement for the world.

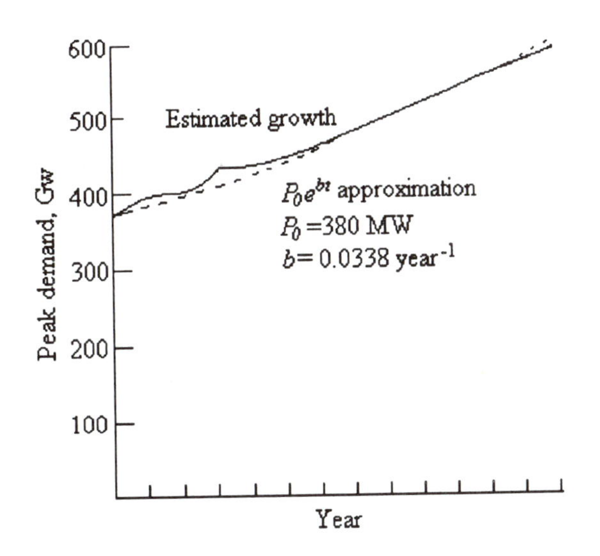

FIGURE 1.3
Peak demand and its approximation by P_0e^{bt}.

$$Q_1 = \int_{-\infty}^{t_1} P_0 e^{bt}\, dt = \frac{P_0}{b} e^{bt_1}$$

and

$$Q_2 = \int_{t_1}^{t_2} P_0 e^{bt}\, dt = \frac{P_0}{b} (e^{bt_d} - 1)\, e^{bt_1}$$

From (1.16), $t_d = (\ln 2)/b$, so Q_2 becomes

$$Q_2 = \frac{P_0}{b} (2 - 1)\, e^{bt_1} = \frac{P_0}{b} e^{bt_1} = Q_1 \tag{1.18}$$

where b is the per-unit growth rate. Fig 1.4 shows the doubling time in years as a function of growth rate in percent per year.

Example 1.6 The consumption of energy in a certain country has a growth rate of 3.3 percent per year. In how many years will the energy consumption be tripled?

Solution

From (1.15) with $Q_2/Q_1 = 3$,

$$3 = e^{0.033t} \text{ or } \ln 3 = 0.033\, t$$

and

$$t = \frac{\ln 3}{0.033} = 32.3 \text{ years}$$

In general, the time T at which the total consumption will equal the reserves, Q_T, may be obtained as follows.

With $t = 0$ at the present time, we have

$$Q_T = \int_0^T P_0 e^{bt}\, dt = \frac{P_0}{b} (e^{bT} - 1)$$

which we may write as

$$e^{bT} = \frac{bQ_T}{P_0} + 1$$

so that

$$T = \frac{1}{b} \ln \left(\frac{bQ_T}{P_0} + 1 \right) \tag{1.19}$$

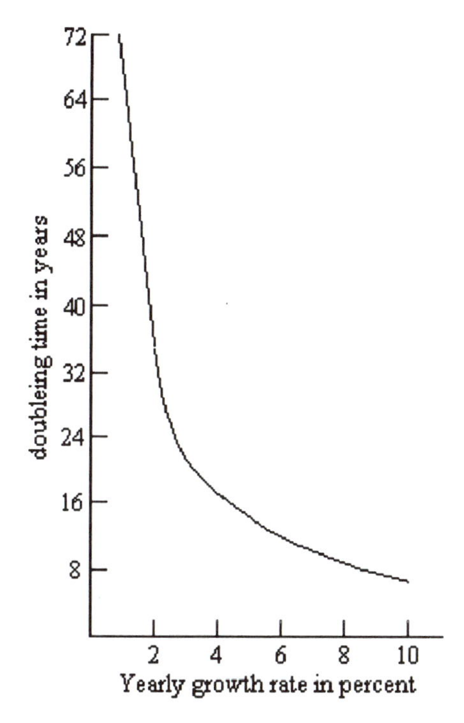

FIGURE 1.4
Doubling time vs. growth rate.

With regard to future growth rates, it is estimated that world-wide electricity demand will grow at an average rate of 3 to 4.3 percent annually throughout the 1990s, reaching 16,000 TWh (16,000 × 10^{12} Wh) by the year 2000. Subsequently, the growth rate is expected to be lower. The global demand is estimated at about 22,000 TWh in the year 2010 and 28,000 TWh [2] in the year 2030.

1.3 MAJOR ENERGY SOURCES

In Section 1.2 we have seen the effect of growth rate on doubling time. The consumption of energy for industrial purposes for the entire world is approximately doubling every 10 years. With such a growth rate, the world's fossil fuel supply—coal, petroleum, and natural gas—will be depleted in a few hundred years unless we exploit other sources of energy. The other major source of energy on the earth is solar radiation. This energy may be used directly as intercepted solar radiation, or indirectly as wind and hydropower. Other significant forms of energy are tidal energy, geothermal energy, and nuclear energy.

TABLE 1.1
Estimate of Fossil-Fuel Reserves

Fuel	Estimated Reserves	Approximate Energy Content (watt-years)
Coal	7.6×10^{12} metric tons	937 per ton
Petroleum	2×10^{12} barrels	168 per barrel
Natural gas	10^{16} ft^3	0.036 per ft^3

Considering fossil fuels first, the estimate of reserves [1] and their approximate energy contents for the world are given in Table 1.1. It may be verified (see Problem 1.16) that if all the energy needs of the world were to be supplied by coal and the doubling time was about 14 years, the entire coal reserve will be depleted in 125 years. Even though the reserves of fossil fuels may be large, their ultimate production depends, in somewhat unpredictable ways, on economic, political, and environmental considerations. Ultimately, we will need alternative energy sources in addition to using fossil fuels more efficiently than we do at present.

We mentioned earlier in this section that solar radiation is a major source of energy. It has been shown that the average worldwide incident power at the earth's surface is 182 W/m^2, which corresponds to a daily average energy of 4.4 kWh/m^2.

The potential for saving fuel by use of solar energy in buildings in the United States was estimated at 0.2 to 0.3 quadrillion Btu in 1985 [3]. The direct use of the solar power is either the active type involving solar cells (or photovoltaic cells) or the passive type utilizing radiation to heat solar collectors. Obviously, the most favorable sites for the production of solar power are desert areas. At present, large-scale utilization of solar energy is limited by the cost of solar cells and solar collector-heat exchanger systems and by the requirement of an adequate energy storage system to smooth out the daily variation. Furthermore, the solar cells generate direct current (dc), and hence inverting and related equipment will be needed to obtain the desired alternating current (ac) for most large-scale operations.

Indirect use of solar energy is manifested in wind energy. Turbine-type wind energy generators transform the kinetic energy of the wind into rotary-shaft motion and, in turn, to electrical energy. The choice of wind as a source of energy depends on various characteristics of wind [4-6].

Among the advantages of wind energy are its widespread availability, limited environmental impact, and the relatively small amount of land that is needed for the plant. The positive environmental impact of wind energy is evident from the fact that wind energy reduces the combustion of fossil fuels and the consequent emissions of pollutants such as carbon dioxide, sulphur dioxide and nitrous oxide. The comparative costs of electric power generation by wind

TABLE 1.2
Relative Cost of Electricity from Different Sources*

Energy source	cents/kWh
Coal	5.2 to 6
Gas-CCGT**	3.4 to 4.2
Nuclear	7.4 to 6.7
Wind	4.3 to 7.7

*In terms of 1991 US dollars
**CCGT-Combined cycle gas turbine

generators and other plants are shown in Table 1.2 [6]. The national target in the United States for installed wind turbine capacity is between 4000 and 8000 MW by the year 2000.

Another indirect means of using solar power is the stream-flow part of the hydrological cycle. In hydropower conversion, the potential energy of a mass of water at a hydraulic head is converted into the kinetic energy of the hydraulic turbine, which drives the electric generator. Since the potential energy is given by (1.6), the potential energy of 1000 kg of water for a head of 100 m is 9.8×10^5 J. Alternatively, a flow rate of 1 m^3/s for a head of 100 m has a hydraulic power of $(9.8 \times 10^3 \times \text{head}) = 9.8 \times 10^3 \times 100 = 9.8 \times 10^5$ W of power. It has been estimated that the total water-power capacity of the world is about 3×10^{12} W, but only 8.5 percent of this capacity has been developed, because many of the regions having the greatest potential have economic problems.

The next form of energy we wish to mention is *tidal energy*, which is obtained by filling a bay, closed by a dam, during periods of high tide, and emptying it during low-tide periods. Tidal power has been in use for centuries. The turbine for use in tidal power generation is reversible, so that the flow of water during both the filling and emptying of the bay may be utilized. Thus, tidal power is available twice during each tidal period of 12h and 25 min. The total worldwide tidal-power potential is estimated at about 64 billion watts. However, large (of the order of gigawatts) tidal-power plants are not feasible, either economically or technically. The high peak powers, produced in two short bursts in every tidal period (of 12 h and 25 min), require expensive turbines. Also, the burst of generated power cannot always be matched with the demand. Because of these factors, and since only a limited number of tidal sites exist, it is not expected that tidal energy can contribute significantly to world energy demand.

It has been found [7,8] that the average outward thermal power density at the earth's surface is 0.063 W/m^2 (or 1.5 $\mu cal/cm^2$-s). *Geothermal power* is obtained by extracting the heat that is stored temporarily in the earth. It is estimated [1] that the total stored thermal energy in major geothermal areas is 4×10^{20} J. With a 25 percent conversion factor, this corresponds to electrical energy of 3×10^6 megawatt-years.

Finally, we consider the two forms of nuclear energy: fission and fusion. In fission, nuclei of a heavy element, such as uranium 235 (^{235}U) split, whereas fusion involves the combination of light nuclei such as deuterium. Among fissionable materials, ^{235}U is the most suitable from an environmental standpoint. But ^{235}U is a rare isotope (each 100,000 atoms of uranium contain 711 atoms of ^{235}U). Thus, naturally available nuclear fuel is in very limited supply. This difficulty may be overcome by the use of breeder reactors, in which uranium 238 is transformed into fissionable plutonium 239 by absorbing neutrons. Similarly, thorium 232 is transformed into uranium 233, which is fissionable. In terms of equivalent energy, with the breeder reactor 1 g of uranium 238 will produce 8.1×10^{10} J of heat, which is approximately equivalent to the heat produced by 2.7 metric tons of coal. The cost of production of electrical energy from nuclear fuels is slightly higher than that from fossil fuels.

Fusion power is scientifically feasible, but the engineering problems have not all been resolved. Thus, no fusion power plants exist in the world. Engineering implementation of a practical fusion-power generating station does not seem feasible in the near future (say, by the year 2000). The hydrogen isotopes deuterium and tritium are suitable as fusion fuels. Deuterium is abundant (one atom to each 6,700 atoms of hydrogen), and the energy cost of separating it would be almost negligible compared with the amount of energy released by fusion. Tritium, however, exists only in tiny amounts in nature and must be bred in a reactor. For fusion to occur, the nuclei must form a plasma (atoms heated to such a high temperature that they are stripped of their electrons). The plasma must be contained or confined in a region of space such that the plasma density is high. Furthermore, the plasma must be contained long enough for the fusion process to take place. Heating and plasma confinement are major engineering problems in fusion power generation [9].

1.4 LIMITATIONS TO POWER GENERATION

From the discussion of Section 1.3, it is clear that some resources for energy are potentially unlimited. However, unlimited resources do not imply that the increase in energy conversion and consumption could be unlimited. The earth cannot sustain physical growth for more than a few tens of successive doublings. Besides constraints on the availability of fuels, and technical limitations, factors limiting energy conversion are biophysical, social, ethical, economic, and political in nature [10].

Biophysical limitation is due to the degradation of all energy to low-temperature heat. All energy obtained from any resource, by an appropriate process of energy conversion, must end up as low-temperature heat added to the surface of the earth, causing the earth's surface temperature to increase. Also, the extreme heat affects climatic events which are rather complex and not fully understood. The burning of fossil fuels produces carbon dioxide and certain solid particles which tend to change the optical properties of air and have intricate effects on the climate. It is generally felt that "global climatic problems could be expected fairly early in the next century if energy use were to continue growing at historical rates." The adverse effects on public health of oxides of sulfur and nitrogen are among the biophysical limitations on energy conversion.

The impact of energy conversion on land, water, wilderness, and ecological systems constitutes a social and ethical limit. Ill effects of energy conversion and utilization include those related to strip mining, water pollution, oil spills, and acid rainfall. Since the Three Mile Island and Chernobyl incidents [11], safe use of nuclear energy poses an ever-increasing ethical problem.

High economic and technical risks also impose a limit on the sustained growth of energy conversion. Complex technology is involved in the utilization of nuclear fuels for power plants. A good portion (about two-thirds) of the nuclear power output may be needed to maintain its own growth, and the cumulative energy deficit may become insurmountable.

Public acceptance of complex and new energy technology—such as the use of nuclear fuels—presents another limit to the growth of energy systems. Because of "the susceptibility of most modern energy technologies to commercial monopoly or technical dependence, all producing inequity," political forces are also considered a constraint to energy growth.

Emissions from thermal power plants include oxides of sulfur and nitrogen (SO_2 and NO_x), carbon monoxide, hydrocarbons, and particulates. Although means are taken to control air-polluting emissions, these emissions and waste-heat removal do present severe problems. In studying the growth of energy conversion processes, we must consider the many limiting factors discussed above [12].

REFERENCES

1. M. K. Hubbert, "The Energy Resources of the Earth," *Scientific American*, September 1971, pp. 31-40.
2. M. Khatib, "Electricity in the Global Energy Scene," *IEE Proc. A*, vol. 140, January 1993, pp. 24-28.
3. *Solar Age*, February 1981, p. 4.
4. D. F. Warne and P. G. Calman, "Generation of Electricity from the Wind," *IEE Reviews* (IEE, London), vol. 124, no. 11R, November 1977, pp. 963-985.
5. B. Sorensen, "Turning to the Wind," *American Scientist*, vol. 69, September-October 1981, pp. 500-506.
6. J. A. Halliday, "Wind Energy: An Option for the UK?" *IEE Proc.-A*, vol. 140, January 1993, pp. 53-62.
7. C. Starr, "Energy and Power," *Scientific American*, September 1971, pp. 3-18.

8. "Electric Power Supply and Demand for the Contiguous United States 1981-1990," *U.S. Dept. of Energy Report DOE/EP-0022*, July 1981.
9. G. H. Miley, *Fusion Energy Conversion*, American Nuclear Society, La Grange Park, IL, 1976.
10. A. Lovins, "Limits to Energy Conversion: The Case for Decentralized Technologies," in *Alternatives to Growth*, vol. 1, D. L. Meadows, ed. Ballinger, Cambridge, MA, 1977, pp. 59-76.
11. *Time*, "Nuclear Nightmare" (cover story), April 9, 1979, pp. 8-19.
12. A. E. Wheldon and C. E. Gregory, "Energy, Electricity and the Environment," *IEE Proc.-A*, vol. 140, January 1993, pp. 2-7.

PROBLEMS

1.1. A certain country consumed 1.6×10^5 GWh energy during one year. What is the equivalent of this energy in quads? (10^{15} Btu = 1 quad).

1.2. The annual energy requirement in a certain region is 2.0 quad. This energy is met by coal having an energy content of 937 Wyr/ton. Calculate the amount of coal required per year to meet this energy need.

1.3. Calculate the annual cost of fuel to run a power station supplying an average daily load of 36 MW. The overall efficiency of the power plant is 18 percent and the price of fuel is 3.1 cents/kWh.

1.4. In an interconnected power system, having an overall efficiency of 15 percent, 12 quads of energy are supplied by coal and 20 quads by natural gas. The heat content of coal is 13,800 Btu/lb and that of natural gas is 1.0 Btu/ft^3. (a) Find the amounts of coal and natural gas consumed. (b) Also calculate the energy available to the consumer.

1.5. A certain amount of fuel contains 15×10^{12} Btu of energy, and converted into electric energy in a power station having a 12 percent overall efficiency. The average daily (24 hr) demand on the station is 5 MW. In how many days will the fuel be totally consumed?

1.6. How many kilocalories of energy are available from a fuel with a 10 quad energy content? In how many days will this fuel be totally consumed if it is used to satisfy a demand of 10^{13} Btu/day at a power plant with an overall efficiency of 20 percent?

1.7. In a power station, 4×10^4 GWh of energy is to be produced in 1 year, half from coal and half from natural gas. The energy content of coal is 900 Wyears/ton, and that of natural gas is 0.03 Wyear/ft^3. How much coal and how much natural gas will be required?

1.8. Rework Problem 1.7 assuming all the energy is to be supplied by (*a*) coal and (*b*) natural gas.

1.9. During a 1-year period, a certain power system consumed energy (in quads) from various sources as follows: coal, 6; oil, 2; gas, 1; and hydro, 0.5. If the overall efficiency of the system is 0.12, how much electric energy (in gigawatthours) could be produced by the system from these sources?

1.10. A 90 percent efficient electric motor operating an elevator lifts a 10-ton load through a height of 60 ft. Calculate the energy required by the motor to do so.

1.11. The load of Problem 1.10 is to be lifted through the entire height of 60 ft within 40s. Determine the minimum horsepower rating of the motor.

1.12. Calculate the energy required for a geared dc motor to lift a 1-ton load through 50 ft in 10s. The motor/gear overall efficiency is 0.51. Also calculate the motor/gear output power.

1.13. Calculate the velocity with which a 200-kg mass must move so that its kinetic energy equals the energy dissipated in a 0.2-Ω resistor through which a 100-A current flows for 2 h.

1.14. Natural gas reserves in a certain country are estimated at 100×10^9 ft^3, with an energy content of 0.025 Wyear/ft^3. If the present peak power demand is 0.5 GW, the power demand growth rate is 5 percent, and all the energy is to be supplied by natural gas, approximately how long will the reserve last?

1.15. The energy consumption in a certain country has a growth rate of 6 percent per year. At this rate, in how many years will the energy consumption be quadrupled?

1.16. If the estimated coal reserves in the United States is 7.6×10^{12} tons, verify that this entire amount will be depleted in 125 years, assuming a fixed growth rate of 5 percent at all times and the present peak demand is 700 GW. The energy content of coal is 937 Wyears/ton.

1.17. One million cubic meters of water is stored in a reservoir feeding a water turbine. If the center of mass of the water is 50 m above the turbine and losses are negligible, how much energy (in megawatthours) will that volume of water produce? The density of water is 933 kg/m^3.

CHAPTER

2

AC CIRCUIT ANALYSIS

Quite often, the study of a power system under certain operating conditions reduces to solving an ac circuit problem. Thus, a degree of proficiency in ac circuit analysis is essential to our study of electric power systems. In this chapter, we briefly review the basic concepts pertaining to ac circuits. Because electrical components of a power system predominantly are three-phase units, we will focus primarily on three-phase systems. In addition, we will pay special attention to power in ac circuits.

2.1 REPRESENTATION OF AC QUANTITIES

In an ac power system under steady-state, alternating voltages (and currents) vary sinusoidally at a fixed frequency (usually 60 Hz in the United States). Thus, an ac voltage v and a current i in a circuit may be represented graphically by the sine wave shown in Figure 2.1. Mathematically, these waveforms may be expressed as:

$$v = V_m \sin \omega t \tag{2.1a}$$

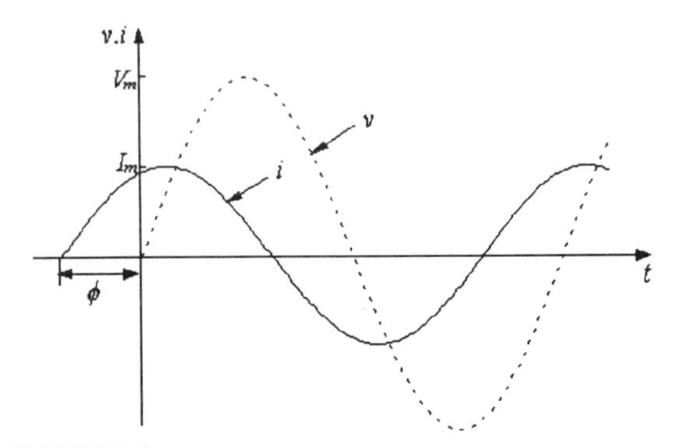

FIGURE 2.1
Representation of an ac voltage and a current.

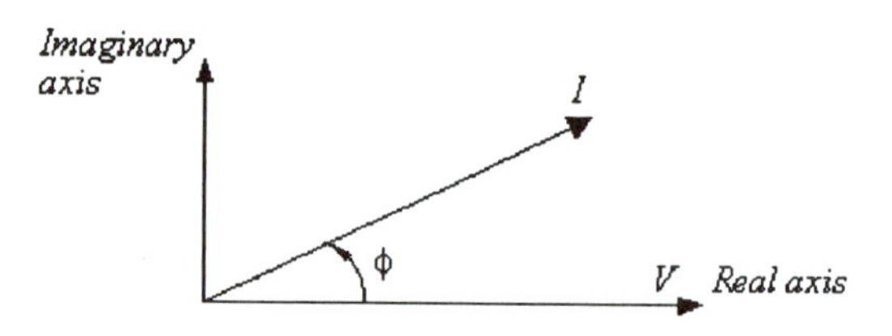

FIGURE 2.2
AC voltage and current represented as phasors.

$$i = I_m \sin(\omega t + \phi) \tag{2.1b}$$

where V_m is the maximum value of v and ω is the angular frequency such that $\omega = 2\pi f$ (f being the frequency in Hz). In (2.1b), ϕ is a phase angle by which i is ahead of v (ahead because the coefficient of ϕ is positive).

Alternating quantities are also represented by phasors. Thus, v and i of Figure 2.1 may be written as phasors

$$V = |V| \angle 0^\circ \tag{2.2a}$$

$$I = |I| \angle \phi \tag{2.2b}$$

where $|V| = V_m/\sqrt{2}$ and $|I| = I_m/\sqrt{2}$ are rms values, and I leads V by an angle ϕ. Pictorially, the phasors (complex numbers written in polar form) V and I are shown in Figure 2.2, where V is the reference phasor. Furthermore, it is clear from Figure 2.2 that

$$V = |V| + j0 \tag{2.3a}$$

and

$$I = |I| \; (\cos\phi \; + \; j\sin\phi) \tag{2.3b}$$

Notice that in Figure 2.2, the horizontal axis is labeled as the real axis, and the vertical axis is the imaginary axis.

Up to this point we have expressed sinusoidal ac quantities graphically (Figure 2.1), analytically as in (2.1), and phasorially as in Figures 2.2 and (2.2) and (2.3). The equivalence of these two forms of representation may be seen by noting that

$$v = \text{Im}\left[V_m e^{j\omega t}\right] = \text{Im}\left[\sqrt{2}\, V e^{j\omega t}\right] \tag{2.4a}$$

and

$$i = \text{Im}\left[I_m e^{j(\omega t + \phi)}\right] = \text{Im}\left[\sqrt{2}\; I e^{j\omega t}\right] \tag{2.4b}$$

where (Im) stands for "imaginary part of" and $V = |V|\angle 0° = (V_m/\sqrt{2})\, e^{j0°}$ and $I = |I|\angle\phi = I_m/\sqrt{2}\, e^{j\phi}$. When the volt-ampere relationships such as $v = Ri$; $v = L(di/dt)$ and $v = 1/C \int idt$ are applied to v and i using the representations given in (2.4a) and (2.4b), the common terms $\sqrt{2}\, e^{j\omega t}$ and "imaginary part of" cancel and the calculus operations are automatically performed when the circuit element is represented by its impedance. Thus, all three volt-ampere relationships are represented by $V = ZI$ and standard phasor analysis for sinusoidal steady-state circuits is applied.

2.2 POWER IN AC CIRCUITS

From (2.1) the instantaneous power in the circuit may be written as

$$p = vi = V_m I_m \sin(\omega t) \sin(\omega t + \phi) \tag{2.5}$$

The circuit absorbs or delivers power depending upon whether the end result of (2.5) is positive or negative. Now, using the trigonometric identity $\sin A \sin B = ½ [\cos (A - B) - \cos (A + B)]$ where $A \equiv \omega t + \phi$ and $B = \omega t$, we may rewrite (2.5) as

$$p = \frac{1}{2} V_m I_m [\cos\phi - \cos(2\omega t + \phi)] \tag{2.6}$$

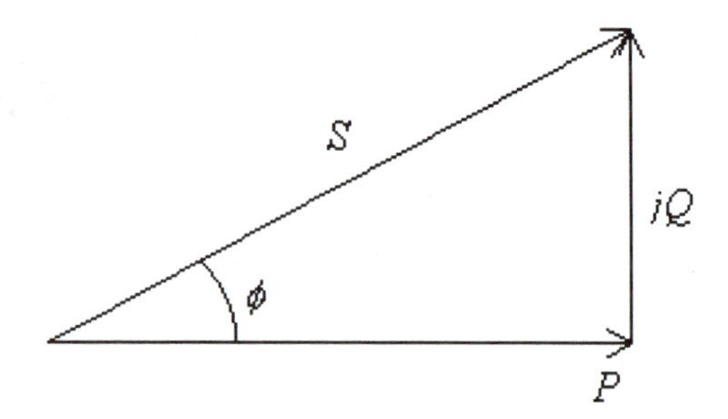

FIGURE 2.3
Power triangle.

The average value of (2.6), which is the average power P, becomes

$$P = \frac{1}{2} V_m I_m \cos\phi = |V|\,|I| \cos \phi \tag{2.7}$$

where $|V| = V_m/\sqrt{2}$ and $|I| = I_m/\sqrt{2}$ are rms values.

In terms of complex exponentials, we may express the voltage and current, respectively, as

$$|V| = |V| \angle 0° = Ve^{j0} \text{ and } I = |I|\angle\phi = |I|e^{j\phi} \tag{2.8}$$

Defining complex power S by

$$S = P + jQ = VI^* \tag{2.9}$$

where I^* is the complex conjugate of I, and (2.8) and (2.9) yield

$$S = P + jQ = |V|\,|I|e^{-j\phi} = |V|\,|I| \cos\phi - j|V|\,|I|\sin\phi \tag{2.10}$$

Now, comparing (2.7) and (2.10), we notice that the first term on the right-hand side of (2.10) corresponds to the average power P. To distinguish it from the second term, we call P the real, true, or active power, measured in W or kW. The second term is

$$Q = -|V|\,|I| \sin \phi \tag{2.11}$$

The quantity $|V|\,|I|\sin \phi$ is called the reactive power, in var or kvar. Notice from (2.11) that the reactive power is negative. On the other hand, it may be easily verified that if we have the current lagging the voltage, that is $I = |I| \angle{-\phi}$ with $V = |V|\angle 0°$, then the reactive power is positive. Consequently, it may be said that an inductive load (current lags voltage) absorbs reactive power whereas a capacitive load (current leads voltage) supplies reactive power. Finally, the magnitude of the complex power $|S|$ is also known as the apparent power, in VA or kVA.

Sometimes it is advantageous to express $S = P + jQ$ graphically as a power triangle as shown in Figure 2.3. In the power triangle, it is conventional to define an angle θ which is equal to the difference between the angles of the voltage and current phasors. In the case where the voltage phasor is at reference angle of 0° and the current phasor is at an angle ϕ, then $\theta = 0° - \phi = -\phi$. The average power from (2.7) then becomes

$$P = |V|\,|I| \cos\theta \tag{2.7a}$$

where $\cos\theta$ is defined as the power factor and θ is the power factor angle. Similarly, the reactive power from (2.11) becomes

$$Q = |V|\,|I| \sin\theta \tag{2.11a}$$

The power factor is defined as leading when θ is negative (current leads the voltage) and lagging when θ is positive (current lags the voltage).

2.3 BALANCED THREE-PHASE SYSTEMS

Most of the commercial electrical power produced in this country is generated in *three-phase* systems. A three-phase system requires the generation of three balanced sinusoidal voltages which by definition will have the same magnitude and frequency, but will be displaced from each other by 120° in phase. These voltages are then interconnected in one of the two types of connections, designated as wye or delta, and shown in Figure 2.4. The result, for either type of connection, is a three-terminal source of power supplying a balanced set of three voltages to a load. The individual generator voltages are called the phase voltages. For the wye connection, these phase voltages are

$$V_{aa'} = |V_p| \angle 0° \tag{2.12}$$

$$V_{bb'} = |V_p| \angle -120° \tag{2.13}$$

$$V_{cc'} = |V_p| \angle 120° \tag{2.14}$$

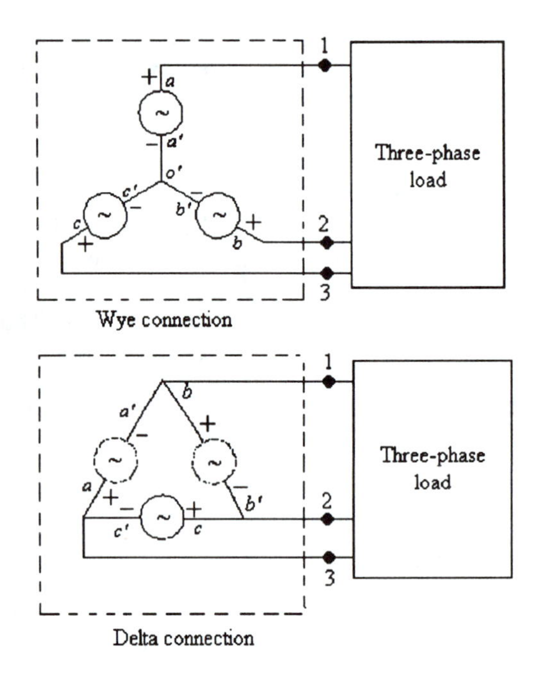

FIGURE 2.4
Three-phase connections.

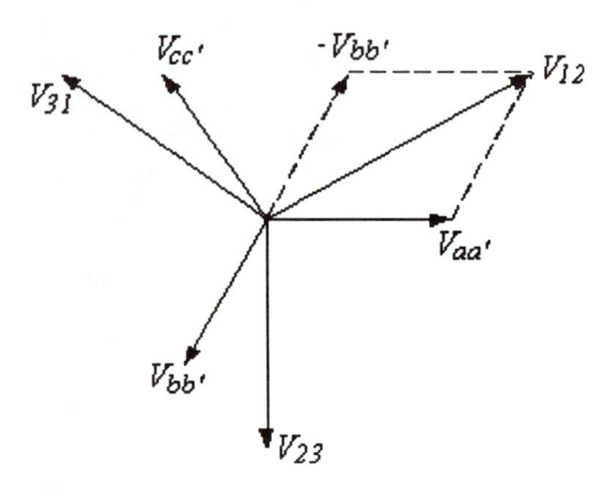

FIGURE 2.5
Phase and line voltages in a wye connection.

Because the origin of the time axis is arbitrary, one of the phasors has been chosen to coincide with the horizontal or real axis of the phasor diagram, as shown in Figure 2.5, which shows all of the individual voltages. The symbol, $|V_p|$ is the rms value of the phase voltage which is the same in all three phases.

In three-phase systems, it is often the line-to-line or simply the line voltages that are of greatest interest. The line voltages V_{12}, V_{23}, and V_{31} in a wye connection can be determined in terms of the phase voltages, as will be demonstrated for V_{12},

$$
\begin{aligned}
V_{12} &= V_{aa'} + V_{b'b} = V_{aa'} - V_{bb'} = |V_p|\angle 0 - |V_p|\angle -120° \\
&= |V_p|(1 + j0) - |V_p|\left(-0.5 - j\,\frac{\sqrt{3}}{2}\right) \\
&= \sqrt{3}\,|V_p|\,\angle 30° = |V_L|\,\angle 30° = (\sqrt{3}\,\angle 30°)\,V_{aa'}
\end{aligned}
\tag{2.15}
$$

where $|V_L| = \sqrt{3}\,|V_p|$ represents the rms value of the *line voltages*. The graphical equivalent of this computation is shown in Figure 2.5. Although we could solve for each of the line voltages in the manner above, symmetry demands that the other two voltages be given by

$$V_{23} = |V_L|\,\angle -90° = (\sqrt{3}\,\angle 30°)\,V_{bb'} \tag{2.16}$$

$$V_{31} = |V_L|\,\angle 150° = (\sqrt{3}\,\angle 30°)\,V_{cc'} \tag{2.17}$$

These are also shown on the phasor diagram of Figure 2.5. For the delta connection, the line voltages are evidently the same as the individual phase voltages, and so no special computations need be made.

It should be noted that the phasors $V_{aa'}$, $V_{bb'}$ and $V_{cc'}$ form what is defined as a balanced positive-sequence set, i.e., the phase order in a clockwise direction is *abc*. If this phase order is reversed such that the sequence is *acb* where $V_{aa'} = |V_p|\angle 0°$, $V_{bb'} = |V_p|\,\angle 120°$ and $V_{cc'} = |V_p|\,\angle -120°$, a similar calculation to that given above will show that $V_{12} = |V_L|\,\angle -30°$, $V_{23} = |V_L|\,\angle 90°$ and $V_{31} = |V_L|\,\angle -150°$. The phase and line voltages in this case form what is defined as a balanced negative-sequence set of phasors. These distinctions will be discussed in greater detail in Chapter 5.

When the load connected to a balanced, three-phase set of voltages is a balanced load, i.e., made up of identical impedances connected in a delta or wye arrangement, then the currents which result form a balanced set of three-phase currents. Thus, suppose the currents in the wye-connected source shown in Figure 2.6 are given by

$$I_{a'a} = |I_p| \angle\phi_p \tag{2.18}$$

$$I_{b'b} = |I_p| \angle(\phi_p - 120°) \tag{2.19}$$

$$I_{c'c} = |I_p| \angle(\phi_p + 120°) \tag{2.20}$$

where $|I_p|$ is the rms value of the current in each phase. The average power delivered by each phase is given by $|V_p||I_p| \cos\theta_p$, where the power factor angle $\theta_p = 0° - \phi_p = -\phi_p$ using phase *a* and the voltage in phase *a* as the references. Similarly, the reactive power delivered by each phase is given by $|V_p||I_p| \sin\theta_p$. The total average and reactive powers absorbed by the load are equal to those delivered by the source and are given (for this *balanced* system) by

$$P_T = 3[|V_p||I_p| \cos\theta_p] \tag{2.21a}$$

and

$$Q_T = 3\left[|V_p||I_p| \sin\theta_p\right] \tag{2.21b}$$

But for the wye connection, the rms phase current is the same as the rms current in the line, $|I_L|$, and we showed earlier that $|V_L| = \sqrt{3}\,|V_p|$, so we can express the total power in terms of line voltages and currents as follows.

$$P_T = 3\left(\frac{|V_L|}{\sqrt{3}}\right)|I_L| \cos\theta_p = \sqrt{3}\,|V_L||I_L| \cos\theta_p \tag{2.22a}$$

and

$$Q_T = 3\left(\frac{|V_L|}{\sqrt{3}}\right)|I_L| \sin\theta_p = \sqrt{3}\,|V_L||I_L| \sin\theta_p \tag{2.22b}$$

It should be noted that the angle θ_p is still the angle between the phase voltage and the phase current.

The expression for the total power in a delta-connected system is the very same. Whereas the line voltage equals the phase voltage for a delta-connected system, the line current is $\sqrt{3}$ times greater than the phase current, as will be demonstrated later in (2.36) - (2.38). Hence, if (2.21) is converted to the line quantities for the delta-connected system, we obtain

$$P_T = 3\,|V_L|\left(\frac{|I_L|}{\sqrt{3}}\right)\cos\theta_p = \sqrt{3}\,|V_L|\,|I_L|\,\cos\theta_p \tag{2.23a}$$

and

$$Q_T = 3\,|V_L|\left(\frac{|I_L|}{\sqrt{3}}\right)\sin\theta_p = \sqrt{3}\,|V_L|\,|I_L|\sin\theta_p \tag{2.23b}$$

which is the same as (2.22).

In working with *balanced* three-phase systems, computations are usually made on a *per-phase* basis, and then the total results for the entire circuit are obtained on the basis of the symmetry and other factors which must apply. Thus, if a set of three identical wye-connected impedances is connected to a wye-connected, three-phase source (as shown in Figure 2.6), because of the symmetry, point 0′ turns out to have the same potential as point 0, and could therefore be connected to it. Thus, each generator seems to supply only its own phase, and the computations on a per-phase basis are therefore legitimate. It is left as an exercise to the reader to show that the connection between points 0 and 0′ will always carry zero current in a balanced system.

Sometimes it is necessary, or desirable, to mathematically convert a set of identical impedances connected in wye to an equivalent set connected in delta, or vice versa (Figure 2.7). In order for this transformation to be valid, the impedances seen between any two of the three terminals must be the same. Thus, getting the impedance Z_{ab} for both the wye and the delta, and equating them, we get,

$$2Z_w = \frac{Z_d(Z_d + Z_d)}{Z_d + Z_d + Z_d} = \frac{2}{3}\,Z_d \tag{2.24}$$

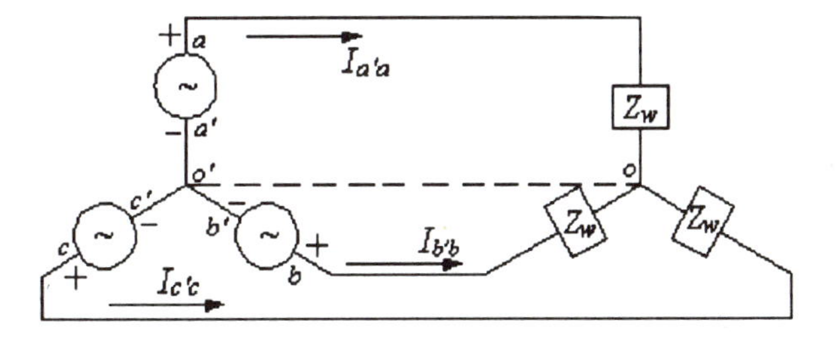

FIGURE 2.6
A three-phase voltage-source supplying a three-phase wye-connected load.

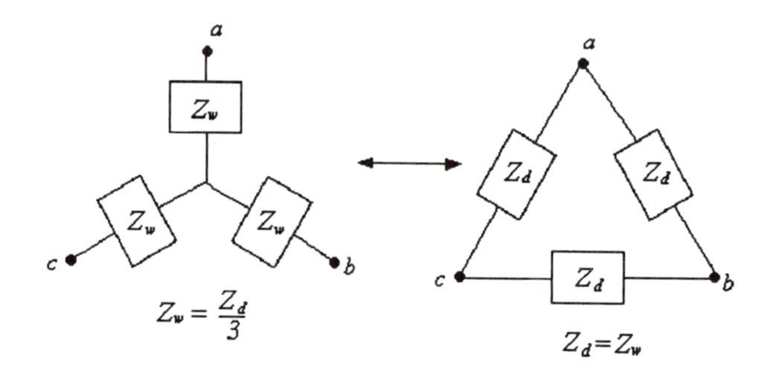

FIGURE 2.7
Wye-delta transformation for a balanced load.

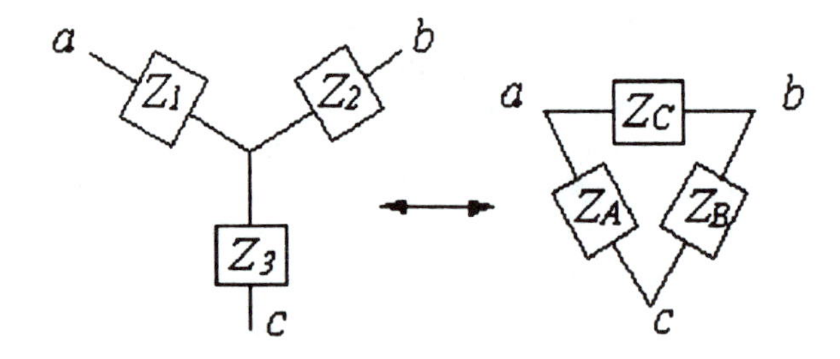

FIGURE 2.8
Wye-delta transformation for an unbalanced load.

From this, we can solve for the wye impedances Z_w in terms of the delta impedances Z_d, or vice versa, such that

$$Z_w = \frac{Z_d}{3} \tag{2.25}$$

and

$$Z_d = 3Z_w \tag{2.26}$$

Hence, we can switch back and forth between balanced wye and delta connected impedances as desirable.

In case we encounter a three-phase load such as that shown in Figure 2.8, for equivalence of impedances at the terminals, we may write

$$Z_{ac} = Z_1 + Z_3 = \frac{Z_A(Z_B + Z_C)}{Z_A + Z_B + Z_C} \tag{2.27}$$

$$Z_{ab} = Z_1 + Z_2 = \frac{Z_C(Z_A + Z_B)}{Z_A + Z_B + Z_C} \tag{2.28}$$

$$Z_{bc} = Z_2 + Z_3 = \frac{Z_B(Z_A + Z_C)}{Z_A + Z_B + Z_C} \tag{2.29}$$

Solving for Z_A, Z_B and Z_C yields

$$Z_A = \frac{1}{Z_2}(Z_1 Z_2 + Z_2 Z_3 + Z_3 Z_1) \tag{2.30}$$

$$Z_B = \frac{1}{Z_1}(Z_1 Z_2 + Z_2 Z_3 + Z_3 Z_1) \tag{2.31}$$

$$Z_C = \frac{1}{Z_3}(Z_1 Z_2 + Z_2 Z_3 + Z_3 Z_1) \tag{2.32}$$

Similarly for Z_1, Z_2, and Z_3 we have,

$$Z_1 = \frac{Z_A Z_C}{Z_A + Z_B + Z_C} \tag{2.33}$$

$$Z_2 = \frac{Z_B Z_C}{Z_A + Z_B + Z_C} \tag{2.34}$$

$$Z_3 = \frac{Z_A Z_B}{Z_A + Z_B + Z_C} \tag{2.35}$$

As an exercise, suppose we were to take the simple network of Figure 2.6 and convert both the source and load to equivalent delta connections. Since the connection between 0 and 0′ carries no current in a balanced system, it may be removed or added as is convenient. With this connection removed, we may apply (2.26) to the load and obtain the delta connected equivalent shown in Figure 2.9. We have also shown previously, (2.15)-(2.17), that for a positive-sequence set of balanced phase voltages with phase a as a reference, $V_{12} = |V_L|\angle 30°$, $V_{23} = |V_L|\angle{-90°}$ and $V_{31} = |V_L|\angle 150°$.

The currents flowing in the equivalent delta-connected load are

$$I_{12} = \frac{V_{12}}{Z_d} = \frac{|V_L|\angle 30°}{Z_d} = |I_p|\angle 30°$$

$$I_{23} = \frac{V_{23}}{Z_d} = \frac{|V_L|\angle{-90°}}{Z_d} = |I_p|\angle{-90°}$$

$$I_{31} = \frac{V_{31}}{Z_d} = \frac{|V_L|\angle 150°}{Z_d} = |I_p|\angle 150°$$

where it has been assumed for simplicity that the angle of the impedance Z_d (and hence Z_w also) is 0° such that Z_w and Z_d are pure resistances. It is left as an exercise for the student to show that the following results are valid for any impedance angle. The quantity $|I_p| = |V_L|/Z_d$ is the rms value of the phase current in the delta-connected load. The rms-phase voltage of the delta connected load is $|V_L| = \sqrt{3}\,|V_p|$ where $|V_p|$ is the phase voltage of the equivalent wye as defined in (2.12)-(2.14).

If we now represent the line currents as I_a, I_b and I_c, Kirchhoff's current law may be applied at node 1 to give

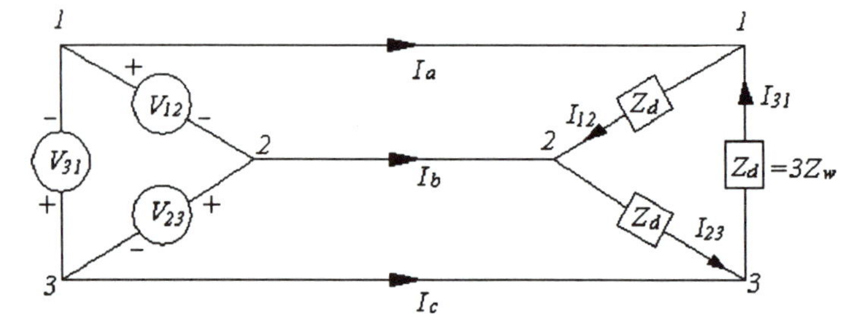

FIGURE 2.9
A balanced three-phase, delta-connected voltage source supplying a balanced delta-connected load.

$$I_a = I_{12} - I_{31} = |I_p|\angle 30° - |I_p|\angle 150°$$

or (2.36)

$$I_a = \sqrt{3}\,|I_p|\angle 0° = |I_L|\ \angle 0° = (\sqrt{3}\angle -30°)\ I_{12}$$

By a similar calculation at nodes 2 and 3

$$I_b = I_{23} - I_{12} = \sqrt{3}\ |I_p|\ \angle -120° = |I_L|\angle -120° = (\sqrt{3}\angle -30°)I_{23} \quad (2.37)$$

$$I_c = I_{31} - I_{23} = \sqrt{3}\ |I_p|\angle 120° = |I_L|\angle 120° = \sqrt{3}\,(-30°)\ I_{31} \quad (2.38)$$

These results are also shown graphically in Figure 2.10

From the above, it may be seen that the rms value of the line current for a balanced delta connection $|I_L| = \sqrt{3}\ |I_p|$ where $|I_p|$ is the rms phase current in the delta. Similarly, the line current I_a lags its corresponding phase current I_{12} by 30° for the positive-sequence situation that has been assumed. Further investigation would show that for a negative-sequence situation, $|I_L|$ is

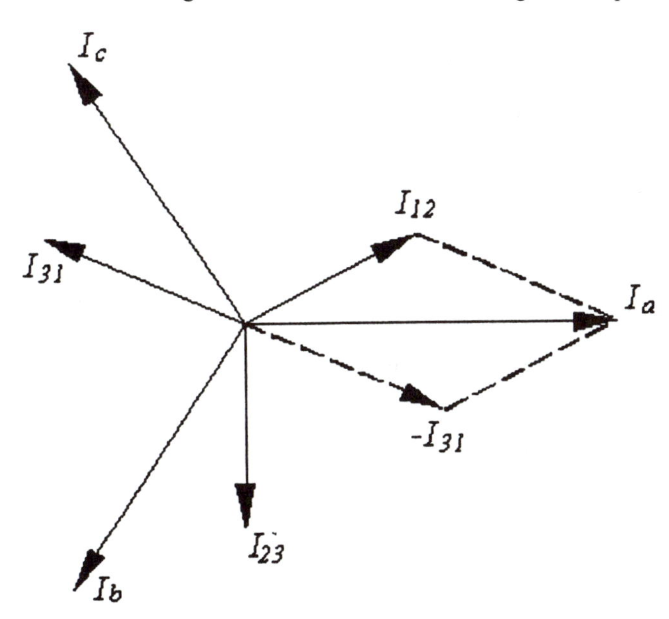

FIGURE 2.10
Phase and line currents for a delta connection.

still equal to $\sqrt{3}\ |I_p|$ but that the phase shift between I_a and I_{12} will be such that I_a would lead I_{12} by 30°. Finally, it should be pointed out that since delta connections equivalent to the wye connections of Figure 2.6 have been used in the proceeding demonstration, it may easily be shown that the line current I_a is equal to the line current $I_{a'a}$ in Figure 2.6.

PROBLEMS

2.1. Evaluate: (a) $10\angle 45° + 10\angle -45°$; (b) $(6.8 + j3.2) + (5 - j1.2)$; (c) $2\angle 60° + 3\angle -30° - 4\angle 45°$; and (d) $2\angle 60° + (5 - j1.232)$.

2.2. Perform the following operations in polar form: (a) $(3 + j3)/(5 + j8)$; (b) $(5\angle 45°)/4\angle 20°)$; (c) $(5\angle 45°)(4\angle -20°)/(3\angle 75°)$.

2.3. The instantaneous voltage and current in an ac circuit are: $v = 155.6 \sin 377t$ V, and $i = 7.07 \sin(377t - 30°)$ A. Represent these as complex exponentials.

2.4. Calculate the apparent, active and reactive powers associated with the circuit of Problem 2.3.

2.5. An inductive coil consumes 500 W of average power at 10 A, 110 V and 60 Hz. Assuming the voltage as reference phasor, obtain an expression for the instantaneous current, expressed as a complex exponential.

2.6. A 110-V 60-Hz inductive load draws $(500 + j500)$ VA complex power. A capacitor C is connected across the load to bring the overall power factor to 0.866 lagging. Determine the value of C, and calculate the new value of complex power for load-capacitor combination.

2.7. For the circuit shown in Figure 2.11, calculate: (a) the current through the 4-Ω resistor; (b) the complex power absorbed by the inductive branch at 60 Hz; (c) the value of C to supply 5 var at 60 Hz; and (d) the total active and reactive powers supplied by the two sources for the value of C found in (c).

2.8. Three impedances $3\angle 0°$ Ω, $4\angle 60°$ Ω and $5\angle 90°$ Ω are connected in wye. This wye-connected load is supplied by a balanced positive sequence 240-V 60-Hz delta-connected three-phase source. Determine the line currents using loop analysis.

2.9. Repeat Problem 2.8 using a wye-delta transformation on the wye-connected load to solve for the line currents.

2.10. A voltage $v = V_m \sin \omega t$ is applied across a resistor R. Show that the instantaneous power in the resistor pulsates at a frequency 2ω.

2.11. Extend the result of Problem 2.10 to a three-phase balanced system. Verify that in this case the total instantaneous power has no pulsating component.

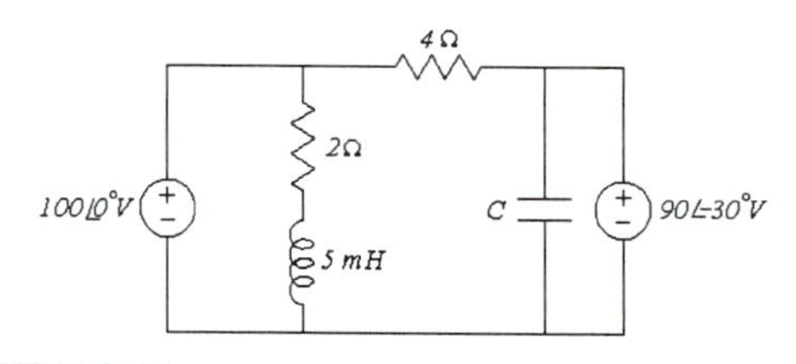

FIGURE 2-11
Problem 2.7.

2.12. For the wye-connected load of Problem 2.8, determine (a) the power drawn by each impedance; (b) the reactive power in each phase; and (c) the overall power factor of the circuit.

2.13. A 460-V three-phase source supplies power to the following three-phase balanced loads: a 200 hp induction motor operating at 94 percent efficiency and 0.88 pf lagging, a 50 kW resistance-heating load, and a combination of other loads totaling 40 kW at 0.70 lagging power factor. Calculate the total complex power.

2.14. A balanced wye-connected three-phase load consisting of three $5\angle 30°\Omega$ impedances is connected to a 230V line. Determine the real and reactive power absorbed in this load.

2.15. Repeat Problem 2.14 for a load impedance of $5\angle -30°\Omega$.

2.16. Assuming that the load in Problem 2.14 is delta connected, determine the current in each phase of the load and use this information to determine the line currents and the real and reactive power absorbed in the load. Compute these powers using both line and phase values.

2.17. Show that the current $I_{00'}$ in Figure 2.6 is always zero in a balanced three-phase system.

2.18. Show that (2.16) and (2.17) are the correct line voltages for the wye connection of Figure 2.4.

2.19. If the voltages $V_{aa'}$, $V_{bb'}$ and $V_{cc'}$ in Figure 2.4 are a balanced negative-sequence set with $V_{aa'}$ as the reference phasor (0° angle), show that $V_{12} = |V_L|\angle -30°$ and $V_{31} = |V_L|\ \angle -150$.

2.20. Show that the currents I_{12}, I_{23} and I_{31} in Figure 2.9 are a balanced set when $Z_d = |Z_d|\angle\theta$.

2.21. If V_{12}, V_{23} and V_{31} in Figure 2.9 are a balanced negative-sequence set with V_{12} as the reference, show that I_a will lead I_{12} by 30°.

CHAPTER 3

COMPONENTS OF ELECTRIC POWER SYSTEMS

Earlier, we discussed energy resources and the fundamentals of electric power systems. We also mentioned that very seldom does a power generating station operate as a separate entity. In fact, power generating stations all over the country are interconnected. In later chapters, we shall study certain aspects of electric power systems. Before we consider the power system as a whole, it is worthwhile to familiarize ourselves with a brief historical development of electric central stations and review qualitatively some of the pertinent components that constitute an electric power system.

3.1 HISTORY OF CENTRAL-STATION ELECTRIC SERVICE

The concept of *central-station* electric power service was first applied at Thomas Edison's Pearl Street Station in New York City in 1882, when the first distribution line was strung along a few city blocks to provide lights in a few homes. The electric power industry in the United States has grown phenomenally from the novelty of one circuit with a few light bulbs to providing the main driving force of the greatest industrial nation on earth. All of this has taken place in the equivalent of one human life span. The high standard of living enjoyed in this country is closely tied to electric laborsaving appliances and tools, and to the high productivity made possible by the electric machinery of industry.

Central-station service, as opposed to individual generators in each home, possessed all the inherent advantages to make it a huge success. Economics of scale, convenience, and relative continuity combined to promote the growth of Edison's idea at a rapid rate. The first limiting factors encountered were voltage drop and resistance losses on the low-voltage *direct-current* (dc) distribution circuits. Distance of the customer from the generating station was severely limited as the problem of voltage regulation became more pronounced out toward the end of the line. The solution to these problems came with the introduction to this country of the *alternating-current* (ac) *transformer* by George Westinghouse in 1885. An ac distribution circuit was placed in service at Great Barrington, Massachusetts, by William Stanley in 1886, proving the feasibility of the technique. Power could be generated at low voltage levels by the simpler ac generator and stepped up to higher voltage for sending over long distances. Because the current required to deliver power at a given voltage is inversely proportional to the voltage, the current requirements and consequently the conductor size could be kept within practical limits and still deliver large amounts of power to distant areas. The first ac transmission line was put in service at Portland, Oregon, in 1890, carrying power 13 miles from a *hydro* generating station on the Willamette River.

During the next decade, *two-phase* and *three-phase* motors and generators were developed and were demonstrated to be superior to *single-phase* machines from the standpoint of size, weight, and efficiency. Three-phase transmission was shown to possess inherent advantages of conductor requirements and losses for a given power need. Consequently, by the turn of the century, it was apparent that three-phase ac transmission systems would become standard. Three frequencies, 25, 50, and 60 Hz, battled for dominance, and it was several decades before the conversion to 60 Hz was complete in the United States.

3.2 ELECTRIC POWER GENERATORS

In order to visualize and discuss an entire functioning power system, it is necessary that we establish the necessary basic components of such a system. The first and most obvious is the three-phase ac synchronous generator or *alternator*. The alternator must be driven mechanically by some sort of *prime mover*. The early prime movers were primarily reciprocating engines and waterwheels. The simplest form of prime mover was the hydro station with a simple waterwheel. Once the original installation was made at a waterfall or dam, the fuel was free forever. For this reason, hydro stations are seldom retired from service. Thousands of tiny hydro stations are still in use today, often unattended and operated by remote control. As the more readily available sources of water power were developed, the emphasis shifted to the *steam*

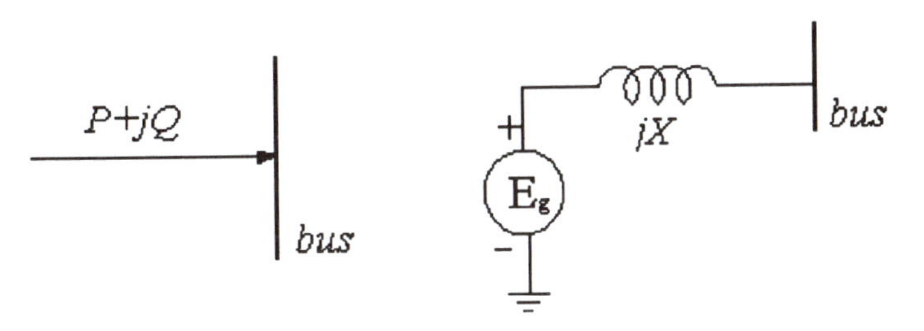

FIGURE 3.1
Synchronous alternator models.

turbine. Fired by fossil fuels such as coal, gas, and oil, steam turbines grew in numbers and unit size until they dominated the power generation industry. Even when nuclear fuels began to command a sizable portion of the market, the nuclear reactor and its complex heat exchanger ultimately served only to make steam for driving a conventional steam turbine. The turbines driven by steam from nuclear reactors must be larger than fossil-fired units if they are to be economically feasible because of the huge fixed costs, which are independent of capability. Direct conversion of energy into its electrical form without the rotating prime mover and alternator offer great promise for the future, but is not yet economically competitive in commercial quantities.

During the course of the discussions in this text, we will be using two entirely different models for the synchronous alternator. The first of these, which will be introduced during our study of power-flow analysis in Chapter 4, is to represent the alternator as a source of real power (P) and reactive power (Q). The second model will be used for fault calculations and system stability analysis in Chapters 5 and 6. It consists of a voltage source connected in series with an impedance (usually a purely inductive reactance). Each of these models for a synchronous generator connected to a power system bus (node) is shown in Figure 3.1. A more detailed discussion of these models may be found in Appendix III.

3.3 TRANSFORMERS

Practical design problems limit the voltage level at the terminals of the alternator to a relatively low value. The transformer is used to *step* the voltage up (as the current is proportionately reduced) to a much higher level so that power can be transmitted up to hundreds of miles while conductor size and losses are kept down within practical limits. Physically, a *power transformer* bank may consist of three single-phase transformers with appropriate electrical connections external to the cases, or a three-phase unit contained in a single tank. The latter is predominant in the larger power ratings for economic reasons. The windings usually are immersed in a special-purpose oil for insulation and cooling.

A detailed discussion of three-phase transformers is presented in Appendix II. Based upon the results of this discussion and the use of the per-unit system in our analyses, it will be seen that it is convenient and practical to model the three-phase transformer by a simple impedance (or admittance). In many applications, the resistive portion of this model is small and the transformer may be represented by a purely inductive reactance (or susceptance). Thus, a transformer connected between two buses (nodes) in a power system may be represented as shown in Figure 3.2 where the impedance (or admittance) is in per unit as will be discussed in Section 3.13.

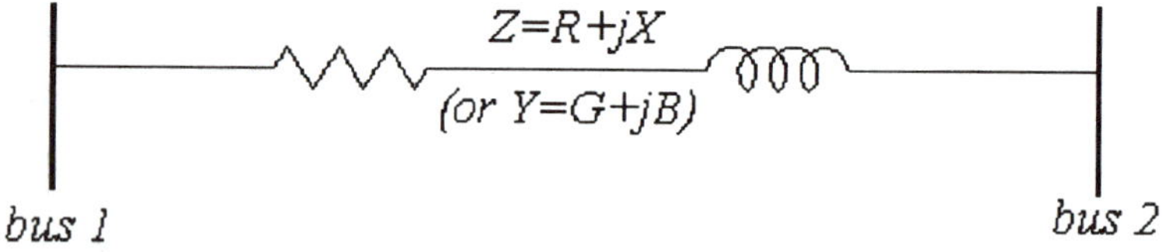

FIGURE 3.2
Transformer model.

FIGURE 3.3
A 345-kV transmission line, insulators, and supporting structure.

3.4 TRANSMISSION LINES

The *transmission line* (Figure 3.3) usually consists of three *conductors* either as three single wires (or as bundles of wires) and one or more neutral conductors, although it is possible sometimes to omit the neutral conductor since it carries only the unbalanced return portion of the line current. A three-phase circuit with perfectly balanced phase currents has no neutral return current. In most instances, the current is accurately balanced among the phases at transmission voltages, so that the neutral conductor may be much smaller. In some locations, the soil conditions permit an effective neutral return current through the earth. The neutral conductors have another equally important function: they are installed above the phase conductors and provide an effective electrostatic shield against lightning. Manufacturers of high-voltage equipment tend to standardize as much as possible on a few *nominal voltage classes*. The most common transmission line-to-line voltage classes in use within the United States are 115, 138, 230, 345, 500, and 765 kV. Developmental work is being done for utilizing voltages up to 2000 kV. Costs of line construction, switchgear, and transformers rise exponentially with voltage, leading to the use of the lowest voltage class capable of carrying the anticipated load over the required distance. However, the cost of energy loss is inversely proportional to the square of transmission voltage. The construction cost and the cost of energy loss are depicted in Figure 3.4. The point of intersection of these two curves gives the theoretical optimum transmission voltage for a total minimum cost.

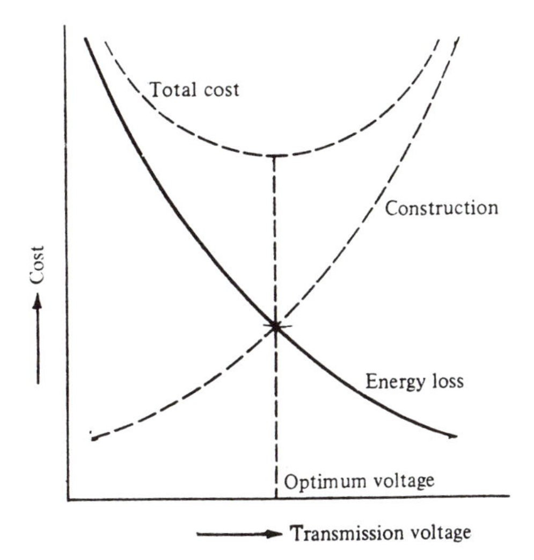

FIGURE 3.4
Determination of approximate optimum transmission voltage.

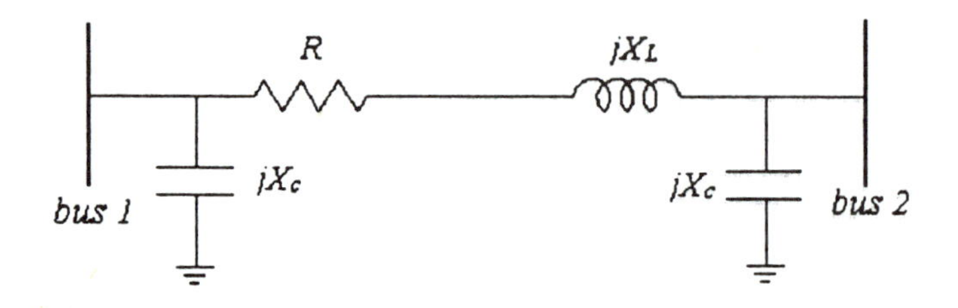

FIGURE 3.5
π-model of a transmission line.

The processes of determining parameters which may be used to represent transmission lines require some discussion and these topics are treated in Appendix IV. For our present purposes, however, it is sufficient to indicate that these lines may be represented by a π (or equivalent T) circuit model as shown in Figure 3.5 for a transmission line connected between two buses in a power system. The shunt branches of this model will normally represent capacitive reactance (or susceptance) between the actual lines and ground, whereas the series path is normally the resistance (or conductance) and inductive reactance (or susceptance) of the transmission line. In many cases, the shunt branches and/or the series resistance in this model are neglected.

3.5 CIRCUIT BREAKERS AND DISCONNECT SWITCHES

Circuit breakers are large three-pole switches located at each end of every transmission-line section, and on either side of large transformers. The primary function of a circuit breaker is to open under the control of automatic *protective relays* in the event of a fault or short circuit in the protected equipment. The relays indicate the severity and probable location of the fault, and may contain sufficient electromechanical or solid-state logic circuitry to decide whether the line or transformer could be re-energized safely, and initiate reclosure of its circuit breakers. In very critical, extra-high-voltage (EHV)* switchyards or substations, a small digital computer may be used to analyze fault conditions and perform logical control functions. If the fault is a transient one from which the

* Commonly accepted high voltage levels are: high voltage (HV), 115 to 230 kV; extra high voltage (EHV), 345 to 765 kV (up to 1000 kV); ultra high voltage (UHV), 1000 kV and up.

FIGURE 3.6
Power circuit breakers.

FIGURE 3.7
A disconnect switch.

system may recover, such as a lightning stroke, the integrity of the network is best served by restoring equipment to service automatically, preferably within a few cycles. If the fault is persistent, such as a conductor on the ground, the relays and circuit breakers will isolate the faulted section and allow the remainder of the system to continue in normal operation. The secondary function of a circuit breaker is that of a switch to be operated manually by a local or remote operator to de-energize an element of the network for maintenance. When a circuit breaker's contacts open under load, there is a strong tendency to arc across the contact gap as it separates. Various methods are used to suppress the arc, including submersion of the contact mechanism in oil or gas such as sulfur hexafluoride (SF_6). The higher voltage classes use a powerful air blast to quench the arc, using several interrupters or contact sets in series for each phase (Figure 3.6).

One side of an open-circuit breaker generally remains energizcd. To completely isolate (or de-energize) a circuit breaker, a disconnect switch is placed in series with the circuit breaker, as shown in Figs. 3.6 and 3.7.

Representation of circuit breakers, fuses and switches in a power system for a power system analysis is a simple process. If the breaker or switch is closed, it is assumed to be a short circuit with zero impedance. If the circuit breaker or switch is open, it is represented by an open circuit or infinite impedance (or zero admittance).

3.6 VOLTAGE REGULATORS

When electric power has been transmitted into the area where it is to be used, it is necessary to transform it back down to a distribution level voltage which can be utilized locally. The step-down transformer bank may be very similar to the step-up bank at the generating station, but of a size to fill the needs only of the immediate area. To provide a constant voltage to the customer, a *voltage regulator* is usually connected to the output side of the step-down transformer. It is a special type of 1:1 transformer with several discrete taps of a fractional percent each over a voltage range of ±10%. A voltage-sensing device and automatic control circuit will position the tap contacts automatically to compensate the low-side voltage for variations in transmission voltage. In many cases, the same effect is accomplished by incorporating the regulator and its control circuitry into the step-down transformer, resulting in a combination device called a *load tap changer* (LTC), and the process is known as tap changing under load (TCUL).

Regulating transformers that are designed to achieve small changes in voltage magnitude may be represented in a power system analysis by a π-equivalent circuit similar to that shown in Figure 3.5. In this case, however, the shunt and series branches may no longer normally be assigned to capacitive and inductive properties, respectively. The need for shunt branches in the model for a regulating transformer arises because the per-unit system (to be discussed later) will not normally allow the regulating transformer to be represented by a single series impedance. Regulating transformers also exist that are designed to modify the phase of the voltage instead of its magnitude. Representation of these devices will normally require the inclusion of a transformer turns ratio in the model. (See Appendix II.)

3.7 SUBTRANSMISSION

Some systems have certain intermediate voltage classes which they consider *subtransmission.* Probably, at the time it was installed, it was considered transmission, but with rapid system growth and a subsequent overlay of higher voltage transmission circuits, the earlier lines were tapped at intervals to serve more load centers and become local feeders. In most systems, 23, 34.5, and 69 kV are considered of subtransmission type, and on some larger systems, 138 kV may also be included in that category, depending on the application. As the frontiers of higher voltages are pushed back inexorably, succeeding higher voltage classes may be relegated to subtransmission service. The π model of Figure 3.5 may also be used to represent subtransmission lines.

3.8 DISTRIBUTION SYSTEMS

A low-voltage *distribution system* is necessary for the practical distribution of power to numerous customers in a local area. A distributing system resembles a transmission system in miniature, having lines, circuit breakers, and transformers, but at lower voltage and power levels. Electrical theory and analytical methods are identical for both, since the distinction is purely arbitrary. Distribution voltages range from 2.3 to 35 kV, with 12.5 and 14.14 kV predominant. Such voltage levels are sometimes referred to as *primary* voltages which are then stepped down to the 240/120 V at which most customers are served. Single-phase distribution circuits are supplied through transformers, balancing the total load on each phase as nearly as possible. Three-phase distribution circuits are erected only to serve large industrial or motor loads. The ultimate transformer which steps voltage down from distribution to customer service level may be mounted on a pole for overhead distribution systems or on a pad or in a vault for underground distribution. Such transformers usually are protected by *fuses* or *fused cutouts.* The π model of Figure 3.5 may also be used to represent distribution lines.

3.9 LOADS

Countless volumes have been written about the systems and techniques necessary for the production and delivery of electric power, but very little has been recorded about *loads*, for which all the other components exist. Perhaps the main reason is that loads are so varied in nature as to defy comprehensive classification. In the simplest concept, any device that utilizes electric power can be said to impose a load on the system. Viewed from the source, all loads can be classified as resistive, inductive, capacitive, or some combination of them. Loads may also be time variant, from a slow random swing to rapid cyclic pulses which cause distracting flicker in the lights of customers nearby. The composite load on a system has a predominant resistive component and a small net inductive component. Inductive loads such as induction motors are far more prevalent than capacitive loads. Consequently, to keep the resultant current as small as possible, *capacitors* are usually installed in quantities adequate to balance most of the inductive current. It has been shown that the power consumed by the composite load on a power system varies with system frequency. This effect is imperceptible to the customer in the range of normal operating frequencies (±0.02 Hz), but can make an important contribution to the control of systems operating in synchronism. System load also varies through daily and annual cycles, creating difficult operating problems. A typical load curve is shown in Figure 3.8.

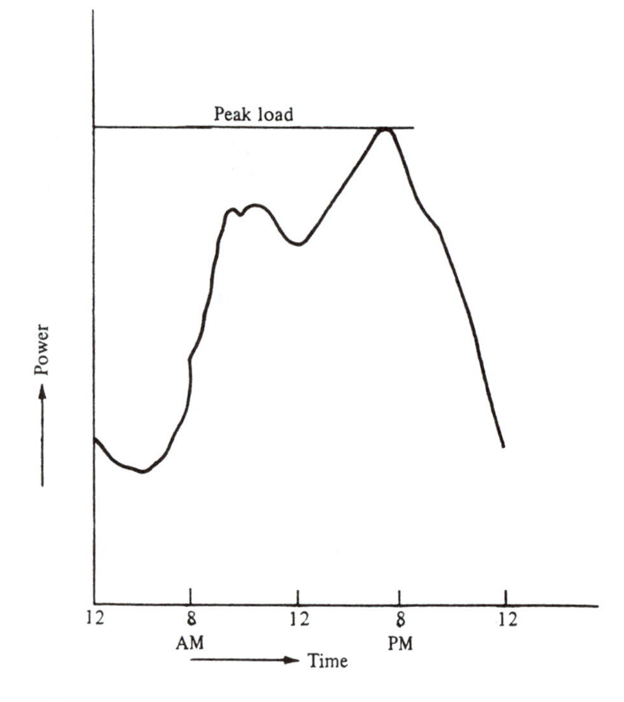

FIGURE 3.8
A typical load curve on a power system.

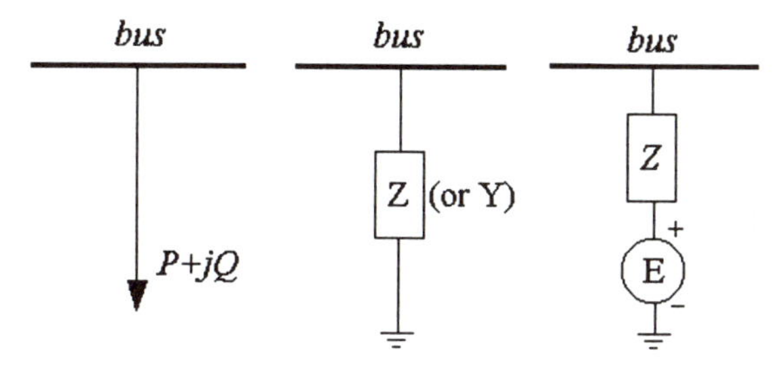

FIGURE 3.9
Power system load models.

Power system loads may be represented as a real power (P) and reactive power (Q) taken from the system (as in power-flow analyses), as an impedance between a system bus and ground, or as a voltage source in series with an impedance in the case of rotating machinery loading. This latter representation is important in cases where a rotating machine will contribute to the system currents during the initial stages of a system fault. These models are shown for a load connected to a power system bus in Figure 3.9.

3.10 CAPACITORS

When applied on a power system for the reduction of inductive current (*power factor correction*), capacitors can be grouped into either transmission or distribution classes. In either case, they should be installed electrically as near to the load as possible for maximum effectiveness. When applied properly, capacitors balance out most of the inductive component of current to the load, leaving essentially a unity power factor load. The result is a reduction in size of the conductor required to serve a given load and a reduction in I^2R losses.

The process of power factor correction may easily be visualized by referring to Figure 3.10. With the capacitor absent from the network, the voltage source supplies current through the resistance R and reactive power is supplied

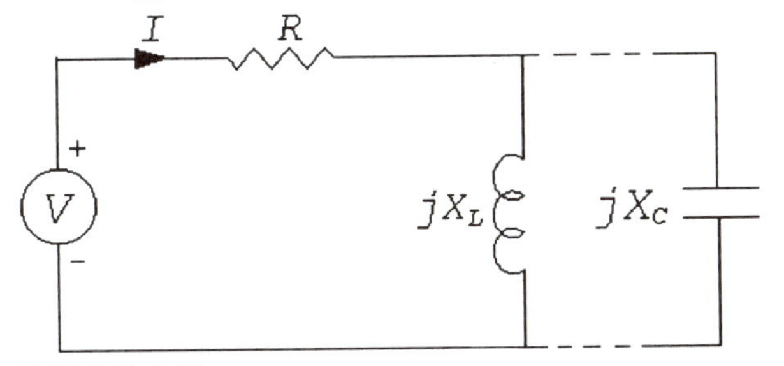

FIGURE 3.10
Power factor correction network.

to the inductive reactance X_L. Suppose now that a capacitor is added to the network with $jX_c = -j/(\omega C) = jX_L = j\omega L$. The impedance of this parallel combination is now infinite and the current which will be drawn from the voltage source is reduced to zero. However, the voltage V appears across both X_L and X_C such that there is a circulating current in this parallel combination. In effect, the power factor has been perfectly corrected and all the reactive power that is being consumed in the inductance is being produced by the capacitor.

Static capacitors may be used at any voltage, but practical considerations impose an upper limit of a few kilovolts per capacitor. Therefore, high-voltage banks must be composed of many capacitors connected in series and parallel. High-capacity transmission capacitor banks should be protected by a high-side circuit breaker and its associated protective relays. Small distribution capacitors may be vault- or pole-top-mounted and protected by fuses.

Industrial loads occasionally require very large amounts of power factor correction, varying with time and the industrial process cycle. The *synchronous condenser* is ideally suited to such an application. Its contribution of either capacitive or inductive current can be controlled very rapidly over a wide range, using automatic controls to vary the excitation current. Physically, it is very similar to a synchronous generator operating at a leading power factor, except that it has no prime mover. The synchronous condenser is started as a motor and has its losses supplied by the system to which it supplies reactive power.

3.11 OVERALL REPRESENTATION OF A POWER SYSTEM

In the preceding sections, we have mentioned the various basic components of an electric power system. In this section we consider the representation of these components interconnected to constitute a power system. First, we review the graphical representation and one-line diagram of a power system. This is followed by the impedance diagrams obtained from the most commonly used equivalent circuits of the components. Finally, because the components have different voltage and kilovoltampere (kVA) ratings, we introduce per-unit quantities as a common basis for analyzing the interconnected components and systems.

GRAPHICAL REPRESENTATION OF COMPONENTS. Figure 3.11 shows the symbols used to represent the typical components of a power system.

ONE-LINE DIAGRAMS. Using the symbols of Figure 3.11, a system consisting of two generating stations interconnected by a transmission line is shown in Figure 3.12, as a one-line diagram. From even such an elementary network as this, it is easy to imagine the confusion that would result in making diagrams showing all three phases. The ratings of all the generators, transformers, and loads are specified and the voltage levels at the buses are assumed to be known.

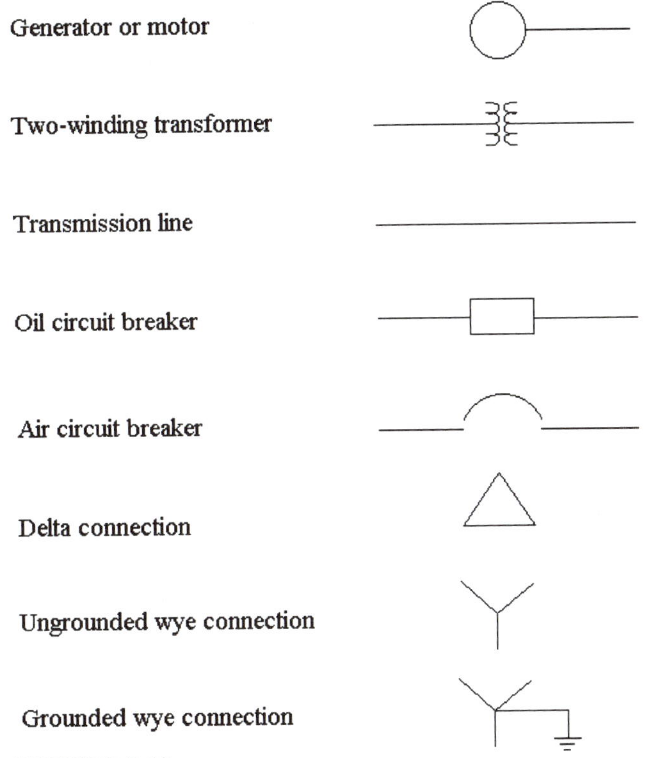

FIGURE 3.11
Symbolic representation of elements of a power system.

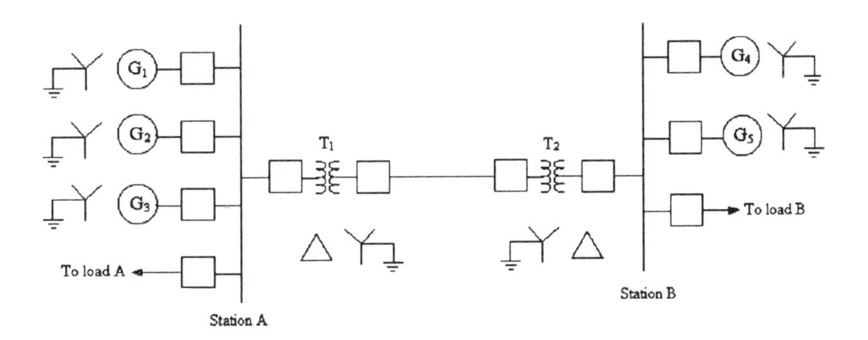

FIGURE 3.12
One-line diagram of a portion of a power system formed by interconnecting two stations.

The advantage of such a one-line representation is rather obvious in that a complicated system can be represented simply. A concerted effort is made to keep the current magnitudes equal in each phase. Consequently, on a balanced system, one phase can represent all three by proper mathematical treatment. From the one-line diagram, the impedance or reactance diagrams can be conveniently developed as shown in the following section. The one-line diagram rather becomes second nature to power system engineers as they attempt to visualize a widespread and complex network.

3.12 EQUIVALENT CIRCUITS AND REACTANCE DIAGRAMS

We note from Figure 3.12 that the power system consists of several major components: generators, transformers, transmission lines, and loads. Equivalent circuits of these components may then be interconnected to obtain a circuit representation for the entire system. In other words, the one-line diagram may be replaced by an *impedance diagram* or a *reactance diagram* (if resistances are neglected). Thus, corresponding to Figure 3.12, the impedance and reactance diagrams are shown in Figure 3.13*(a)* and *(b)*, respectively, on a per phase basis. In the equivalent circuits of the components in Figure 3.13*(a)*, we have made the following assumptions.

1. Rotating machines are represented by a voltage source in series with an inductive reactance. The internal resistance is negligible compared to the reactance.

2. The loads are lagging power factor (resistance and inductance) and are represented as an impedance.

3. The transformer core is assumed ideal, and the transformer can be represented by a reactance connected in series with the winding resistances.

4. The transmission line is of medium length and can be denoted by a T circuit. An alternate representation, such as by a π circuit is equally applicable. (See Appendix IV.)

5. The delta/wye-connected transformer, T_1, is replaced by an equivalent wye/wye-connected transformer (by a delta-to-wye transformation) so that the impedance diagram may be drawn on a per phase basis. The exact nature and values of the impedances (or reactances) are determined in later chapters. Similar comments apply to the transformer T_2.

The reactance diagram, Figure 3.13*(b)*, is drawn by neglecting all the resistances. In this reactance diagram, impedances to ground (the passive loads and transmission line capacitance) have also been neglected. This type of system

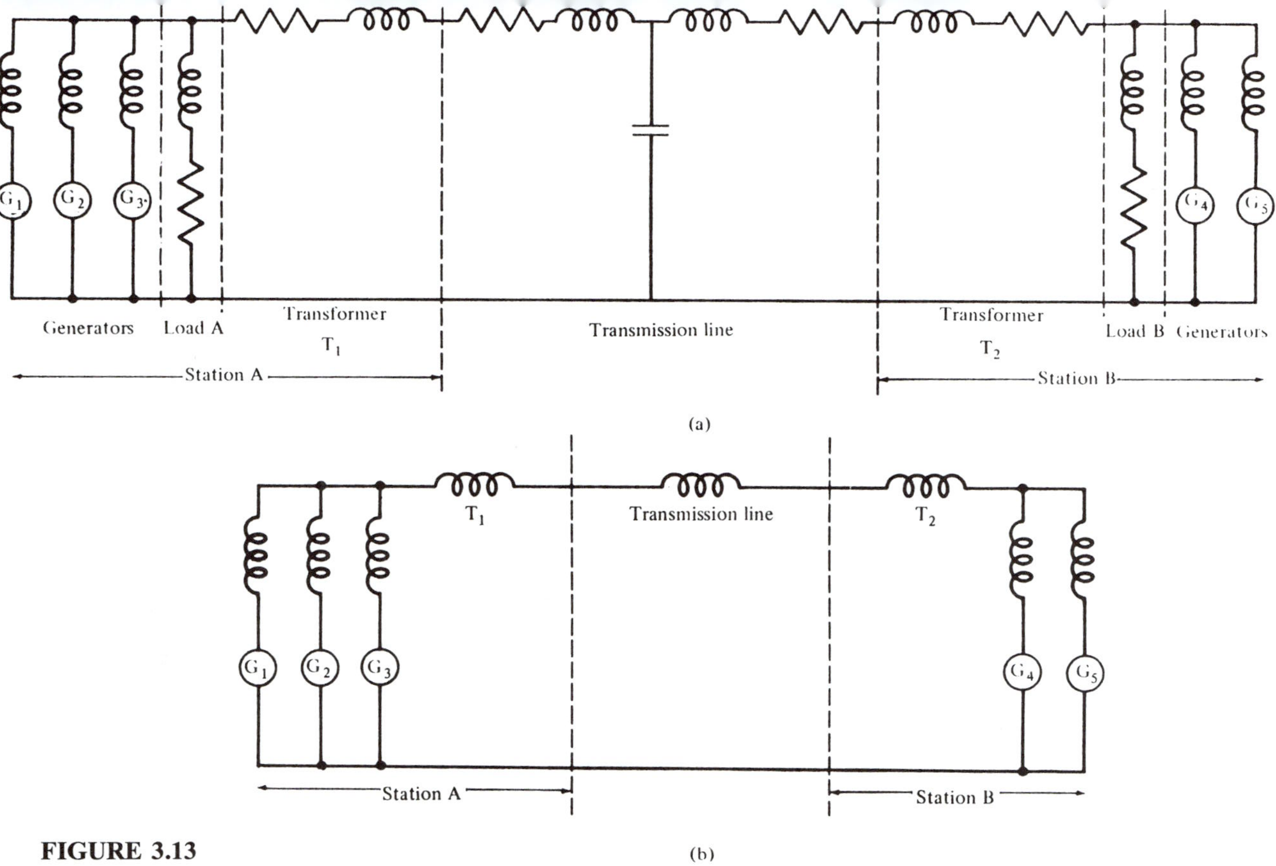

FIGURE 3.13
(*a*) Impedance diagram for Fig. 3.12(*b*); (*b*) corresponding reactance diagram, neglecting shunt connections.

representation is common when performing fault analyses as will be seen in Chapter 5.

3.13 PER-UNIT REPRESENTATION

When making computations on a power system network having two or more voltage levels, it is very cumbersome to convert currents to a different voltage level at each point where they flow through a transformer, the change in current being inversely proportional to the transformer turns ratio. A much simplified system has been devised whereby a set of *base quantities* are assumed for each voltage level, and each parameter is expressed as a fraction of its respective base. For convenience, base quantities are chosen such that they correspond rather closely to the range of values normally expected in the parameter. For instance, if a voltage base of 345 kV is chosen and under certain operating conditions the actual system voltage is 334 kV, the ratio of actual to base voltage is 0.97. This is expressed as 0.97 *per-unit* volts or simply 0.97 pu as all per-unit quantities are dimensionless. An equally common practice is to multiply each result by 100, under which the preceding would be expressed as 97 *percent.* With experience, this technique can give an excellent grasp of or feeling for the system conditions, whether normal or abnormal. It works equally well for single-phase or polyphase circuits and can even eliminate the need to consider whether a three-phase component is connected in wye or delta. Per-unit or percent quantities and their bases must obey the same relationships (such as Ohm's law and Kirchhoff's laws, etc.) to each other as with other systems of units. A minimum of four base quantities are required to define completely a per-unit system: voltage, current, power, and impedance (or admittance). If two of these are chosen arbitrarily, for example, voltage and power, the others are fixed. Therefore, on a per-phase basis the following relationships hold:

$$\text{base current} = \frac{\text{base voltamperes}}{\text{base voltage}} \quad \text{amperes} \tag{3.1}$$

$$\text{base impedance} = \frac{\text{base voltage}}{\text{base current}} \quad \text{ohms} \tag{3.2}$$

$$\text{per unit voltage} = \frac{\text{actual voltage}}{\text{base voltage}} \quad \text{pu} \tag{3.3}$$

$$\text{per unit current} = \frac{\text{actual current}}{\text{base current}} \quad \text{pu} \tag{3.4}$$

$$\text{per unit impedance} = \frac{\text{actual impedance}}{\text{base impedance}} \quad \text{pu} \qquad (3.5)$$

The preceding equations are now applied to the following simple example.

Example 3.1 A 0.8 lagging power factor load drawing 10 kVA at 1800 V is connected to a generator by a transmission line having an impedance of $(12 + j20)$ ohms. Assume 10 kVA and 2000 V as base quantities, and calculate the per-unit voltage at the load, per-unit current, and per-unit impedance of the line.

Solution

The following table identifies the per-unit (pu) values and the base quantities:

Quantity	Per-unit value	Unit value or base Quantity
Voltamperes	1.0	10 kVA (assumed)
Voltage	0.9	2000 V (assumed)
Current	1.11	10,000/2000 = 5A
Impedance	$0.03 + j0.05$	2000/5 = 400 Ω

In a three-phase system, the base kVA may either be chosen as the three-phase kVA and the base voltage as line-to-line voltage, or the base values may be taken as the per phase quantities. In either case, the per-unit three-phase kVA and voltage on the three-phase kVA base and the per-unit per phase kVA and voltage on the kVA per phase base are the same. This point is illustrated by the next example,

Example 3.2 Consider a three-phase wye-connected 50,000-kVA 120-kV system. Express 40,000-kVA three-phase apparent power and 100 kV line-to-line voltage in per-unit values on *(a)* the three-phase base and *(b)* the per phase base.

Solution

(a) Three-phase basis:

base kVA = 50,000 kVA (three phase)

base kV = 120 kV (line-to-line)

$$\text{pu kVA} = \frac{40{,}000}{50{,}000} = 0.8 \text{ pu}$$

$$\text{pu kV} = \frac{100}{120} = 0.83 \text{ pu}$$

(b) Per phase basis:

$$\text{base kVA} = \frac{1}{3} \times 50{,}000 = 16{,}667 \text{ kVA}$$

$$\text{base kV} = \frac{120}{\sqrt{3}} = 69.28 \text{ kV (for a wye connection)}$$

$$\text{actual kVA/phase} = 40{,}000/3 = 13{,}333 \text{ kVA}$$

$$\text{pu kVA} = \frac{13{,}333}{16{,}667} = 0.8 \text{ pu as before}$$

$$\text{actual kV/phase} = \frac{100}{\sqrt{3}} = 57.73 \text{ kV (for a wye connection)}$$

$$\text{pu kV} = \frac{57.73}{69.28} = 0.83 \text{ pu as before}$$

We stated earlier that a power system consists of generators, transformers, transmission lines, and loads. The per-unit impedances of generators and transformers, as supplied from tests by manufacturers, are generally based on their own ratings. However, these per-unit values could be referred to a new base according to the following equation:

$$(\text{pu impedance})_{\text{new base}} = \frac{[(\text{VA})_{\text{new base}}][(\text{V})^2_{\text{old base}}]}{[(\text{VA})_{\text{old base}}][(\text{V})^2_{\text{new base}}]} (\text{pu impedance})_{\text{old base}} \tag{3.6}$$

If the old base and the new base voltages are the same, (3.6) simplifies to

$$(\text{pu impedance})_{\text{new base}} = \frac{\text{new base voltamperes}}{\text{old base voltamperes}} (\text{pu impedance})_{\text{old base}} \tag{3.7}$$

The impedances of transmission lines are normally expressed in ohms, which can be easily converted to the pu value using the appropriate impedance base.

Example 3.3 A three-phase wye-connected 6.25-kVA 220-V synchronous generator has a reactance of 8.4 Ω per phase. Choose the rated kVA and voltage as base values, and determine the per-unit reactance. Convert this per-unit value to a 230-V 7.5-kVA base.

Solution

$$\text{base VA} = 6250 \text{ VA (three phase)}$$

$$\text{base V} = 220 \text{ V (line-to-line)}$$

In a three-phase system, the total volt-amperes is $\sqrt{3}$ × line voltage × line (wye) current so that

$$\text{base A} = \frac{6250}{\sqrt{3} \times 220} = 16.4 \text{ A}$$

$$\text{base V (wye)} = \frac{220}{\sqrt{3}} = 127.0 \text{ V}$$

$$\text{base reactance (wye)} = \frac{127.0}{16.4} = 7.74\ \Omega$$

$$\text{pu reactance} = \frac{8.4}{7.74} = 1.09 \text{ pu}$$

On the 230 V 7.5 kVA base, from (3.6) we obtain

$$\text{pu reactance} = \left(\frac{7500}{6250}\right)\left(\frac{220}{230}\right)^2 1.09 = 1.2 \text{ pu}$$

It should be noted that the base current and base impedance in this problem could have been calculated using the following general formulas. Referring to the original base values

$$\text{base A} = \frac{\text{base volt amperes (three phase)}}{\sqrt{3} \times \text{base line voltage}} \tag{3.8}$$

and

$$= \frac{6250}{\sqrt{3}\ (220)} = 16.4 \text{ A}$$

$$\text{base impedance} = \frac{(\text{base line voltage})^2}{\text{base volt amperes (three phase)}} \tag{3.9}$$

$$= \frac{(220)^2}{6250} = 7.74\ \Omega$$

Example 3.4 A three-phase 13-kV transmission line delivers 8 MVA of load at rated current. The per phase impedance of the line is $(0.01 + j0.05)$ pu on a 13-kV 8-MVA base. How much voltage will be dropped across the line?

$$\text{base kVA} = 8000 \text{ kVA}$$

$$\text{base kV} = 13 \text{ kV}$$

$$\text{base A} = \frac{8000}{\sqrt{3}\,13} = 355.3 \text{ A}$$

$$\text{base Z} = \frac{13{,}000^2}{8{,}000{,}000} = 21.125 \ \Omega$$

$$\text{impedance} = 21.125\,(0.01 + j0.05)$$

$$= 0.211 + j1.06 \ \Omega$$

$$\text{voltage drop} = 355.3\,(0.211 + j1.06)$$

$$= 74.97 + j376.62 \text{ or } 384 \text{ V}$$

The per-unit system gives us a better feeling of relative magnitudes of various quantities, such as voltage, current, power, and impedance. We will have a better appreciation of the usefulness of the per-unit concept in later chapters, particularly in fault calculations. Conventionally, we choose one specific VA base value for the entire system under consideration, and the ratio of the line voltage bases on either side of a transformer is chosen to be the same as the transformer line-to-line turns ratio. Consequently, per-unit reactances (or impedances) remain unchanged whether referred to the high-voltage side or to the low-voltage side of the transformer, as illustrated by the next example.

Example 3.5 A 25 kVA 220/110 V three-phase transformer has a 0.04 Ω leakage reactance referred to the low-voltage side. Calculate the per-unit leakage impedance referred to the low-voltage and the high-voltage sides of the transformer. Use the transformer ratings as base values. Note that three-phase transformers are rated as three-phase kVA and line voltage for either wye or delta connections.

Solution

$$(\text{VA})_{\text{base}} = 25 \text{ kVA (three-phase)}$$

$$V_{\text{base LV}} = 110 \text{ V (line-to-line)}$$

$$V_{\text{base HV}} = 220 \text{ V (line-to-line)}$$

From (3.9)

$$Z_{\text{base}} = \frac{(V_{\text{base}})^2}{VA_{\text{base}}} \tag{3.10}$$

Thus,

$$Z_{\text{base LV}} = \frac{110^2}{25{,}000} = 0.484\ \Omega$$
$$Z_{\text{pu LV}} = \frac{j0.04}{0.484} = j0.0826\ \text{pu} \tag{3.11}$$

Referred to the HV-side, with the turns ratio = $\frac{220}{110}$ = 2, we have

$$Z_{\text{HV}} = a^2 Z_{\text{LV}} = 2^2 \times j0.04 = j0.16\ \Omega$$

$$Z_{\text{base HV}} = \frac{220^2}{25{,}000} = 1.936\ \Omega$$

$$Z_{\text{pu HV}} = \frac{Z_{\text{HV}}}{Z_{\text{base HV}}} = \frac{j0.16}{1.936} = j0.0826\ \text{pu} \tag{3.12}$$

Notice that (3.11) and (3.12) give the same per-unit value, as expected. Thus the turns ratio is not included in our representation of the transformer of Figure 3.2.

We finally consider the next example which involves the basic components of a power system and illustrates the effects of changes in base voltages because of the presence of transformers.

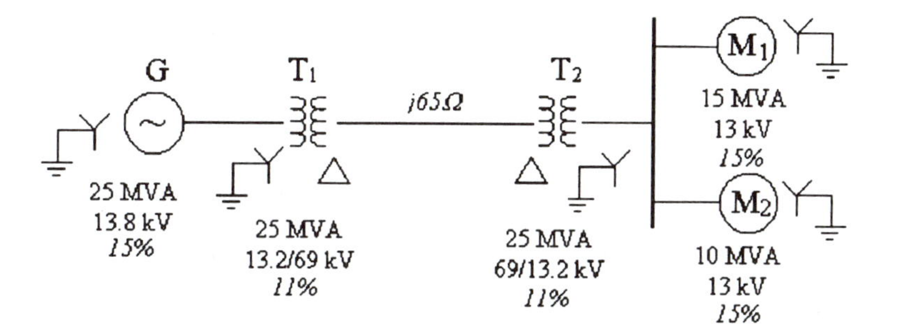

FIGURE 3.14
Single-line diagram of a power system.

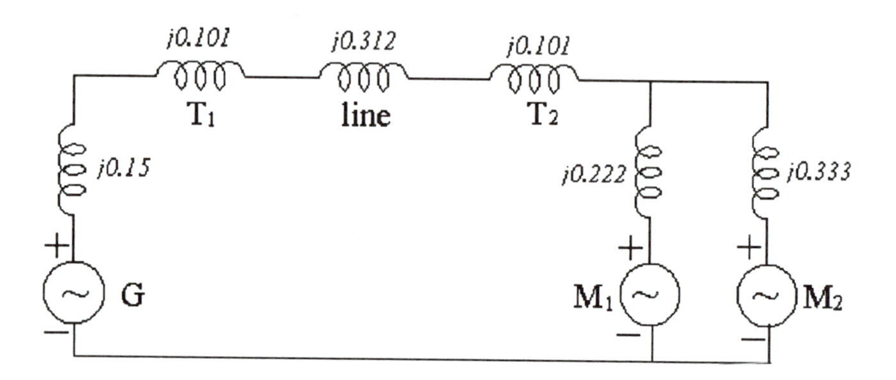

FIGURE 3.15
Reactance diagram for the system of Figure 3.14.

Example 3.6 A one-line diagram of a three-phase power system is shown in Figure 3.14. Choose 13.8 kV, the generator voltage as the base voltage and 25 MVA as the base VA and draw a reactance diagram. Label all values of reactances as per unit. Note that rotating machine and transformer impedances are given based upon their own rated three-phase volt-amperes and line voltages.

Solution

The chosen base values are 25 MVA (three-phase) and 13.8 kV (line-to-line). Thus, the generator reactance is 0.15 per unit. From (3.6), the transformer reactance becomes $(25/25)\ (13.2/13.8)^2 0.11 = 0.101$ per unit. The base voltage at the transmission line is $13.8 \times (69/13.2) = 72.136$ kV. Since the ohmic value of the line reactance is 65Ω the per-unit line reactance becomes $65(25)/(72.136)^2 = 0.312$ per unit. Finally applying (3.6) to the motor reactances, we obtain: $X_{M1} = 0.15(25/15)(13/13.8)^2 = 0.222$ pu, and $X_{M2} = 0.15(25/10)(13.0/13.8)^2 = 0.333$ pu.

Having obtained these numerical values, we draw the reactance diagram shown in Figure 3.15.

PROBLEMS

3.1. A single-phase system is rated at 300 kVA and 11 kV. Using these as base values, find the base current and base impedance.

3.2. Using 10 MVA and 345 kV as base values for a single-phase system, express 138 kV, 60 MVA, 250 A, and 50Ω in (*a*) per-unit and (*b*) percent values.

3.3. For a single-phase system, use 50 Ω and 250 A as base impedance and base current, respectively. What is the base kVA and base voltage?

3.4. The per-unit values of impedance, current and voltage of a single-phase system are 0.9, 0.3 and 0.27, respectively. The base impedance is 35 Ω and the base current is 80 A. Determine the actual values of impedance, current, voltage, and voltamperes.

3.5. A single-phase, 10-kVA, 200-V generator has an internal impedance Z_g of $j2\Omega$. Using the ratings of the generator as base values, determine the generated per-unit voltage (E_g in Figure 3.1) that is required to produce full-load current under short-circuit conditions.

3.6. Let a 5-kVA, 400/200-V single-phase transformer be approximately represented by a 2-Ω reactance referred to the low-voltage side. Considering the rated values as base quantities, express the transformer reactance as a per-unit quantity.

3.7. The per-unit impedance of a single-phase system is 0.7. The base kVA is 300 kVA and the base voltage is 11 kV. What is the ohmic value of the impedance? Will this ohmic value change if 400 kVA and 38 kV are chosen as base values? What is the per-unit impedance for the 400 kVA and 38 kV base values?

3.8. A 100-kVA 20/5-kV single-phase transformer has an equivalent impedance of 10 percent. Calculate the impedance of the transformer referred to *(a)* the 20-kV side and *(b)* the 5-kV side.

3.9. Express the per-unit admittance Y_{pu} of a single-phase power system in terms of the base voltage V_{base} and the base voltamperes $(VA)_{base}$.

3.10. A three-phase, wye-connected system is rated at 50 MVA and 120 kV. Express 40,000 kVA of three-phase apparent power as a per-unit value referred to *(a)* the three-phase system kVA as base and *(b)* the per-phase system kVA as base.

3.11. A three-phase 345-kV transmission line has a series impedance of $(4 + j60)$ Ω and a shunt admittance of $j2 \times 10^{-3}$ S as shown in Figure 3.5. Using 100 MVA and the line voltage as base values, calculate the per-unit impedance and per-unit admittance of the line.

3.12. A three-phase, wye-connected, 6.25-kVA, 220-V synchronous generator has a reactance of 8.4 Ω per phase. Using the rated kVA and voltage as base values, determine the per-unit reactance. Then refer this per-unit value to a 230-V, 7.5-kVA base.

3.13. A three-phase, 13-kV transmission line delivers 8 MVA of load. The per-phase series impedance of the line is $(0.01 + j0.05)$ pu, referred to a 13-kV, 8-MVA base. What is the voltage drop across the line?

3.14. A portion of a three-phase power system consists of two generators in parallel, connected to a step-up transformer that links them with a 230-kV transmission line. The ratings of these components are

Generator G_1: 10 MVA, 4160V 12 percent reactance

Generator G_2: 5 MVA, 4160V 8 percent reactance

Transformer: 15 MVA, 4.16:230 kV 6 percent reactance

Transmission line: $(4 + j60)$ Ω, 230 kV

where the percent reactances are computed on the basis of the individual component ratings. Express the reactances and the transmission line impedance in percent with 15 MVA as the base value.

3.15. A three-phase transmission line supplies a reactive load at a lagging power factor. The load draws 1.2 pu current at 0.6 pu voltage while taking 0.5 pu (true) power. If the base voltage is 20 kV and the base current is 160 A, calculate the power factor and the ohmic value of the resistance of the load.

3.16. A one-line diagram of a two-generator three-phase system is shown in Figure 3.16. Redraw the diagram to show all values in per unit on a 7000-kVA base.

3.17. Redraw the diagram of Figure 3.16 showing all values in ohms.

3.18. Draw an impedance diagram for the three-phase system shown in Figure 3.17, expressing all values as per-unit quantities on a 50 kVA base.

3.19. Draw a per-unit reactance diagram for the three-phase system shown in Figure 3.18. Choose a 20 MVA, 66 kV base at the transmission line.

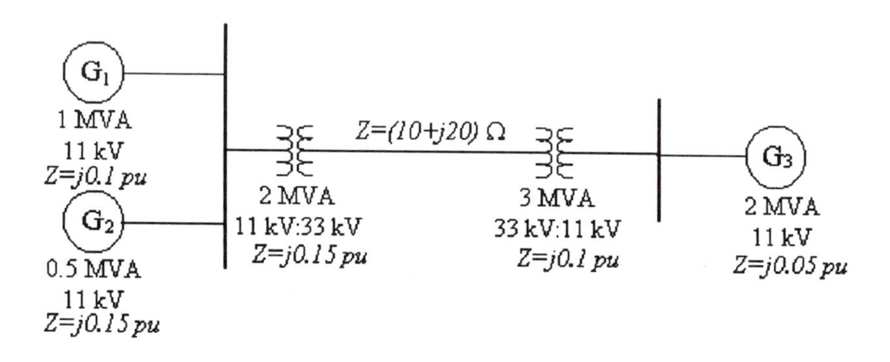

FIGURE 3.16
Problems 3.16 and 3.17.

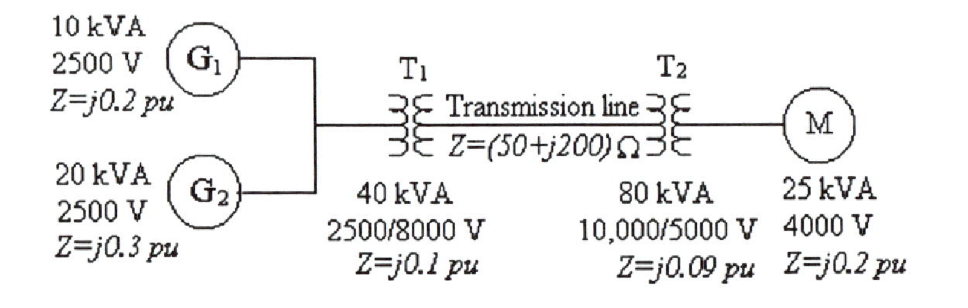

FIGURE 3.17
Problem 3.18.

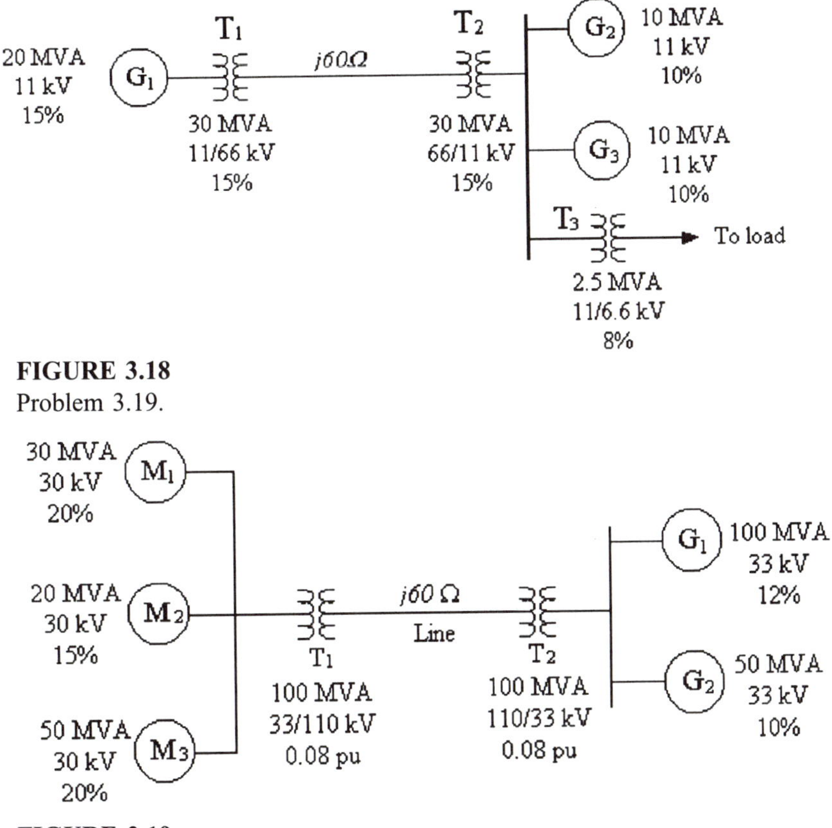

FIGURE 3.18
Problem 3.19.

FIGURE 3.19
Problem 3.20

3.20. Three-phase generators G_1 and G_2 supply motor loads M_1, M_2, and M_3, as shown in Figure 3.19. Transformers T_1 and T_2 are rated at 100 MVA and 33/110 kV, and each has a reactance of 0.08 per unit. Assuming 100 MVA and 33 kV are used as base values at the generators, obtain all the reactances as per-unit values.

3.21. Use the results of Problem 3.20 to draw a reactance diagram for the system shown in Figure 3.19.

CHAPTER 4

POWER FLOW STUDIES

Power flow studies, also known as load flow studies, are extremely important in evaluating the operation of power systems, controlling them, and planning for future expansions. For specified bus conditions, a power flow study yields mainly the real and reactive power flows and phasor voltage at each bus on the system, although much additional information is available from the computer printouts of typical power flow studies.

The principles involved in power flow studies are straightforward, but a study relating to a real power system can be carried out only with a digital computer. Then the required numerical computations are performed systematically by means of an iterative procedure, such as the Gauss-Seidel method or the Newton-Raphson method. These methods are facilitated by using the concept of the bus admittance matrix, which is briefly reviewed in this chapter. We also consider the fast decoupled power method which is a modified Newton-Raphson method especially applicable to conditions existing in a power system. Before considering these numerical methods, we illustrate the concept of power flow by obtaining explicit expressions for the power flow in a lossless short transmission line. Control of reactive power flow is discussed toward the end of this chapter. Examples are presented to illustrate computer applications to power flow studies. In all cases, we assume a balanced system to obtain a single-phase, single-line representation of the system.

4.1 POWER FLOW IN A SHORT TRANSMISSION LINE

We assume the short transmission line shown in Figure 4.1*(a)* has negligible resistance and a series impedance of jX ohms per phase. The per-phase sending-end and receiving-end voltages are V_S and V_R, respectively. We wish to determine the real and reactive power at the sending end and at the receiving end, given that V_S leads V_R by an angle δ.

Complex power S, in volt amperes, is given by

$$S = P + jQ = VI^* \quad \text{VA} \tag{4.1}$$

where I^* is the complex conjugate of I. Thus, on a per-phase basis, at the sending end we have

$$S_S = P_S + jQ_S = V_S I^* \quad \text{VA} \tag{4.2}$$

From Figure 4.1*(a)*, I is given by

$$I = \frac{1}{jX}(V_S - V_R)$$

so

$$I^* = \frac{1}{-jX}(V_S^* - V_R^*) \tag{4.3}$$

Substituting (4.3) in (4.2) yields

$$S_S = \frac{V_S}{-jX}(V_S^* - V_R^*) \tag{4.4}$$

Now, from the phasor diagram of Figure 4.1*(b)*,

$$V_R = |V_R| \angle 0° \qquad \text{so} \qquad V_R = V_R^* = |V_R|$$

and

$$V_S = |V_S| \angle \delta = |V_S| \, e^{j\delta}$$

Hence, (4.4) becomes

$$S_S = \frac{|V_S|^2 - |V_R|\,|V_S|\,e^{j\delta}}{-jX}$$

$$= \frac{|V_S|\,|V_R|}{X}\sin\delta + j\frac{1}{X}(|V_S|^2 - |V_S|\,|V_R|\cos\delta)$$

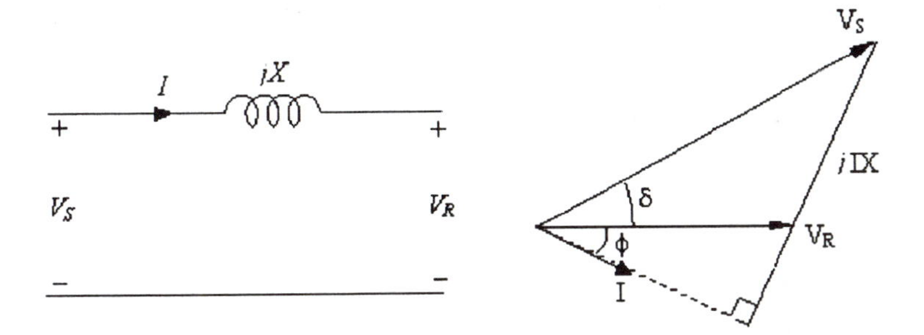

FIGURE 4.1

(a) A short lossless transmission line; (b) phasor diagram.

(The latter equation requires some manipulation.) Finally, since $S_S = P_S + jQ_S$, we may write

$$P_S = \frac{1}{X}(|V_S|\,|V_R|\sin\delta) \quad \text{W} \tag{4.5}$$

and

$$Q_S = \frac{1}{X}(|V_S|^2 - |V_S|\,|V_R|\cos\delta) \quad \text{var} \tag{4.6}$$

Similarly, at the receiving end we have

$$S_R = P_R + jQ_R = V_R I^*$$

Proceeding as above, we finally obtain

$$P_R = \frac{1}{X}(|V_S|\,|V_R|\sin\delta) \quad \text{W} \tag{4.7}$$

and

$$Q_R = \frac{1}{X}(|V_S|\,|V_R|\cos\delta - |V_R|^2) \quad \text{var} \tag{4.8}$$

From this simple example, a number of significant conclusions may be derived. First, if it is assumed that the magnitude of the sending and receiving end voltages are maintained at approximately 1.0 per unit, the transfer of real power depends primarily on the angle δ, which is known as the *power angle*. This is in contrast to a dc system where the power flow depends entirely on the relative magnitudes of the sending end and receiving end voltages. Moreover, the transmitted power varies approximately as the product of the voltage levels. The maximum power transfer occurs when δ = 90° and

$$P_{R(\max)} = P_{S(\max)} = \frac{|V_S|\,|V_R|}{X} \tag{4.9}$$

Finally, from (4.6) and (4.8), it is clear that reactive power will flow in the direction of the lower voltage. If the system operates with $\delta \approx 0$, then the average reactive power flow over the line is given by

$$Q_{\text{av}} = \frac{1}{2}(Q_S + Q_R) = \frac{1}{2X}(|V_S|^2 - |V_R|^2) \quad \text{var} \tag{4.10}$$

This equation shows the strong dependence of the reactive power flow on the voltage difference.

Up to this point we have neglected the I^2R loss in the line. If we now assume that R is the resistance of the line per phase, then the line loss is given by

$$P_{\text{line}} = |I|^2 R \quad \text{W} \tag{4.11}$$

From (4.1), we have

$$I^* = \frac{P + jQ}{V}$$

and

$$I = \frac{P - jQ}{V^*}$$

Thus,

$$II^* = |I|^2 = \frac{P^2 + Q^2}{|V|^2}$$

and (4.11) becomes

$$P_{\text{line}} = \frac{(P^2 + Q^2)\,R}{|V|^2} \quad \text{W} \tag{4.12}$$

indicating that both real and reactive power contribute to the line losses. Thus, it is important to reduce reactive power flow to reduce line losses.

Example 4.1 For the system shown in Figure 4.2 (where complex numbers denote per-unit complex power), it is desired that $|V_R| = |V_S| = 1$ pu. The loads, as shown, are $S_{L1} = 6 + j10$ pu and $S_{L2} = 14 + j8$ pu. The line impedance is $j0.05$ pu. If the real power input at each bus is 10 pu, calculate the real and reactive powers at both ends of the line.

Solution

Let

$$V_R = 1\ \angle 0^\circ \quad \text{and} \quad V_S = 1\ \angle \delta$$

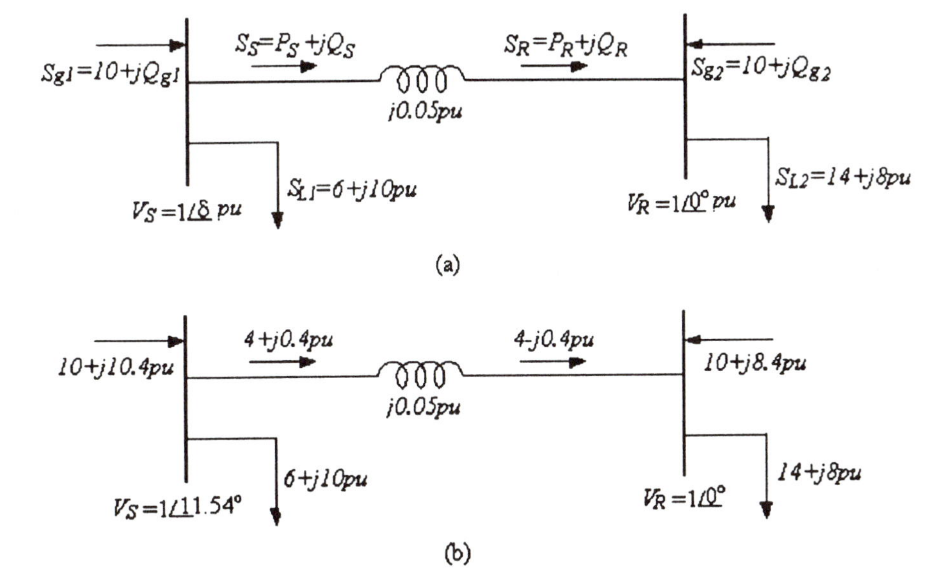

FIGURE 4.2
(*a*) Diagram for Example 4.1. (*b*) Completed power flow for Example 4.1.

By conservation of energy, $P_R = P_S = (10 - 6) = 4.0$ pu (remember that the line has no real power losses). From (4.5) or (4.7)

$$4 = \frac{(1)(1)}{0.05}\sin\delta$$

From which $\delta = 11.54°$ and $V_S = 1\ \angle 11.54°$. From (4.6) and (4.8):

$$Q_S = \frac{1}{0.05}\left((1)^2 - (1)(1)\cos 11.54°\right) = 0.4 \text{ pu}$$

$$Q_R = \frac{1}{0.05}\left((1)(1)\cos 11.54° - (1)^2\right) = -0.4 \text{ pu}$$

Therefore, $Q_{g2} = 8 + 0.4 = 8.4$ pu and $Q_{g1} = 10 + 0.4 = 10.4$ pu and the completed power flow is as shown in Figure 4.2(b). Note that the reactive power loss in the line is equal to $0.4 + 0.4 = 0.8$ pu. The student should also realize that it was necessary to generate a reactive power of 8.4 pu at the receiving end of the line in order to maintain a voltage magnitude of 1 pu at this end of the line.

4.2 AN ITERATIVE PROCEDURE

We are able to obtain an analytical expression for the power flow in our idealized case; however, in an actual power system, explicit analytical solutions are not forthcoming because of load fluctuations, system complexity, and because the receiving-end voltage may not be known. Then, numerical methods must be used to solve for unknown quantities—generally via an iterative procedure.

Figure 4.3 shows a two-bus system, with the real power represented by solid arrows and the reactive power by dashed arrows. The governing equations for the system are (on a per-phase basis)

$$S_2 = V_2 I^*$$

$$V_1 = V_2 + Z_l I$$

with the symbols as defined in Figure 4.3. Solving for V_2 and eliminating I from these equations, we obtain

$$V_2 = V_1 - Z_l I = V_1 - Z_l \frac{S_2^*}{V_2^*} \tag{4.13}$$

To solve (4.13) iteratively, assuming that V_1, Z_l and S_2 are specified, we would assume a value for V_2 and call it $V_2^{(0)}$. We would substitute this in the right-hand side of (4.13) and solve for V_2, calling the new value of V_2, obtained in this first iteration, $V_2^{(1)}$. We would then substitute $(V_2^{(1)})^*$ in the right-hand side of (4.13) and obtain a new value $V_2^{(2)}$. This procedure would be repeated until convergence, i.e., new value is equal to the substituted value within a desired precision. The iterative process we would use is thus given by the general equation, or *algorithm*,

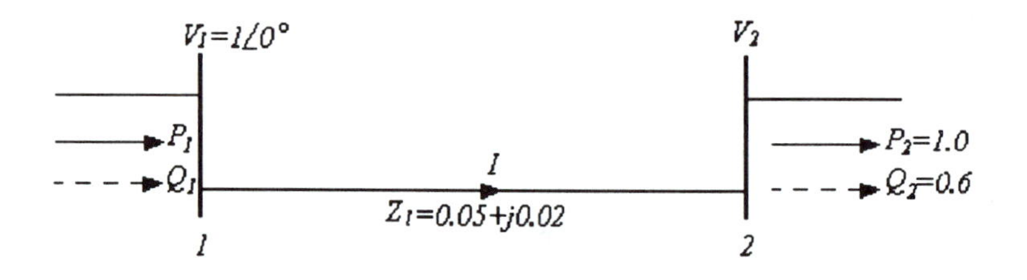

FIGURE 4.3
A two-bus system.

$$V_2^{(k)} = V_1 - \frac{Z_l S_2^*}{\left(V_2^{(k-1)}\right)^*} \tag{4.14}$$

We now illustrate the application of this approach by the following example.

Example 4.2 In Figure 4.3, let $V_1 = 1\ \angle 0°$, $Z_l = 0.05 + j0.02$, and $P_2 + jQ_2 = 1.0 + j0.6$ (all per unit). Determine V_2 and $P_1 + jQ_1$.

Solution

Based on the given numerical values, we make the initial assumption $V_2 = 1\ \angle 0°$ and use (4.14) iteratively to obtain the following values:

Iteration	V_2, pu
0	$1.0 + j0$
1	$0.938 + j0.01$
2	$0.934 + j0.01$
3	$0.933 + j0.01$
4	$0.933 + j0.01$

Now, with $V_2 = 0.933 + j0.01$ pu $\approx 0.933\ \angle 0°$ pu, we have

$$I = \frac{S_2^*}{V_2^*} = \frac{1.0 - j0.6}{0.934 - j0.01} = 1.079 - j0.632 \quad \text{pu}$$

Then

$$I^* = 1.079 + j0.632$$

and since

$$V_1 = 1\ \angle 0°$$

we have

$$P_1 + jQ_1 = S_1 = V_1 I^* = (1\ \angle 0°)\ (1.079 + j0.632)$$

$$= 1.079 + j0.632$$

In this example, convergence is achieved in just three iterations. Different data, such as a greater load, might require more iterations for convergence to the solution. Or, convergence may not be achieved at all if a solution does not exist or if the starting point of the iteration process is not appropriate.

As a further illustration of the application of this procedure, we consider the following example.

Example 4.3 For the system of Example 4.2, it is desired to have $|V_1| = |V_2| = 1.0$ pu by supplying reactive power at bus 2. Determine the reactive power that must be supplied at bus 2 to raise $|V_2|$ from 0.933 pu calculated in Example 4.2 to 1.0 pu.

Solution

From (4.1) we obtain,

$$I = \frac{S_2^* + jQ_2'}{V_2^*}$$

which, when substituted in (4.13), yields

$$V_1 = V_2 + \frac{Z_l}{V_2^*}(S_2^* + jQ_2')$$

where Q_2' represents the added reactive power at bus 2. We now substitute in the above (note that the reference angle of 0° has been changed from bus 1 to bus 2)

$$|V_1| = 1 \qquad V_2 = 1\ \angle 0° \qquad Z_l = 0.05 + j0.02 \qquad S_2^* = 1 - j0.6$$

and obtain by equating magnitudes

$$1 = |1 + (0.05 + j0.02)\ [1 + j(Q_2' - 0.6)]|$$

Hence,

$$Q_2' = 4.02 \text{ pu}$$

Therefore, injecting 4.02 pu of reactive power into bus 2 will raise the magnitude of V_2 from 0.933 pu to 1.0 pu. This example demonstrates how the injection of reactive power into a bus will raise the bus voltage. In other words, injecting 4.02 pu reactive power at bus 2 will make the voltages at the two buses equal.

4.3 BUS ADMITTANCE MATRIX

Up to this point we considered only two-bus systems. Clearly, a power system has a large number of buses. In this case, a systematic formulation of voltage-current relationships in terms of network admittances facilitates the solution of power flow problems. Before obtaining the general results, we consider a four-bus system represented by the one-line diagram of Figure 4.4*(a)*. This system may be represented by the network shown in Figure 4.4*(b)* where the *y*s are the complex admittances and the *V*s are phasor values of the node voltages measured with respect to the reference. The currents I_1, I_2, I_3 and I_4 are injected into each node by a current source. In terms of the node voltages V_1, V_2, V_3 and V_4 and the given admittances, Kirchhoff's current law yields

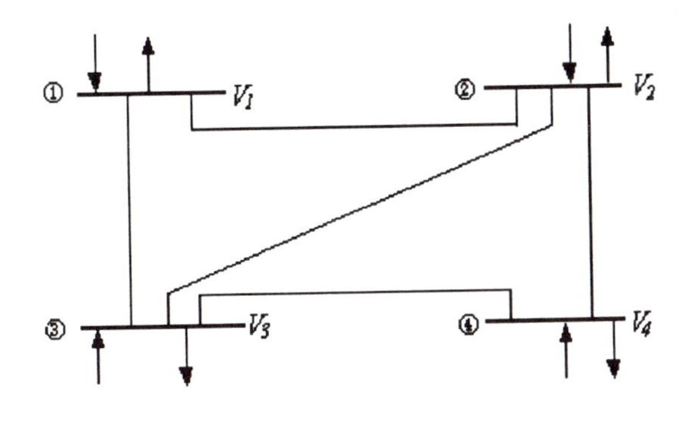

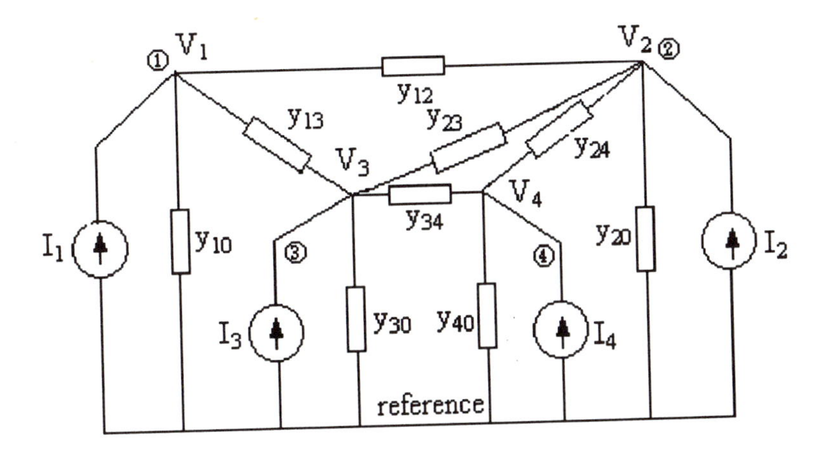

FIGURE 4.4
A four-bus system.

$$I_1 = V_1 y_{10} + (V_1 - V_2) y_{12} + (V_1 - V_3) y_{13}$$

$$I_2 = V_2 y_{20} + (V_2 - V_1) y_{12} + (V_2 - V_3) y_{23} + (V_2 - V_4) y_{24}$$

$$I_3 = V_3 y_{30} + (V_3 - V_1) y_{13} + (V_3 - V_2) y_{23} + (V_3 - V_4) y_{34}$$

$$I_4 = V_4 y_{40} + (V_4 - V_2) y_{24} + (V_4 - V_3) y_{34}$$

Rearranging these equations and rewriting them in matrix form, we obtain

$$\begin{bmatrix} I_1 \\ I_2 \\ I_3 \\ I_4 \end{bmatrix} = \begin{bmatrix} y_{10}+y_{12}+y_{13} & -y_{12} & -y_{13} & 0 \\ -y_{12} & y_{20}+y_{12}+y_{23}+y_{24} & -y_{23} & -y_{24} \\ -y_{13} & -y_{23} & y_{30}+y_{13}+y_{23}+y_{34} & -y_{34} \\ 0 & -y_{24} & -y_{34} & y_{40}+y_{24}+y_{34} \end{bmatrix} \begin{bmatrix} V_1 \\ V_2 \\ V_3 \\ V_4 \end{bmatrix} \tag{4.15}$$

Equation (4.15) may be written as

$$\begin{bmatrix} I_1 \\ I_2 \\ I_3 \\ I_4 \end{bmatrix} = \begin{bmatrix} Y_{11} & Y_{12} & Y_{13} & Y_{14} \\ Y_{21} & Y_{22} & Y_{23} & Y_{24} \\ Y_{31} & Y_{32} & Y_{33} & Y_{34} \\ Y_{41} & Y_{42} & Y_{43} & Y_{44} \end{bmatrix} \begin{bmatrix} V_1 \\ V_2 \\ V_3 \\ V_4 \end{bmatrix} \tag{4.16}$$

where

$$\begin{aligned} Y_{11} &= y_{10} + y_{12} + y_{13} \\ Y_{22} &= y_{20} + y_{12} + y_{23} + y_{24} \\ Y_{33} &= y_{30} + y_{13} + y_{23} + y_{34} \\ Y_{44} &= y_{40} + y_{24} + y_{34} \\ Y_{12} &= Y_{21} = -y_{12} \\ Y_{13} &= Y_{31} = -y_{13} \\ Y_{14} &= Y_{41} = -y_{14} = 0 \\ Y_{23} &= Y_{32} = -y_{23} \\ Y_{24} &= Y_{42} = -y_{24} \\ Y_{34} &= Y_{43} = -y_{34} \end{aligned}$$

Each admittance Y_{ii} (i = 1, 2, 3, 4) is called the *self-admittance* (or *driving-point admittance*) of node i and is equal to the algebraic sum of all the admittances terminating on the node. Each off-diagonal term Y_{ik} (i, k = 1, 2, 3, 4) is called the *mutual admittance* (or *transfer admittance*) between nodes i and k and is equal to the negative of the sum of all admittances connected directly between those nodes. Further, $Y_{ik} = Y_{ki}$.

For a general network with N nodes, therefore, Kirchhoff's current law in terms of node voltages may be written in matrix form as

$$\boldsymbol{I} = \boldsymbol{Y}_{\text{bus}} \boldsymbol{V} \tag{4.17}$$

where

$$\boldsymbol{Y}_{\text{bus}} = \begin{bmatrix} Y_{11} & Y_{12} & \cdots & Y_{1N} \\ Y_{21} & Y_{22} & \cdots & Y_{2N} \\ \cdots & \cdots & \cdots & \cdots \\ Y_{N1} & Y_{N2} & \cdots & Y_{NN} \end{bmatrix} \tag{4.18}$$

is called the *bus admittance matrix*, and $\boldsymbol{V}$ and $\boldsymbol{I}$ are the N-element *node voltage matrix* and *node current matrix*, respectively.

In (4.18), the first subscript on each Y indicates the node at which the current is applied, and the second subscript indicates the node whose voltage is responsible for a particular component of the current. Further, the admittances along the diagonal are the self-admittances, and the off-diagonal admittances are the mutual admittances. It follows from (4.17) and (4.18) that the current entering a node k is given by

$$I_k = \sum_{n=1}^{N} Y_{kn} V_n \tag{4.19}$$

Having introduced the concept of bus admittance matrix, we now return to power flow studies.

4.4 THE POWER FLOW EQUATIONS

As noted in the preceding section, the bus admittance matrix is useful in a systematic approach to the solution of power flow problems. Before discussing this approach, we need to define the following special buses:

1. A *load bus* is a bus for which the active and reactive powers P and Q (generation and/or load) are known, and $|V|$ and δ are to be found.

2. A *controlled voltage* (or *generator*) *bus* is a bus for which the magnitude of the bus voltage $|V|$ and the corresponding generated and load power P are known, and Q and δ are to be obtained.

3. A *swing bus* (or *slack bus*) is a generator bus at which $|V|$ and δ are specified, and the total P and Q are to be determined. For convenience, we normally choose $V\angle\delta = 1\angle 0°$ per unit. There is one swing bus in any power flow, and the real and reactive power flows into this bus will include the system real and reactive power losses. This approach is necessary in order to maintain the system power balance.

From (4.19), we may write the kth injected nodal current as

$$I_k = \sum_{n=1}^{N} Y_{kn} V_n \tag{4.20}$$

which may also be written as

$$I_k = Y_{kk} V_k + \sum_{\substack{n=1 \\ n \neq k}}^{N} Y_{kn} V_n \tag{4.21}$$

Solving for V_k yields

$$V_k = \frac{I_k}{Y_{kk}} - \frac{1}{Y_{kk}} \sum_{\substack{n=1 \\ n \neq k}}^{N} Y_{kn} V_n \tag{4.22}$$

Now, let $P_k + jQ_k$ be the power injected into bus k. Then

$$V_k^* I_k = P_k - jQ_k \tag{4.23}$$

and we have

$$I_k = \frac{P_k - jQ_k}{V_k^*} \tag{4.24}$$

Finally, (4.22) and (4.24) give, for N nodes,

$$V_k = \frac{1}{Y_{kk}} \left(\frac{P_k - jQ_k}{V_k^*} - \sum_{\substack{n=1 \\ n \neq k}}^{N} Y_{kn} V_n \right) \qquad \text{for } k = 2, 3, \ldots, N \tag{4.25}$$

This set of $(N - 1)$ equations constitutes the power flow equations. These equations are solved iteratively beginning with an educated guess of magnitudes and angles of load bus voltages and of voltage angles at the controlled voltage buses. Since power systems are normally operated near rated voltage, this guess is often an assumption of 1.0 pu voltage magnitude and 0° voltage angle at the system buses. Note that (4.25) does not include an equation for bus 1 ($k = 1$). This is necessary because one swing (or slack) bus must be designated and the voltage at this bus is specified.

4.5 THE GAUSS AND GAUSS-SEIDEL METHODS

The Gauss and Gauss-Seidel methods are iterative procedures for solving simultaneous (nonlinear) equations, such as (4.25). We illustrate the Gauss method with the following example.

Example 4.4 Solve for x and y in the system

$$y - 3x + 1.9 = 0$$
$$y + x^2 - 1.8 = 0$$

Solution

To solve with the Gauss method, we rewrite the given equations as

$$x = \frac{y}{3} + 0.633 \tag{4.26}$$

$$y = 1.8 - x^2 \tag{4.27}$$

We now make an initial guess of $x^{(0)} = 1$ and $y^{(0)} = 1$, update x with (4.26), and update y with (4.27). That is, we compute

$$x^{(1)} = \frac{y^{(0)}}{3} + 0.633 = \frac{1}{3} + 0.633 = 0.966 \tag{4.28}$$

and

$$y^{(1)} = 1.8 - (x^{(0)})^2 = 1.8 - 1 = 0.8 \tag{4.29}$$

In succeeding iterations we compute, more generally,

$$x^{(l+1)} = \frac{y^{(l)}}{3} + 0.633 \tag{4.30}$$

and

$$y^{(l+1)} = 1.8 - (x^{(l)})^2 \tag{4.31}$$

After several iterations, we obtain $x = 0.938$ and $y = 0.917$. A few more iterations would bring us very close to the exact results: $x = 0.93926$ and $y = 0.9178$. However, it must be pointed out that an "uneducated guess" of the initial values (such as $x^{(0)} = y^{(0)} = 100$) would have caused the solution to diverge.

If we were to use the Gauss-Seidel method in the above example, we would still use (4.30) to compute $x^{(l+1)}$, but we would then use the just-computed $x^{(l+1)}$ to find $y^{(l+1)}$. Instead of (4.30) and (4.31), the algorithm for the Gauss-Seidel method would be

$$x^{(l+1)} = \frac{y^{(l)}}{3} + 0.633$$
$$y^{(l+1)} = 1.8 - (x^{(l+1)})^2$$

Extrapolating the above results, we find that the Gauss-Seidel algorithm for the power flow equations (4.25) is

$$V_k^{(l+1)} = \frac{1}{Y_{kk}}\left[\frac{P_k - jQ_k}{(V_k^{(l)})^*} - \sum_{n=1}^{k-1} Y_{kn}V_n^{(l+1)} - \sum_{n=k+1}^{N} Y_{kn}V_n^{(l)}\right] \quad \text{for } k = 2, 3, \ldots, N \tag{4.32}$$

In formulating (4.32), we have assumed that bus 1 is the swing bus ($V_1 = |V_1| \angle\delta_1$ is specified) and hence the computations begin with bus 2. This approach is necessary in order to allow a generator at bus 1 to supply the real and reactive

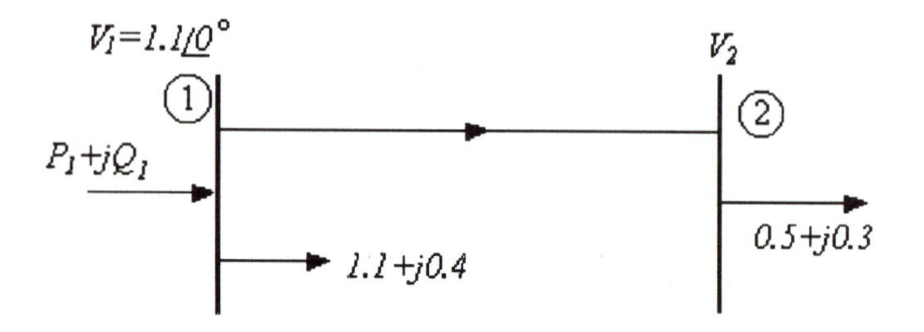

FIGURE 4.5
Example 4.5.

system losses. The superscript $l + 1$ indicates voltage values computed during the current iterations whereas l refers to voltage values from the previous iteration.

Example 4.5 For the two-bus system of Figure 4.5, with the data as shown and with $Y_{11} = Y_{22} - 1.6\ \angle{-80°}$ pu and $Y_{21} = Y_{12} = 1.6\ \angle 100°$ pu determine the per-unit voltage at bus 2 by the Gauss-Seidel method. Note that bus 1 is the swing bus and bus 2 is a load bus.

Solution

The respective power into the two buses is (remembering that power into a bus [generator power] is positive while power out of a bus [load power] is negative)

$$S_1 = (P_1 - 1.1) + j(Q_1 - 0.4) \text{ pu}$$

$$S_2 = -0.5 - j0.3 \text{ pu}$$

From (4.32), we have the Gauss-Seidel algorithm

$$V_2^{(l+1)} = \frac{1}{Y_{22}}\left[\frac{P_2 - jQ_2}{(V_2^{(l)})^*} - Y_{21}V_1\right] \tag{4.33}$$

With the given numerical values, (4.33) becomes

$$\begin{aligned} V_2^{(l+1)} &= \frac{1}{1.6\ \angle{-80°}}\left[\frac{-0.5 + j0.3}{(V_2^{(l)})^*} - (1.6\ \angle 100°)(1.1\ \angle 0°)\right] \\ &= (0.625\ \angle 80°)\left[\frac{0.583\ \angle 149°}{(V_2^{(l)})^*} - 1.76\ \angle 100°)\right] \end{aligned} \tag{4.34}$$

To begin the iterations, we let $V_2^{(0)} = 1.0\ \angle{-10°}$ pu. Then (4.34) yields

$$V_2^{(1)} = (0.625) \angle 80° \left[(0.583 \angle 139°) - (1.76 \angle 100°)\right]$$

$$= 0.8485 \angle -15.6° \text{ pu}$$

The next iteration yields

$$V_2^{(2)} = (0.625) \angle 80° \left(\frac{0.583 \angle 149°}{0.848 \angle 15.6°} - (1.76 \angle 100°) \right)$$

$$= 0.778 \angle -17.67° \text{ pu}$$

The results of subsequent iterations are:

$$V_2^{(3)} = 0.74 \angle -19.2° \; ; \; V_2^{(4)} = 0.72 \angle -20° \; ;$$

$$V_2^{(5)} = 0.70 \angle -20.5° \; ; \; V_2^{(6)} = 0.69 \angle -21.1° \; ;$$

$$V_2^{(7)} = 0.67 \angle -21.5° \; ; \; V_2^{(8)} = 0.67 \angle -22.1° \; .$$

Further iteration is unnecessary. Notice the slow convergence of Gauss-Seidel process. This convergence may be enhanced with the use of an acceleration factor α where $1 \leq \alpha \leq 2$ and

$$V_{acc}^{(l)} = V^{(l-1)} + \alpha \left[V^{(l)} - V^{(l-1)}\right]$$

However, even the accelerated Gauss-Seidel algorithm is slowly converging and we must look towards a better approach.

4.6 THE NEWTON-RAPHSON METHOD

The Newton-Raphson iterative procedure converges much more quickly than the Gauss-Seidel process and is generally the method of choice for solving power flow problems. To see how this algorithm may be applied to the solution of a system of equations, consider two functions of two variables x_1 and x_2, such that

$$f(x_1, x_2) = C_1 \tag{4.35}$$

$$f_2(x_1, x_2) = C_2 \tag{4.36}$$

where C_1 and C_2 are constants. Let $x_1^{(0)}$ and $x_2^{(0)}$ be initial estimates of solutions to (4.35) and (4.36), and let $\Delta x_1^{(0)}$ and $\Delta x_2^{(0)}$ be the values by which the initial estimates differ from the correct solutions. That is,

$$f_1(x_1^{(0)} + \Delta x_1^{(0)}, x_2^{(0)} + \Delta x_2^{(0)}) = C_1 \tag{4.37}$$

$$f_2(x_1^{(0)} + \Delta x_1^{(0)},\ x_2^{(0)} + \Delta x_2^{(0)}) = C_2 \tag{4.38}$$

Expanding the left-hand side of each of these equations in a Taylor's series, we obtain

$$f_1(x_1^{(0)},\ x_2^{(0)}) + \Delta x_1^{(0)} \left.\frac{\partial f_1}{\partial x_1}\right|_{x_1^{(0)},x_2^{(0)}} + \Delta x_2^{(0)} \left.\frac{\partial f_1}{\partial x_2}\right|_{x_2^{(0)},x_2^{(0)}} + \text{higher order terms} = C_1 \tag{4.39}$$

$$f_2(x_1^{(0)},\ x_2^{(0)}) + \Delta x_1^{(0)} \left.\frac{\partial f_2}{\partial x_1}\right|_{x_1^{(0)},x_2^{(0)}} + \Delta x_2^{(0)} \left.\frac{\partial f_2}{\partial x_2}\right|_{x_2^{(0)},x_2^{(0)}} + \text{higher order terms} = C_2 \tag{4.40}$$

Neglecting derivatives of order greater than one and writing the result in matrix form yields

$$\begin{bmatrix} C_1 - f_1(x_1^{(0)},\ x_2^{(0)}) \\ C_2 - f_2(x_1^{(0)},\ x_2^{(0)}) \end{bmatrix} = \begin{bmatrix} \dfrac{\partial f_1}{\partial x_1} & \dfrac{\partial f_1}{\partial x_2} \\ \dfrac{\partial f_2}{\partial x_1} & \dfrac{\partial f_2}{\partial x_2} \end{bmatrix}_{x_1^{(0)},\ x_2^{(0)}} \begin{bmatrix} \Delta x_1^{(0)} \\ \Delta x_2^{(0)} \end{bmatrix} \tag{4.41}$$

where the derivatives are evaluated at $x_1^{(0)}$ and $x_2^{(0)}$. Equation (4.41) may be abbreviated as

$$\begin{bmatrix} \Delta C_1^{(0)} \\ \Delta C_2^{(0)} \end{bmatrix} = \boldsymbol{J}^{(0)} \begin{bmatrix} \Delta x_1^{(0)} \\ \Delta x_2^{(0)} \end{bmatrix} \tag{4.42}$$

where the matrix $\boldsymbol{J}^{(0)}$ is called the *Jacobian matrix* (of the initial estimates), and $\Delta C_1^{(0)}$ and $\Delta C_2^{(0)}$ are the differences specified on the left side of (4.41).

Solution of the matrix equation (4.42) gives $\Delta x_1^{(0)}$ and $\Delta x_2^{(0)}$. Then a better estimate of the solution is

$$x_1^{(1)} = x_1^{(0)} + \Delta x_1^{(0)} \tag{4.43}$$

$$x_2^{(1)} = x_2^{(0)} + \Delta x_2^{(0)} \tag{4.44}$$

Note that this is not the exact solution since higher order terms were neglected in (4.39) and (4.40). Repeating the process using $x_1^{(1)}$ and $x_2^{(1)}$ gives a still better estimate, and the iterations are continued until $\Delta x_1^{(l)}$ and $\Delta x_2^{(l)}$ become smaller than a predetermined value.

Example 4.6 Solve the following equations by the Newton-Raphson method:

$$x_1^2 - 4x_2 - 4 = 0$$

$$2x_1 - x_2 - 2 = 0$$

Solution

Let $x_1^{(0)} = 1$ and $x_2^{(0)} = -1$ be the starting point for the first iteration. Then

$$f_1(x_1^{(0)}, x_2^{(0)}) = 1 + 4 - 4 = 1$$

$$f_2(x_1^{(0)}, x_2^{(0)}) = 2 + 1 - 2 = 1$$

and the partial derivatives evaluated at $x_1^{(0)}$ and $x_2^{(0)}$ are

$$\frac{\partial f_1}{\partial x_1} = 2x_1 = 2 \qquad \frac{\partial f_2}{\partial x_1} = 2$$

$$\frac{\partial f_1}{\partial x_2} = -4 \qquad \frac{\partial f_2}{\partial x_2} = -1$$

Now (4.41) yields the equations

$$\begin{bmatrix} 0 - f_1(x_1^{(0)}, x_2^{(0)}) \\ 0 - f_2(x_1^{(0)}, x_2^{(0)}) \end{bmatrix} = \begin{bmatrix} \dfrac{\partial f_1}{\partial x_1} & \dfrac{\partial f_1}{\partial x_2} \\ \dfrac{\partial f_2}{\partial x_1} & \dfrac{\partial f_2}{\partial x_2} \end{bmatrix} \begin{bmatrix} \Delta x_1^{(0)} \\ \Delta x_2^{(0)} \end{bmatrix}$$

and substitution yields

$$\begin{bmatrix} -1 \\ -1 \end{bmatrix} = \begin{bmatrix} 2 & -4 \\ 2 & -1 \end{bmatrix} \begin{bmatrix} \Delta x_1^{(0)} \\ \Delta x_2^{(0)} \end{bmatrix}$$

Simultaneous solution gives us $\Delta x_1^{(0)} = -0.5$ and $\Delta x_2^{(0)} = 0$. Thus, better estimates of x_1 and x_2 are

$$x_1^{(1)} = x_1^{(0)} + \Delta x_1 = 1 - 0.5 = 0.5$$

and

$$x_2^{(1)} = x_2^{(0)} + \Delta x_2 = -1 + 0 = -1.0$$

Proceeding as above with these new estimates, we find that a second and third iteration yield

$$x_1^{(2)} = 0.5357 \quad \text{and} \quad x_2^{(2)} = -0.9286$$

$$x_1^{(3)} = 0.5359 \quad \text{and} \quad x_2^{(3)} = -0.9282$$

Clearly, such problems are solved most conveniently with a digital computer.

To apply the Newton-Raphson method to a power flow problem, for the kth and nth buses we let

$$V_k = |V_k| \angle\delta_k \qquad V_n = |V_n| \angle\delta_n \qquad Y_{kn} = |Y_{kn}| \angle\theta_{kn}$$

Then, from (4.20) and (4.24),

$$P_k - jQ_k = \sum_{n=1}^{N} |V_k V_n Y_{kn}| \angle(\theta_{kn} + \delta_n - \delta_k) \tag{4.45}$$

so that

$$P_k = \sum_{n=1}^{N} |V_k V_n Y_{kn}| \cos(\theta_{kn} + \delta_n - \delta_k) \tag{4.46}$$

and

$$Q_k = -\sum_{n=1}^{N} |V_k V_n Y_{kn}| \sin(\theta_{kn} + \delta_n - \delta_k) \tag{4.47}$$

For the moment, let us assume that our system consists of only load buses plus the one necessary swing bus. For all the load buses, P_k and Q_k are specified and we will designate these quantities as P_{ks} and Q_{ks}. These correspond to knowing C_1 and C_2 in (4.41). The next step is to estimate $|V|$ and δ at each bus except the swing bus (bus 1) where the voltage is known.

We then substitute these estimated values (which correspond to the estimated values for x_1 and x_2) in (4.46) and (4.47) to calculate Ps and Qs that correspond to $f_1(x_1^{(0)}, x_2^{(0)})$ and $f_2(x_1^{(0)}, x_2^{(0)})$ in (4.41). These will be designated as $P_{kc}^{(0)}$ and $Q_{kc}^{(0)}$. Next we compute

$$\Delta P_k^{(0)} = P_{ks} - P_{kc}^{(0)} \tag{4.48}$$

$$\Delta Q_k^{(0)} = Q_{ks} - Q_{kc}^{(0)} \tag{4.49}$$

where the subscripts s and c mean, respectively, specified and calculated values. These correspond to the values on the left side of (4.42) and are called the power mismatches.

Corresponding to (4.41) and (4.42), the matrix equation for a three-bus system (with bus 1 as the swing bus and hence omitted) is

$$\begin{bmatrix} \Delta P_2^{(0)} \\ \Delta P_3^{(0)} \\ \Delta Q_2^{(0)} \\ \Delta Q_3^{(0)} \end{bmatrix} = \begin{bmatrix} \frac{\partial P_2}{\partial \delta_2} & \frac{\partial P_2}{\partial \delta_3} & \frac{\partial P_2}{\partial |V_2|} & \frac{\partial P_2}{\partial |V_3|} \\ \frac{\partial P_3}{\partial \delta_2} & \frac{\partial P_3}{\partial \delta_3} & \frac{\partial P_3}{\partial |V_2|} & \frac{\partial P_3}{\partial |V_3|} \\ \frac{\partial Q_2}{\partial \delta_2} & \frac{\partial Q_2}{\partial \delta_3} & \frac{\partial Q_2}{\partial |V_2|} & \frac{\partial Q_2}{\partial |V_3|} \\ \frac{\partial Q_3}{\partial \delta_2} & \frac{\partial Q_3}{\partial \delta_3} & \frac{\partial Q_3}{\partial |V_2|} & \frac{\partial Q_3}{\partial |V_3|} \end{bmatrix}_{\substack{\delta_2^{(0)},\, \delta_3^{(0)} \\ |V_2|^{(0)},\, |V_2|^{(0)}}} \begin{bmatrix} \Delta \delta_2^{(0)} \\ \Delta \delta_3^{(0)} \\ \Delta |V_2^{(0)}| \\ \Delta |V_3^{(0)}| \end{bmatrix} \qquad (4.50)$$

or

mismatches (power)	=	Jacobian	×	corrections (voltage)

Equation (4.50) is solved by inverting the Jacobian. The values determined for $\Delta\delta_k^{(0)}$ and $\Delta V_k^{(0)}$ are added to the previous estimates of V_k and δ_k to obtain new estimates with which to start the next iteration. The process is repeated until the voltage corrections are as small as desired.

If controlled voltage buses are contained in the power system, the Newton-Raphson iterative solution process is actually simplified because the order of the Jacobian is reduced by one for each controlled voltage bus. This can be seen by referring to (4.50) where we will now assume that bus 2 is a voltage controlled bus instead of a load bus. Then Q_{2S} and hence $\Delta Q_2^{(0)}$ are unknown and $\Delta|V_2|$ = 0 since $|V_2|$ is specified. As a result, the third column of the Jacobian matrix $\left(\frac{\partial(P,Q)}{\partial|V_2|}\right)$ is zero and $\Delta|V_2|$ in the correction matrix is zero. This effectively eliminates the third equation of the Newton-Raphson procedure and result is as shown in (4.50*a*).

$$\begin{bmatrix} \Delta P_2^{(0)} \\ \Delta P_3^{(0)} \\ \Delta Q_3^{(0)} \end{bmatrix} = \begin{bmatrix} \frac{\partial P_2}{\partial \delta_2} & \frac{\partial P_2}{\partial \delta_3} & \frac{\partial P_2}{\partial |V_3|} \\ \frac{\partial P_3}{\partial \delta_2} & \frac{\partial P_3}{\partial \delta_3} & \frac{\partial P_3}{\partial |V_3|} \\ \frac{\partial Q_3}{\partial \delta_2} & \frac{\partial Q_3}{\partial \delta_3} & \frac{\partial Q_3}{\partial |V_3|} \end{bmatrix} \begin{bmatrix} \Delta \delta_2^{(0)} \\ \Delta \delta_3^{(0)} \\ \Delta |V_3^{(0)}| \end{bmatrix} \qquad (4.50a)$$

To illustrate the Newton-Raphson method as applied to power flow analysis, we now consider the following examples.

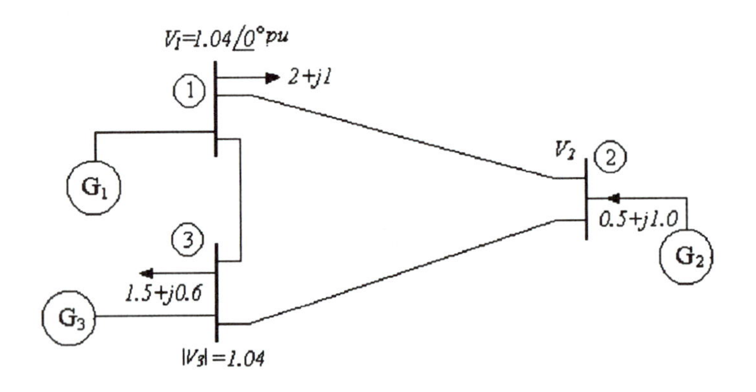

FIGURE 4.6
Example 4.7.

Example 4.7 For the system shown in Figure 4.6

$$\boldsymbol{Y}_{\text{bus}} = \begin{bmatrix} 24.23\ \angle -75.95^\circ & 12.13\ \angle 104.04^\circ & 12.13\ \angle 104.04^\circ \\ 12.13\ \angle 104.04^\circ & 24.23\ \angle -75.95^\circ & 12.13\ \angle 104.04^\circ \\ 12.13\ \angle 104.04^\circ & 12.13\ \angle 104.04^\circ & 24.23\ \angle -75.95^\circ \end{bmatrix} \text{pu}$$

Given the per-unit voltages and power as shown, determine V_2 by the Newton-Raphson method. Note that bus 1 is the swing bus, bus 2 is a load bus (even though there is only specified generated power at this bus) and bus 3 is a voltage controlled bus. Reactive power produced in generator G_3 is used to maintain $|V_3|$ at 1.04 pu.

Solution

Estimate $V_2^{(0)} = 1\angle 0^\circ$ pu and $\delta_3^{(0)} = 0$. This is our educated initial guess for the bus voltages. Then from (4.46)

$$\begin{aligned} P_{2C}^{(0)} &= |V_2^{(0)}|\,|V_1^{(0)}|\,|Y_{21}|\cos(\theta_{21} + \delta_1^{(0)} - \delta_2^{(0)}) + |V_2^{(0)}|^2\,|Y_{22}|\cos\theta_{22} \\ &\quad + |V_2^{(0)}|\,|V_3^{(0)}|\,|Y_{23}|\cos(\theta_{23} + \delta_3^{(0)} - \delta_2^{(0)}) \\ &= (1)(1.04)(12.13)\cos 104.04^\circ + (1)(24.23)\cos(-75.95^\circ) \\ &\quad + (1)(1.04)(12.13)\cos 104.04^\circ \\ &= -0.2386 \text{ pu} \end{aligned}$$

Similarly,

$$P_{3C}^{(0)} = |V_3^{(0)}||V_1^{(0)}||Y_{31}|\cos(\theta_{31} + \delta_1^{(0)} - \delta_3^{(0)}) + |V_3^{(0)}||V_2^{(0)}||Y_{32}|$$

$$\text{x } \cos(\theta_{32} + \delta_2^{(0)} - \delta_3^{(0)}) + |V_3^{(0)}|^2|Y_{33}|\cos\theta_{33}$$

$$= (1.04)(1.04)(12.13)\cos 104.04° + (1.04)(12.13)\cos 104.04°$$

$$+ (1.04)^2(24.23)\cos(-75.95°)$$

$$= 0.119 \text{ pu}$$

Also, from (4.47),

$$Q_{2C}^{(0)} = -|V_2^{(0)}||V_1^{(0)}||Y_{21}|\sin(\theta_{21} + \delta_1^{(0)} - \delta_2^{(0)}) - |V_2^{(0)}|^2|Y_{22}|\sin\theta_{22}$$

$$-|V_2^{(0)}||V_3^{(0)}|\sin(\theta_{23} + \delta_2^{(0)} - \delta_3^{(0)})$$

$$= -(1)(1.04)(12.13)\sin 104.04° - (1)(24.23)\sin(-75.95°)$$

$$-(1)(1.04)(12.13)\sin 104.04°$$

$$= -0.972 \text{ pu}$$

Now, from (4.48)

$$\Delta P_2^{(0)} = 0.5 - (-0.2386) = 0.7386 \text{ pu}$$

$$\Delta P_3^{(0)} = -1.5 - 0.119 = -1.619 \text{ pu}$$

Similarly, from (4.49)

$$\Delta Q_2^{(0)} = 1 - (-0.972) = 1.972 \text{ pu}$$

For the given three-bus system (with $|V_3|$ known), (4.50) becomes

$$\begin{bmatrix} \Delta P_2^{(0)} \\ \Delta P_3^{(0)} \\ \Delta Q_2^{(0)} \end{bmatrix} = \begin{bmatrix} \frac{\partial P_2}{\partial \delta_2} & \frac{\partial P_2}{\partial \delta_3} & \frac{\partial P_2}{\partial |V_2|} \\ \frac{\partial P_3}{\partial \delta_2} & \frac{\partial P_3}{\partial \delta_3} & \frac{\partial P_3}{\partial |V_2|} \\ \frac{\partial Q_2}{\partial \delta_2} & \frac{\partial Q_2}{\partial \delta_3} & \frac{\partial Q_2}{\partial |V_2|} \end{bmatrix} \begin{bmatrix} \Delta\delta_2^{(0)} \\ \Delta\delta_3^{(0)} \\ \Delta|V_2^{(0)}| \end{bmatrix} \tag{4.51}$$

Differentiating (4.46) and (4.47) and substituting the numerical values yields, from (4.51) above,

$$\frac{\partial P_k}{\partial \delta_2} = \frac{\partial}{\partial \delta_2}\left(\sum_{n=1}^{N} |V_k V_n Y_{\text{ln}}|\cos(\theta_{lcn} + \delta_n - s_k)\right)$$

where $k = 2$ for the first Jacobian element and

$$\frac{\partial P_2}{\partial \delta_2} = \frac{\partial}{\partial \delta_2}\left[|V_2V_1Y_{21}|\cos(\theta_{21} + \delta_1 - \delta_2)\right.$$

$$+ |V_2V_2Y_{22}|\cos(\theta_{22} + \delta_2 - \delta_2)$$

$$\left.+ |V_2V_3Y_{23}|\cos(\theta_{23} + \delta_3 - \delta_2)\right]$$

$$\frac{\partial P_2}{\partial \delta_2} = |V_2V_1Y_{21}|\sin(\theta_{21} + \delta_1 - \delta_2)$$

$$+ |V_2V_3Y_{23}|\sin(\theta_{23} + \delta_3 - \delta_2)$$

$$\left.\frac{\partial P_2}{\partial \delta_2}\right|_{(0)} = (1)(1.04)(12.13)\sin(104.04° + 0° - 0°)$$

$$+ (1)(1.04)(12.13)\sin(104.04° + 0° - 0°)$$

$$\left|\frac{\partial P_2}{\partial \delta_2}\right|_{(0)} = 12.24 + 12.24 = 24.48$$

and by similar calculations

$$\begin{bmatrix} 0.7386 \\ -1.619 \\ 1.972 \end{bmatrix} = \begin{bmatrix} 24.48 & -12.23 & 5.64 \\ -12.23 & 24.95 & -3.06 \\ -6.11 & 3.06 & 22.54 \end{bmatrix} \begin{bmatrix} \Delta\delta_2^{(0)} \\ \Delta\delta_3^{(0)} \\ \Delta|V_2^{(0)}| \end{bmatrix} \tag{4.52}$$

or

$$\begin{bmatrix} \Delta\delta_2^{(0)} \\ \Delta\delta_3^{(0)} \\ \Delta|V_2^{(0)}| \end{bmatrix} = \begin{bmatrix} 24.48 & -12.23 & 5.64 \\ -12.23 & 24.95 & -3.06 \\ -6.11 & 3.06 & 22.54 \end{bmatrix}^{-1} \begin{bmatrix} 0.7386 \\ -1.619 \\ 1.972 \end{bmatrix}$$

$$= \begin{bmatrix} 0.05179 & 0.02653 & -0.00937 \\ 0.02666 & 0.05309 & 0.00051 \\ 0.01043 & -0.00001 & 0.04176 \end{bmatrix} \begin{bmatrix} 0.7386 \\ -1.619 \\ 1.972 \end{bmatrix}$$

Solving for the corrections in (4.52) gives

$\Delta\,\delta_2^{(0)} = (0.05179)(0.7386) + (0.02653)(-1.619) - (0.00937)(1.972) = -0.023$ rad

$\Delta\,\delta_3^{(0)} = (0.02666)(0.7386) + (0.05309)(-1.619) + (0.00051)(1.972) = -0.065$ rad

$\Delta|V_2^{(0)}| = (0.01043)\,(0.7386) + (0.00001)\,(1.619) + (0.04176)\,(1.972) = 0.09$ pu

It should be noted that the voltage angle corrections are in radians.

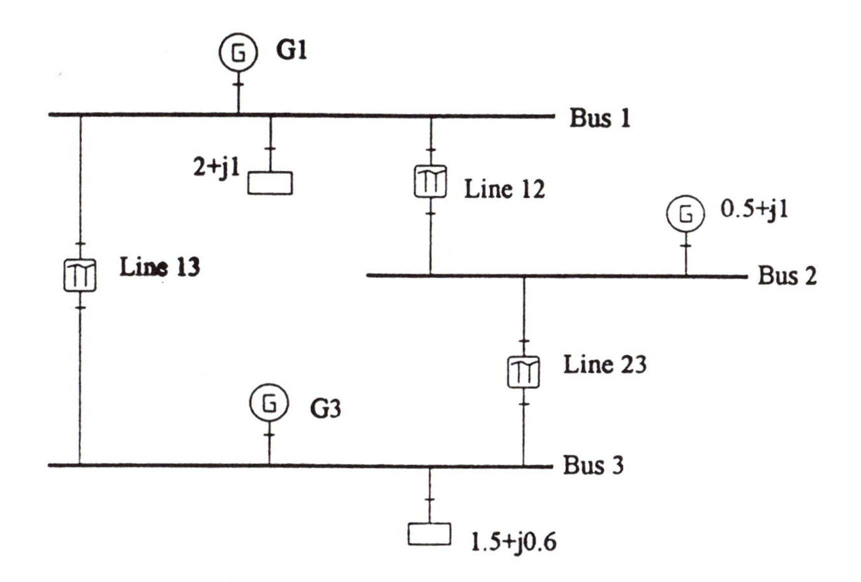

FIGURE 4.7
Computer generated single-line diagram for Example 4.7

Thus,

$$\delta_2^{(1)} = 0 - 0.023 = -0.023 \text{ rad or } -1.31°$$

$$|V_2|^{(1)} = 1 + 0.09 = 1.09 \text{ pu}$$

$$\delta_3^{(1)} = 0 - 0.065\ (180/\pi) = -3.73°$$

This procedure is repeated until, upon convergence, we obtain $V_2 = 1.081\ \angle -1.37°$ pu.

It seems clear that power flow calculations using hand-held calculators become quite cumbersome, even for this simplified 3-bus example. Fortunately, software is available for personal computers which will ease this burden and will allow for power flow computations on real-world systems of a size which would be impossible to analyze without these tools.

As a simple example, SKM Systems Analysis Power* Tools for Windows was applied to this 3-bus problem. The single-line diagram of this system that was drawn with the computer is shown in Figure 4.7 and a portion of the printed output of this power flow analysis is given in Table 4.1. Printed output of the power flows and system voltages imposed on the single-line diagram is shown

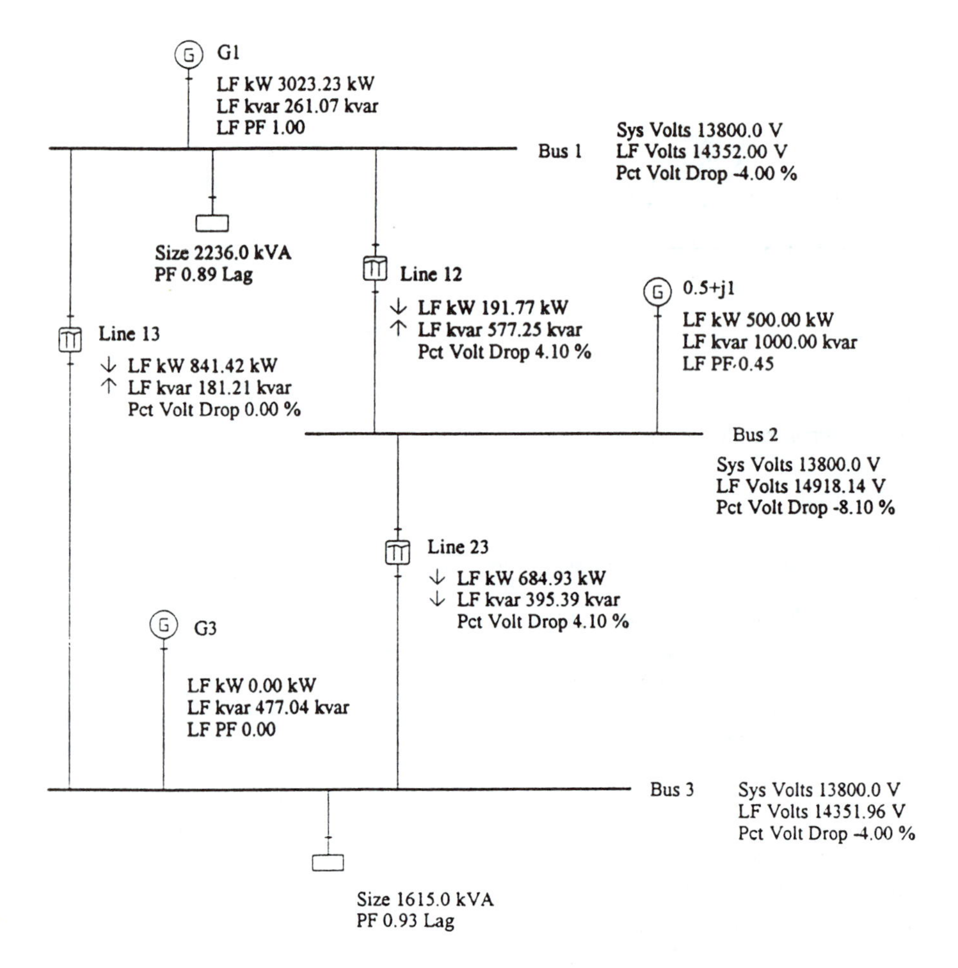

FIGURE 4.8
Computer generated results on the single-line diagram of Example 4.7.

TABLE 4.1. Computer-generated output for Example 4.7.

BALANCED VOLTAGE DROP AND LOAD FLOW ANALYSIS (SWING GENERATORS)

SOURCE	VOLTAGE	ANGLE	KW	KVAR	VD% (UTILITY IMPEDANCE)
G1	1.040	.00	3023.23	261.07	.00

BALANCED VOLTAGE DROP AND LOAD FLOW ANALYSIS (PV GENERATOR SCHEDULE REPORT)

	---VOLTAGE---		-KVAR LIMITS-		---ACTUAL----	
PV SOURCE NAME	SCHED.	ACTUAL	MIN	MAX	KW	KVAR
G3	1.040	1.040	0.	5000.	0.	477.

BALANCED VOLTAGE DROP AND LOAD FLOW BUS DATA SUMMARY

BUS NAME	BASE VOLT	PU VOLT	BUS NAME	BASE VOLT	PU VOLT
Bus 1	13800.00	1.0400	Bus 2	13800.00	1.0810
Bus 3	13800.00	1.0400			

BALANCED VOLTAGE DROP AND LOAD FLOW BRANCH DATA SUMMARY

FROM NAME	TO NAME	TYPE	VD%	AMPS	KVA	RATING%
Bus 1	Bus 2	FDR	-4.10	24.47	608.27	UNKNOWN
Bus 1	Bus 3	FDR	.00	34.62	860.71	UNKNOWN
Bus 2	Bus 3	FDR	4.10	30.61	790.86	UNKNOWN

NOTE: FDR RATING% = % AMPS RATING BASED ON LIBRARY FLA OR BRANCH INPUT FLA
TX2 RATING% = % KVA RATING BASED ON TRANSFORMER FL KVA

3 BUSES

*** T O T A L S Y S T E M L O S S E S ***
31. KW 125. KVAR

```
                 BALANCED VOLTAGE DROP AND LOAD FLOW ANALYSIS
***************************************************************************
VOLTAGE EFFECT ON LOADS MODELED
VOLTAGE DROP CRITERIA: BRANCH =  3.00 %   BUS =  5.00


==== BUS:  Bus 1           DESIGN VOLTS:   13800 BUS VOLTS:   14352 %VD: -4.00
========================= PU BUS VOLTAGE:  1.040       ANGLE:    .0 DEGREES
             PROJECTED BUS LOAD:  1990.0 KW  1019.5 KVAR
             *** SWING  GENERATOR: G1                 3023.2 KW   261.1 KVAR

LOAD   TO:  Bus 2           FEEDER AMPS:  24.4 VOLTAGE DROP:  -566. %VD: -4.10
PROJECTED POWER FLOW:    191.8 KW    -577.2 KVAR    608.3 KVA  PF:    .32 LEADING
LOSSES THRU FEEDER:        6.8 KW     27.4 KVAR     28.2 KVA

LOAD   TO:  Bus 3           FEEDER AMPS:  34.6 VOLTAGE DROP:     0. %VD:    .00
PROJECTED POWER FLOW:    841.4 KW    -181.2 KVAR    860.7 KVA  PF:    .98 LEADING
LOSSES THRU FEEDER:       13.7 KW     54.8 KVAR     56.5 KVA


==== BUS:  Bus 2           DESIGN VOLTS:   13800 BUS VOLTS:   14918 %VD: -8.10
========================= PU BUS VOLTAGE:  1.081       ANGLE:  -1.4 DEGREES
             **** PQ TYPE MACHINE:0.5+j1              500.0 KW  1000.0 KVAR

LOAD FROM:  Bus 1           FEEDER AMPS:  24.4 VOLTAGE DROP:  -566. %VD: -4.10
PROJECTED POWER FLOW:    184.9 KW    -604.6 KVAR    632.3 KVA  PF:    .29 LEADING
LOSSES THRU FEEDER:        6.8 KW     27.4 KVAR     28.2 KVA

LOAD   TO:  Bus 3           FEEDER AMPS:  30.6 VOLTAGE DROP:   566. %VD:  4.10$
PROJECTED POWER FLOW:    684.9 KW     395.4 KVAR    790.9 KVA  PF:    .87 LAGGING
LOSSES THRU FEEDER:       10.7 KW     42.8 KVAR     44.1 KVA


==== BUS:  Bus 3           DESIGN VOLTS:   13800 BUS VOLTS:   14352 %VD: -4.00
========================= PU BUS VOLTAGE:  1.040       ANGLE:  -3.8 DEGREES
             NET BRANCH DIVERSITY LOAD:  1501.9 KW    593.6 KVAR
             ** PV TYPE GENERATOR: G3                     .0 KW   477.0 KVAR

LOAD FROM:  Bus 1           FEEDER AMPS:  34.6 VOLTAGE DROP:     0. %VD:    .00
PROJECTED POWER FLOW:    827.7 KW    -236.0 KVAR    860.7 KVA  PF:    .96 LEADING
LOSSES THRU FEEDER:       13.7 KW     54.8 KVAR     56.5 KVA

LOAD FROM:  Bus 2           FEEDER AMPS:  30.6 VOLTAGE DROP:   566. %VD:  4.10$
PROJECTED POWER FLOW:    674.2 KW     352.6 KVAR    760.8 KVA  PF:    .89 LAGGING
LOSSES THRU FEEDER:       10.7 KW     42.8 KVAR     44.1 KVA
```

in Figure 4.8. Base values for this problem were assumed to be 1000 kVA three-phase and 13,800 V line-to-line. Inspection of the computer output reveals additional information on the state of this system while the voltage at bus 2 is seen to agree with that calculated previously. Notice that $\delta_3 = -3.8°$ and the generator G_3 is producing 0.477 pu reactive power in order to maintain $|V_3| =$ 1.04 pu. The system losses of 31 kW and 125 kvar are supplied by the generator G_1 at the swing bus. Note also that the system loads are specified as a total kVA and a power factor rounded to two decimal places. This results in some minor loading differences from the hand-held calculator example. Thus, the load on bus 1 is specified as $1.99 + j1.0195$ pu instead of $2.0 + j1.0$ pu. Actual load values are given in the printed output. Line power flows are given at the beginning (top) of the lines in Figure 4.8.

4.7 ALTERNATIVE NEWTON-RAPHSON METHODS

From the discussion in the preceding section, it may be seen that the Newton-Raphson method converges very rapidly and is therefore the method of choice over Gauss-Seidel iteration for a system with N buses. However, the computation intensity per iteration is large in the Newton-Raphson procedure as the Jacobian matrix of order $2N-2$ (reduced by one for each voltage controlled bus) must be inverted during each iteration. The subject of how to reduce computation time per iteration in this procedure without significantly affecting the rate of convergence has been of considerable study. Among the methods which have been investigated are as follows:

1. Using a number of Gauss-Seidel iterations prior to switching to the Newton-Raphson procedure.

2. Using a calculated Jacobian inverse for more than one iteration before updating the Jacobian.

3. Using sparse matrix techniques to invert the Jacobian.

4. Decoupling the power flow equations.

The reductions in computations associated with 1 and 2 seem evident. The Gauss-Seidel method converges fairly well initially but slows down as the solution is approached. Newton-Raphson converges much more quickly when the solution is close. The Gauss-Seidel procedure may also be helpful in smoothing out a poor initial guess of the system voltages.

The advantages of 3 are not apparent from the simple examples discussed in this introductory text. However, for a power system with hundreds or thousands of buses, each bus will normally be connected only to several other buses in the system. Thus, for example, consider a 1000 bus system where bus 2 is only connected to buses 1, 3 and 4. In this case, row 2 and column 2 of the bus admittance matrix will each have 4 non-zero elements and 996 elements that are equal to zero. This sparcity or predominance of zero entries carries over to the Jacobian matrix where mathematical techniques for quickly inverting sparse matrices may be applied. This results in a very significant reduction in computation time for large systems.

Finally, the fast decoupled power method is essentially a modified Newton-Raphson method, and involves relatively less computation time. To develop the procedure (or algorithm) of the fast decoupled method, we refer to Example 4.7. We consider the Jacobian of the first iteration as given by (4.52) and observe that in the third column the terms in the first and second rows are much smaller than the term in the third row. It may be verified that the preceding observation holds for subsequent iterations. Similarly, in the third row, the elements in the first and second columns are considerably smaller than that in the third column. In the Jacobian, these smaller terms are associated with the following operations:

$$\frac{\partial P_2}{\partial |V_2|}; \quad \frac{\partial P_3}{\partial |V_3|}; \quad \frac{\partial Q_2}{\partial \delta_2}; \quad \text{and} \quad \frac{\partial Q_2}{\partial \delta_3}.$$

The resulting numerical terms are small because a change in voltage magnitude (say $|V_2|$) does not substantially change the power (P_2) as discussed in Section 4.1. Similarly, a change in the voltage angle (δ_2) has little effect on the reactive power (Q_2). We conclude that P and Q are decoupled and the corresponding method of obtaining the power flow on a system is known as the decoupled power flow method.

In formulating an algorithm based on the above discussion, we assume a complete PQ decoupling. This requires that all derivatives of real power (P) with respect to voltage ($|V|$) and all derivatives of reactive power (Q) with respect to voltage angle (δ) in the Jacobian be set to zero. Consequently, the Jacobian of the kth iteration may be written as:

$$\boldsymbol{J}^{(k)} = \begin{bmatrix} J_{11}^{(k)} & 0 \\ 0 & J_{22}^{(k)} \end{bmatrix} \tag{4.53}$$

Comparing (4.53) and (4.50), it follows that the elements of $\boldsymbol{J}_{11}{}^{(k)}$ are derivatives of power with respect to voltage angles. Similarly, the elements of $\boldsymbol{J}_{22}{}^{(k)}$ are derivatives of reactive power with respect to the magnitudes of bus voltages. It is to be noted that voltage angles and bus voltage magnitudes are independent variables, which constitute the state vector of the system.

To simplify further, we express (4.50) in terms of the Jacobian (4.53). Thus,

$$\begin{bmatrix} \Delta \boldsymbol{P} \\ \Delta \boldsymbol{Q} \end{bmatrix}^{(k)} = \begin{bmatrix} \boldsymbol{J}_{11} & 0 \\ 0 & J_{22} \end{bmatrix}^{(k)} \begin{bmatrix} \Delta\delta \\ \Delta|\boldsymbol{V}| \end{bmatrix}^{(k)} \tag{4.54}$$

Notice that (4.54) represents decoupled vector equations:

$$\Delta \boldsymbol{P}^{(k)} = \boldsymbol{J}_{11}^{(k)} \, \Delta\delta^{(k)} \tag{4.55}$$

$$\Delta \boldsymbol{Q}^{(k)} = \boldsymbol{J}_{22}^{(k)} \, \Delta|\boldsymbol{V}|^{(k)} \tag{4.56}$$

The terms on the left-hand side of these equations represent mismatches. On the right-hand side, the first terms in (4.55) and (4.56) are the Jacobians and the second terms denote corrections. The elements of $\boldsymbol{J}_{11}^{(k)}$ and $\boldsymbol{J}_{22}^{(k)}$ are the submatrices along the diagonal of the Jacobian of (4.51); that is,

$$\boldsymbol{J}_{11}^{(k)} = \begin{bmatrix} \dfrac{\partial P_2}{\partial \delta_2} & \dfrac{\partial P_2}{\partial \delta_3} \\ \dfrac{\partial P_3}{\partial \delta_2} & \dfrac{\partial P_3}{\partial \delta_3} \end{bmatrix}^{(k)} \tag{4.57}$$

and

$$\boldsymbol{J}_{22}^{(k)} = \begin{bmatrix} \dfrac{\partial Q_2}{\partial |V_2|} \end{bmatrix}^{(k)} \tag{4.48}$$

In order to obtain an algorithm, we consider the various elements in the above two Jacobians. Thus, in conjunction with (4.46) and (4.47) we have

$$\begin{aligned} \frac{\partial P_2}{\partial \delta_2} &= - |V_1 V_2 Y_{12}| \sin(\theta_{21} + \delta_1 - \delta_2) - |V_2 V_3 Y_{23}| \sin(\theta_{23} + \delta_2 - \delta_3) \\ &= \left\{ |V_1 V_2 Y_{12}| \left[\cos\theta_{21} \sin(\delta_2 - \delta_1) + \sin\theta_{21} \cos(\delta_1 - \delta_2) \right] \right\} \\ &\quad + \left\{ |V_2 V_3 Y_{23}| \left[\cos\theta_{23} \sin(\delta_2 - \delta_3) + \sin\theta_{23} \cos(\delta_3 - \delta_2) \right] \right\} \end{aligned} \tag{4.59}$$

Next, we define conductances

$$G_{21} = Y_{21} \cos\theta_{21} \; ; \quad G_{23} = Y_{23} \cos\theta_{23} \tag{4.60}$$

and susceptances

$$B_{21} = Y_{21} \sin\theta_{21} \; ; \quad B_{23} = Y_{23} \sin\theta_{23} \tag{4.61}$$

We further make the following (reasonable) approximations, since $(\delta_i - \delta_j)$ is very small,

$$G_{ij} \sin(\delta_i - \delta_j) \simeq 0 \tag{4.62}$$

$$B_{ij} \cos(\delta_i - \delta_j) \simeq B_{ij} \tag{4.63}$$

Combining (4.59) through (4.63) yields

$$\frac{\partial P_2}{\partial \delta_2} = |V_1||V_2| B_{21} + |V_2||V_3| B_{23} = \sum_{\substack{j=1 \\ j \neq 2}}^{3} |V_2||V_j| B_{2j} \tag{4.64}$$

Equation (4.64) may be generalized to

$$\frac{\partial P_i}{\partial \delta_i} = |V_i| \sum_{\substack{j=1 \\ j \neq 1}}^{n} |V_j| B_{ij} = |V_i| \left(\sum_{j=1}^{n} |V_j| B_{ij} - |V_i| B_{ii} \right) \tag{4.65}$$

Now, for a flat voltage profile, $|V_i| = |V_j| = 1.0$ pu and the term under summation sign

$$\sum_{j=1}^{n} B_{ij} = 0 \tag{4.66}$$

This fact may be verified for the special case of $\boldsymbol{Y}_{\text{bus}}$ given in Example 4.7 by adding the terms of a particular row. Under these conditions, (4.65) simplifies to

$$\frac{\partial P_i}{\partial \delta_i} = -|V_i|^2 B_{ii} \tag{4.67}$$

Proceeding as above, we have

$$\begin{aligned} \frac{\partial P_2}{\partial \delta_3} &= |V_2||V_3||Y_{23}| \sin(\delta_2 - \delta_3 - \theta_{23}) \\ &= |V_2||V_3||Y_{23}| \sin(\delta_2 - \delta_3) \cos\theta_{23} - |V_2||V_3||Y_{23}| \cos(\delta_2 - \delta_3) \sin\theta_{23} \end{aligned} \tag{4.68}$$

Combining (4.60), (4.61), (4.63) and (4.68) yields

$$\frac{\partial P_2}{\partial \delta_3} = -|V_2||V_3| B_{23} \tag{4.69}$$

Again, for a flat voltage profile $|V_2| = |V_3| = 1$ pu, and (4.69) reduces to

$$\frac{\partial P_2}{\partial \delta_3} = -B_{23} \tag{4.70}$$

Equation (4.69) may be expressed in the general form

$$\frac{\partial P_i}{\partial \delta_j} = -|V_i||V_j|B_{ij} \tag{4.71}$$

Consequently, for a flat voltage profile the general **B** matrix becomes

$$\boldsymbol{B} = \begin{bmatrix} B_{22} & B_{23} & \cdots & B_{2n} \\ B_{32} & \cdot & \cdot & \cdot \\ \cdot & \cdot & \cdot & \cdot \\ \cdot & \cdot & \cdot & \cdot \\ B_{n2} & \cdots & \cdots & B_{nn} \end{bmatrix} \tag{4.72}$$

which is related to the $\boldsymbol{J}_{11}^{(k)}$ matrix of (4.57) by

$$\boldsymbol{J}_{11}^{(k)} = -\boldsymbol{B} \tag{4.73}$$

In order to express $\boldsymbol{J}_{11}^{(k)}$ in the form consistent with (4.66) and (4.71), we may write

$$\boldsymbol{J}_{11}^{(k)} = -[\boldsymbol{V}^{(k)}]\,\boldsymbol{B}\,[\boldsymbol{V}^{(k)}] \tag{4.74}$$

where

$$[\boldsymbol{V}^{(k)}] = \begin{bmatrix} |V_2^k| & 0 & \cdots & 0 \\ 0 & & & \cdot \\ \cdot & & & \cdot \\ \cdot & & & \cdot \\ 0 & \cdots & \cdots & |V_n^{(k)}| \end{bmatrix} \tag{4.75}$$

In terms of voltages given by (4.75), (4.45) may then be written as

$$\Delta \boldsymbol{P}^{(k)} = -[\boldsymbol{V}^{(k)}]\boldsymbol{B}[\boldsymbol{V}^{(k)}]\,\Delta\delta^{(k)} \tag{4.76}$$

Now, from (4.75) we observe that the elements of the $[\mathbf{V}^{(k)}]^{-1}$ matrix are $1/|V_j^{(k)}|$ along the diagonal. Thus, premultiplying both sides of (4.76) by $[\mathbf{V}^{(k)}]^{-1}$ yields

$$\begin{bmatrix} \dfrac{\Delta P_2}{|V_2|} \\ \dfrac{\Delta P_3}{|V_3|} \\ \vdots \\ \dfrac{\Delta P_n}{|V_n|} \end{bmatrix}^{(k)} = -\,\boldsymbol{B}[\boldsymbol{V}^{(k)}]\ \Delta\delta^{(k)} \tag{4.77}$$

We may now choose $[\mathbf{V}^{(k)}]$ to be the identity matrix and solve for $\Delta\delta^{(k)}$ from (4.77) to obtain

$$\begin{bmatrix} \Delta\delta_2 \\ \Delta\delta_3 \\ \vdots \\ \Delta\delta_n \end{bmatrix}^{(k)} = -\begin{bmatrix} B_{22} & \cdots & B_{2n} \\ B_{32} & & \vdots \\ \vdots & & \vdots \\ B_{n2} & \cdots & B_{nn} \end{bmatrix}^{-1} \begin{bmatrix} \dfrac{\Delta P_2}{|V_2^k|} \\ \vdots \\ \dfrac{\Delta P_n}{|V_n^k|} \end{bmatrix}^{(k)} \tag{4.78}$$

Or,

$$\Delta\delta^{(k)} = -\boldsymbol{B}^{-1}\left[\frac{\Delta \boldsymbol{P}}{|\boldsymbol{V}|}\right]^{(k)} \tag{4.79}$$

Finally, we consider the Jacobian $\boldsymbol{J}_{22}^{(k)}$, as given by (4.58). From (4.47), for a three-bus system we have

$$\begin{aligned} Q_2 = {} & |V_2V_1Y_{21}|\sin(\theta_{21} + \delta_1 - \delta_2) + |V_2^2Y_{22}|\sin\theta_{22} \\ & + |V_2V_3Y_{23}|\sin(\theta_{23} + \delta_3 - \delta_2) \end{aligned} \tag{4.80}$$

Differentiating (4.80) with respect to $|V_2|$ yields

$$\begin{aligned} \frac{\partial Q_2}{\partial|V_2|} = {} & |V_1Y_{21}|\sin(\theta_{21} + \delta_1 - \delta_2) + 2|V_2Y_{22}|\sin\theta_{22} \\ & + |V_3Y_{23}|\sin(\theta_{23} + \delta_3 - \delta_2) \end{aligned} \tag{4.81}$$

Combining (4.81) with (4.60) through (4.63) yields

$$\frac{\partial Q_2}{\partial |V_2|} = -|V_1|B_{21} - |V_3|B_{23} - 2|V_2|\, B_{22} \tag{4.82}$$

But,

$$|V_1|B_{21} + |V_2|B_{22} + |V_3|B_{23} = 0 \tag{4.83}$$

Or,

$$- |V_1|\, B_{21} - |V_3|B_{23} = |V_2|B_{22} \tag{4.84}$$

Substituting (4.84) in (4.82) yields

$$\boldsymbol{J}_{22}^{(k)} = \frac{\partial Q_2}{\partial |V_2|} = -2|V_2|B_{22} + |V_2|B_{22} = -\,|V_2|B_{22} \tag{4.85}$$

In a general form, (4.85) becomes

$$\boldsymbol{J}_{22}^{(k)} = -\left[\boldsymbol{V}^{(k)}\right] \boldsymbol{B} \tag{4.86}$$

Substituting (4.86) in (4.56) gives

$$\Delta \boldsymbol{Q}^{(k)} - -\left[\boldsymbol{V}^{(k)}\right]\boldsymbol{B}\ \Delta \boldsymbol{V}^{(k)} \tag{4.87}$$

Proceeding as in (4.77) through (4.79), we finally obtain

$$\begin{bmatrix} \Delta|V_2| \\ \Delta|V_3| \\ \vdots \\ \Delta|V_n| \end{bmatrix}^{(k)} = -\begin{bmatrix} B_{22} & \cdots & B_{2n} \\ B_{32} & & \vdots \\ \vdots & & \vdots \\ B_{n2} & \cdots & B_{nn} \end{bmatrix}^{-1} \begin{bmatrix} \dfrac{\Delta Q_2}{|V_2|} \\ \vdots \\ \dfrac{\Delta Q_n}{|V_n|} \end{bmatrix}^{(k)} \tag{4.88}$$

Or,

$$\Delta \boldsymbol{V}^{(k)} = -\,\boldsymbol{B}^{-1} \left[\frac{\Delta \boldsymbol{Q}}{|\boldsymbol{V}|}\right]^{(k)} \tag{4.89}$$

Equations (4.79) and (4.89) give the complete algorithm for the fast decoupled power method. It is to be noted that terms relating to voltage-controlled buses do not appear in the Jacobian $\boldsymbol{J}_{22}$.

We now solve the three-bus problem by the method presented in this Section.

Example 4.8 Solve for the voltage $|V_2|$ of the system given in Example 4.7.

Solution

From the $\boldsymbol{Y}_{\text{bus}}$ of Example 4.7 we form the $\boldsymbol{B}$ matrix by eliminating the first row and first column and retaining only the imaginary parts. Thus,

$$\boldsymbol{B} = \begin{bmatrix} -23.5 & 11.77 \\ 11.77 & -23.5 \end{bmatrix}$$

And (4.78) may then be written as

$$\begin{bmatrix} \Delta\delta_2 \\ \Delta\delta_3 \end{bmatrix}^{(k)} = \begin{bmatrix} 23.5 & -11.77 \\ -11.77 & 23.5 \end{bmatrix}^{-1} \begin{bmatrix} \dfrac{\Delta P_2}{V_2} \\ \dfrac{\Delta P_3}{1.04} \end{bmatrix}^{(k)}$$

$$= \begin{bmatrix} 0.05676 & 0.02838 \\ 0.02838 & 0.05676 \end{bmatrix} \begin{bmatrix} \dfrac{\Delta P_2}{V_2} \\ \dfrac{\Delta P_3}{1.04} \end{bmatrix}^{(k)}$$

Expressing the above as two equations, we obtain

$$\Delta\delta_2^{(k)} = 0.05676 \frac{\Delta P_2^{(k)}}{|V_2|^{(k)}} + 0.02729\ \Delta P_3^{(k)} \tag{4.91}$$

$$\Delta\delta_3^{(k)} = 0.02838 \frac{\Delta P_2^{(k)}}{|V_2|^{(k)}} + 0.05458\ \Delta P_3^{(k)} \tag{4.92}$$

From Figure 4.6 the voltage on bus 3 is fixed at $|V_3| = 1.04$. So, all elements pertinent to $|V_3|$ in the $\boldsymbol{B}$ matrix in (4.87) are removed. Thus, we have

$$\Delta\left|V_2^{(k)}\right| = \frac{1}{23.5} \frac{\Delta Q_2^{(k)}}{\left|V_2^{(k)}\right|} = 0.04255 \frac{\Delta Q_2^{(k)}}{\left|V_2^{(k)}\right|} \tag{4.93}$$

Equations (4.91) through (4.93) constitute the recursive equations of the state vector, the elements of which are δ_2, δ_3 and $|V_2|$. However, in the problem at hand we are required to solve for $|V_3|$, which is obtained directly from (4.93). Consequently, we initially have

$$\Delta\left|V_2^{(0)}\right| = 0.04255 \frac{\Delta Q_2^{(0)}}{\left|V_2^{(0)}\right|} \tag{4.94}$$

From Example 4.7 we have $\Delta Q_2^{(0)}$=2.33 pu and $|V_2|$ = 1.0 was initially assumed. Substituting these in (4.94) yields

$$\Delta\left|V_2^{(0)}\right| = 0.04255 \times \frac{2.33}{1.0} = 0.099 \text{ pu}$$

And

$$\left|V_2^{(1)}\right| = \left|V_2^{(0)}\right| + \Delta\left|V_2^{(0)}\right| = 1 + 0.099 = 1.099 \text{ pu}$$

Successive iterations using this value of $|V_2|$ in (4.91), (4.92) and the power flow equations (4.46) and (4.47) yields $|V_2|$ = 1.081 pu which is the same as in Example 4.7. However, it may be verified that it takes more iterations to converge compared to the number of iterations required in the Newton-Raphson method. The principal advantage of the decoupled power method is the reduced computer power required to obtain a solution.

Finally, it is reiterated that regardless of the iterative procedure chosen, a problem of any practical significance can only be solved on a computer. Software programs are commercially available to solve power flow problems involving thousands of buses.

4.8 CONTROL OF REACTIVE POWER FLOW

In Section 4.4 we observed that the magnitude of the bus voltage is specified at a voltage controlled bus. We recall from Sections 4.1 and 4.2 that reactive power controls bus voltage magnitudes. Thus, controlling reactive power will result in maintaining a bus voltage magnitude at some specified level.

Four methods of controlling reactive power on a bus using components connected to the bus are:

(i) By adjusting synchronous generator or motor field excitation
(ii) By using shunt capacitors
(iii) By regulating transformers
(iv) By static var compensators.

Effects of changing the field excitation on a synchronous generator or motor may be understood by referring to Figure 4.9 where we represent the synchronous machine as a voltage source E_g in series with a reactance as given in Figure 3.1. The power system bus to which the machine is connected is assumed to be operating at a voltage of 1.0 $\angle 0°$ pu.

Remembering now that $|E_g|$ increases as the field excitation in the synchronous machine is increased, it is a simple matter to raise or lower $|E_g|$ by raising or lowering the generator or motor field current. If we now confine our discussion to reactive power flow only and neglect any real power flow, δ

must be zero according to (4.7) and average reactive power flow is proportional to $(|E_g|^2 - 1^2)$ in accordance with (4.10). Therefore, overexciting (applying a large field current) in either a synchronous generator or an unloaded synchronous motor will result in a reactive power flow into the system bus. The effect of this reactive power flow is to raise the bus voltage (see Example 4.3). A synchronous motor applied in this manner is often called a synchronous condenser.

Use of *shunt capacitors* to control the reactive power and thereby a bus voltage involves installing shunt capacitor banks at the buses. Each capacitor bank supplies reactive power at the point at which it is placed. Hence, it reduces the line current necessary to supply reactive power to a load and reduces the voltage drop in the line via power-factor improvement.

A capacitor bank installed at a particular system bus having an assumed voltage of 1.0 $\angle 0°$ pu is shown in Figure 4.10. In this case, the reactive power flow Q into the bus is

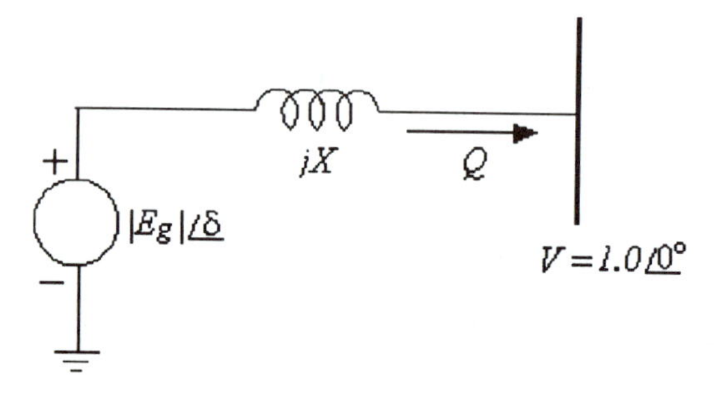

FIGURE 4.9
Synchronous machine connected to a power system bus.

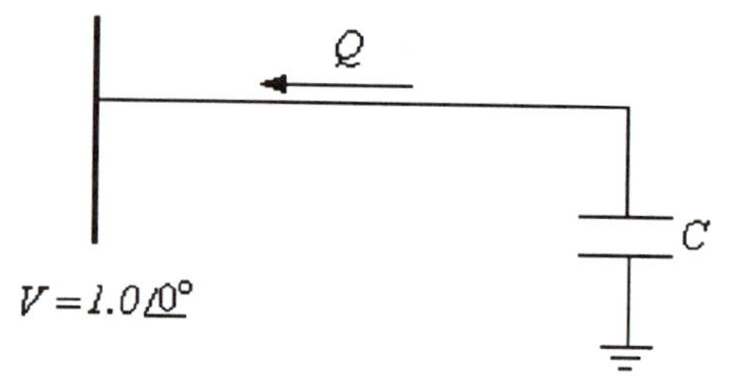

FIGURE 4.10
Capacitor bank connected to a power system bus.

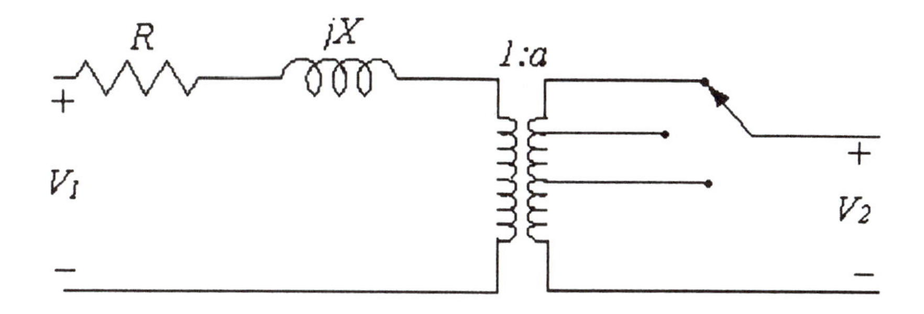

FIGURE 4.11
Per phase representation of a tap-changing or regulating transformer.

$$Q = -|I|^2 X_C = -\left|\frac{1.0}{X_C}\right|^2 X_C = \frac{1}{|X_C|}$$

Since $X_C = -1/\omega C$, then $Q = \omega C$ into the bus which will once again tend to raise the bus voltage above 1.0 pu.

Turning now to the third method, *regulating transformers* may be used to control the flow of real as well as reactive power, since regulating transformers can regulate the magnitudes and phase angles of line voltages. Such transformers usually provide an adjustment of voltage magnitude in the range up to ±10%. These transformers have taps on their windings to vary the transformation ratio. Tap changing may be done on no-load or on load. Load-tap-changing (LTC) transformers operate in conjunction with motors, responding to relays, for automatic tap changing.

On a per phase basis, a tap-changing or regulating transformer is shown in Figure 4.11. The magnitude of a is usually within ±10% of unity and a may be real or complex. When a is real, the regulating transformer is primarily used to control reactive power flow between buses by increasing or decreasing the voltage magnitude on the tap-changing side of the transformer. When a is complex, the angle associated with a modifies the power angle δ and real power flow is controlled in this manner. For a discussion of how such so-called phase shifting transformers may be constructed, the student is referred to more advanced texts on power systems. When the transformation ratio a is real, the regulating transformer may be represented in a power flow analysis by a π-model as discussed in Section II.9.

Finally, *static var compensators* consist of electronically switched inductors and capacitors, and operate on the same principles as do shunt capacitors and reactors.

4.9 COMPUTER SOLUTION OF POWER FLOW PROBLEMS

As has been mentioned previously, computer methods are the only practical approach to power flow analysis. In Section 4.6, one commercially available software package, SKM Systems Analysis Power* Tools for Windows was applied to a simple 3-bus system of the type that would normally be associated with the operation of an electric utility. Power flow analysis is also of interest in the design and operation of industrial power systems as well as utility power systems and the programs that are available have the capability to analyze both situations. In using such software, the wide variety of components that constitute either type of power system are easily specified and connected to define the system. Calculations such as power flow analysis, fault analysis (to be discussed in detail in Chapter 5), system protection, stability, and others may then be conducted with the click of a mouse button or with a single keystroke.

As a final example in this chapter, another commercially available software package, ETAP Power Station from Operation Technology, Inc., will be utilized on an industrially oriented power system. This example, as well as the utility-oriented example given previously, are applied to small systems in order to illustrate the capabilities. It should be remembered, however, that these programs are capable of performing analyses on both utility and industrial power systems consisting of hundreds or even a thousand or more buses.

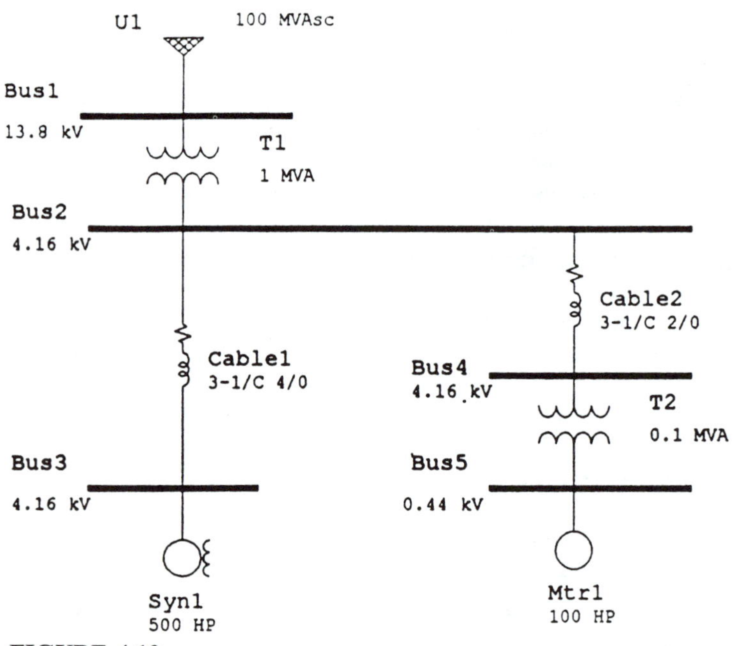

FIGURE 4.12
Single-line diagram for the power system of Example 4.9 showing rated values.

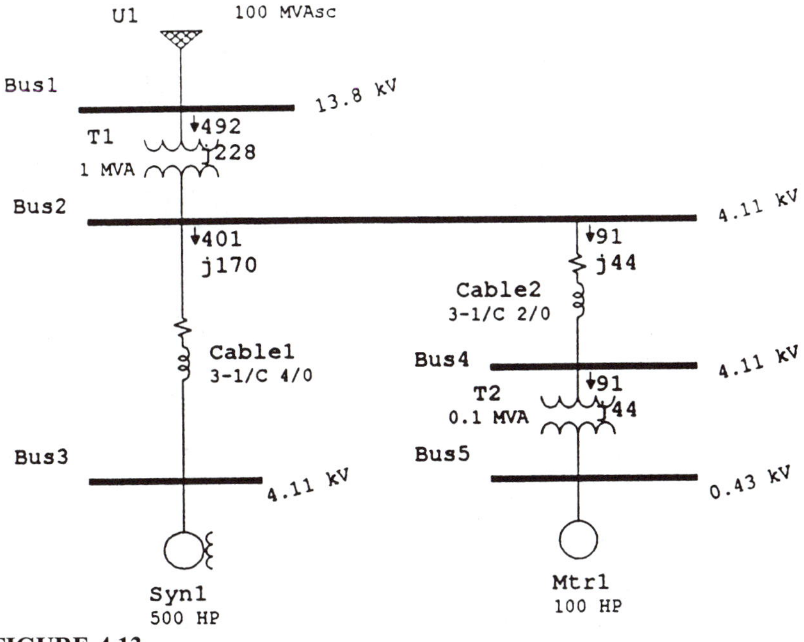

FIGURE 4.13
Single-line diagram for the power system of Example 4.9 showing actual power flows and system voltages.

There are 5 buses in the industrial power system of Example 4.9 with bus 1 connected to an electrical utility having a short-circuit capacity of 100 MVA as shown in the computer-generated single-line diagram of Figure 4.12. Transformer T1 is used to step down the rated utility distribution voltage of 13.8 kV to 4160 V and 500 ft. of 4/0 cable connects the output of this transformer to a fully loaded 500 hp synchronous motor. A second branch circuit from the output of T1 is routed through 1000 ft. of 2/0 cable to a 4160:440 V step-down transformer which supplies a fully loaded 100 hp induction motor. Nominal ratings of all elements of the system are as shown in Figure 4.12.

The computer-generated output of the power flow study on this system is shown in Figure 4.13. Actual bus voltages in kV and the power flows in kW and kvar are as indicated. Contingencies relating to system additions, design changes, etc. may easily be investigated using this tool.

PROBLEMS

4.1. For a system of the type shown in Figure 4.2, $|V_S| = 1.0$ pu and $|V_R| = 1.1$ pu. The loads are $S_{L1} = 3 + j4$ pu, $S_{L2} = 7 + j6$ pu, and the real power supplied by each generator is 5.0 pu. If the line reactance is 0.08 pu, calculate the reactive input power required at both ends of the line.

4.2. In the system of Figure 4.3, $V_1 = 1\angle 0°$ pu, $Z_l = (0.10 + j0.10)$ pu, and $P_2 + jQ_2 = (0.1 + j0.1)$ pu. Calculate V_2 using an iterative approach.

4.3. Determine the real and reactive power supplied to bus 1 in Problem 4.2.

4.4. How much reactive power must be supplied at bus 2 in Problem 4.2 so that $|V_2| = 1.1$ pu?

4.5. For the system shown in Figure 4.3, $Z_l = (0.2 + j0.6)$ pu, $V_1 = 1.1\angle 0°$ pu, and $P_2 + jQ_2 = (0.059 + j0.176)$ pu. Determine the complex power on bus 1.

4.6. Two buses are interconnected by a transmission line of impedance $(0.03 + j1.2)$ pu. The voltage on one bus is $1\angle 0°$, and the load on the other bus is $(1 + j0.4)$ pu. Determine the per-unit voltage on this second bus. Also calculate the per-unit real and reactive power into the first bus.

4.7. The voltages on the two buses of Problem 4.6 are to be made equal in magnitude by supplying reactive power at the second bus. How much reactive power must be supplied?

4.8. The per-unit impedance of a short transmission line is $j0.06$. The per-unit load on the line is $(1 + j0.6)$ pu at a receiving-end voltage of $1\angle 0°$ pu. Calculate the average reactive power flow over the line.

4.9. Obtain the bus admittance matrix for the network shown in Figure 4.14. The values shown are per-unit admittances.

4.10. A generator is connected to a system as shown by the equivalent circuit of Figure 4.15. If $V_t = 0.97\angle 0°$ pu, calculate the complex power delivered by the generator. Also determine E_g.

4.11. Solve the following system of equations by the Gauss-Seidel method, starting with $x(0) = y(0) = 0$:

$$10x + 5y = 6$$
$$2x + 9y = 3$$

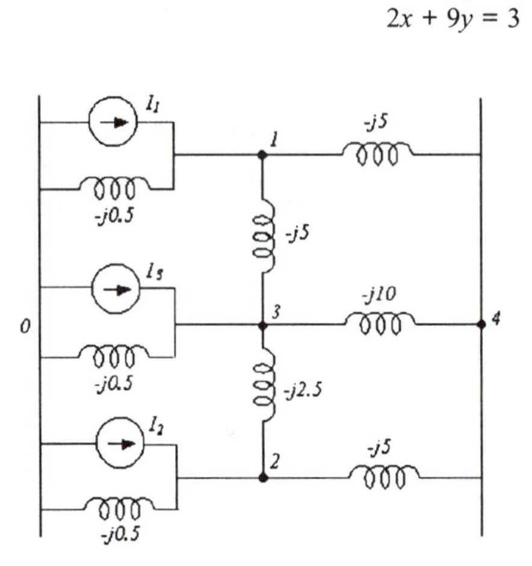

FIGURE 4.14
Problem 4.9.

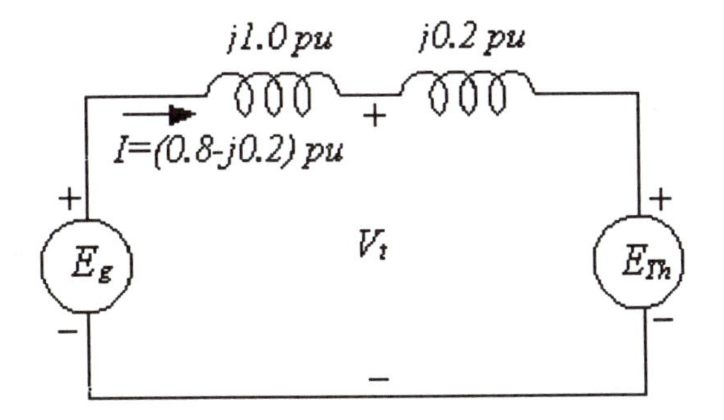

FIGURE 4.15
Problem 4.10.

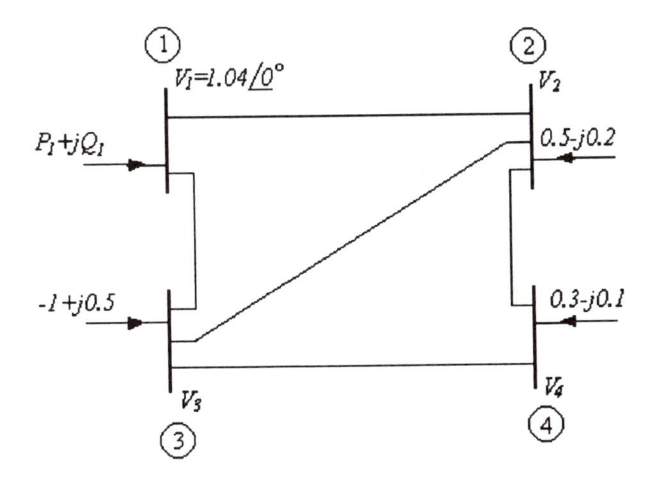

FIGURE 4.16
Problems 4.12, 4.13 and 4.14.

4.12. For the system shown in Figure 4.16, the bus admittance matrix is

$$Y_{bus} = \begin{bmatrix} 3 - j9 & -2+j6 & -1 + j3 & 0 \\ -2 + j6 & 3.666 - j11 & -0.666 + j2 & -1 + j3 \\ -1 + j3 & 0.666 - j2 & 3.666 - j11 & -2 + j6 \\ 0 & -1 + j3 & -2+j6 & 3 - j9 \end{bmatrix} \text{pu}$$

With the complex power on buses 2, 3, and 4 as shown in the figure, determine the value for V_2 that is produced by the first iteration of the Gauss-Seidel procedure. Use an initial guess of $V_2 = V_3 = V_4 = 1.0 \angle 0°$

4.13. Determine the values for V_3 and V_4 in the system of Figure 4.16 and Problem 4.12, as produced by the first iteration of the Gauss-Seidel procedure.

4.14. Determine the value for V_2 of Problem 4.12 as produced by the second iteration of the Gauss-Seidel procedure.

4.15. Evaluate x in the equation $x + \sin x = 2$ by the Gauss-Seidel method. Start with $x(0) = 0$ rad.

4.16. Rework Problem 4.15 using the Newton-Raphson method, and compare the number of iterations required to achieve convergence by the two methods.

4.17. In a five-bus system

$$Y_{21} = Y_{23} = 0 \text{ pu}$$

$$Y_{22} = 7.146 \angle -84.6° \text{ pu}$$

$$Y_{24} = 2.490 \angle 95.1° \text{ pu}$$

$$Y_{25} = 4.980 \angle 95.1° \text{ pu}$$

Determine $V_2^{(1)}$ by the Gauss-Seidel method if $P_2 - jQ_2 = -2 + j0.7$. Begin with $V_1^{(0)} = V_2^{(0)} = V_3^{(0)} = V_4^{(0)} = V_5^{(0)} = 1\angle 0°$ pu.

4.18. For a two-bus system with bus 1 as the swing bus, the bus impedance matrix is

$$Z_{\text{bus}} = \begin{bmatrix} j1.33 & j1 \\ j1 & j1.5 \end{bmatrix} \text{ pu}$$

Start with $V_1 = V_2^{(0)} = 1.05\angle 0°$, and solve for V_2 by the Gauss-Seidel method if the load on bus 2 is $0.1 + j0$ pu.

4.19. Rework Problem 4.18 using one iteration of the Newton-Raphson method.

4.20. For a three-bus system

$$Y_{\text{bus}} = \begin{bmatrix} 1.6 - j8 & -0.8 + j4 & -0.8 + j4 \\ -0.8 + j4 & 1.6 - j8 & -0.8 + j4 \\ -0.8 + j4 & -0.8 + j4 & 1.6 - j8 \end{bmatrix} \text{ pu}$$

Bus 1 is the swing bus and the load on bus 2 is $0.8 + j0.6$ pu. Using an initial voltage of $1.0 \angle 0°$ at all the system buses, perform one iteration in the Gauss-Seidel procedure.

4.21. If Problem 4.20 were to be solved using the Newton-Raphson procedure, determine the power mismatch at bus 2 for the first iteration.

4.22. Show that the entries in the second and third rows of the first column of the Jacobian matrix in (4.52) of Example 4.7 are the correct values.

4.23. Show that the entries in the second column of the Jacobian matrix in (4.52) of Example 4.7 are the correct values.

4.24. Show that the entries in the third column of the Jacobian matrix in (4.52) of Example 4.7 are the correct values.

4.25. Remove the generator from bus 3 of Example 4.7 and repeat the power flow study for this system. Note that bus 3 now becomes a load bus and $|V_3|$ can no longer be maintained at 1.04 pu. [Note: This problem is intended for solution on a digital computer].

4.26. Repeat Example 4.5 using an acceleration factor of $\alpha = 1.6$. Compare your results with those of the example.

4.27. Perform the second and third iterations for Example 4.6.

4.28. Obtain a closed form solution to the equations of Example 4.6 and compare your result with that of Problem 4.27.

4.29. Solve equation 2 of Example 4.6 for x_1 and solve equation 1 for x_2. Perform three Gauss-Seidel iterations with the given initial guess and no acceleration factor. Compare your results with those of Problems 4.27 and 4.28.

4.30. Repeat Problem 4.29 with an acceleration factor of 1.5.

4.31. Verify that real and reactive power balances are maintained in the system output given in Figure 4.8 and Table 4.1.

4.32. Calculate the rms current flowing between buses 1 and 2 of Example 4.7 in amperes and compare with that given in Table 4.1.

4.33. Calculate the real and reactive line losses between buses 1 and 2 of Example 4.7 and compare with those given in Table 4.1.

4.34. Repeat Example 4.5 using the Newton-Raphson solution procedure with the same initial guess of $V_2 = 1.0 \angle{-10°}$ pu. Perform two iterations and compare with the Gauss-Seidel solution.

4.35. Using $V_2 = 1.081 \angle{-1.37°}$ pu and $\delta_3 = -3.8°$, calculate the reactive power that must be supplied by the generator G3 in Example 4.7 so that $|V_3| = 1.04$ pu. (Note that your answer should differ slightly from the computer-generated result due to roundoff).

CHAPTER

5

FAULT CALCULATIONS AND POWER SYSTEM PROTECTION

Under normal conditions, a power system operates as a balanced three-phase ac system. A significant departure from this condition is often caused by a fault. A fault may occur on a power system for a number of reasons, some of the common ones being lightning, high winds, snow, ice, and frost. Faults give rise to abnormal operating conditions, usually excessive currents and voltages at certain points on the system. Protective equipment is used on the system to prevent continuing abnormal conditions which could lead to the destruction of system components. The magnitudes of fault currents determine the interrupting capacity of the circuit breakers and settings of protective relays. Steps must be taken to remove any fault as quickly as possible. Faults may occur within or at the terminals of a generator, transformer, transmission line or any other system component.

5.1 TYPES OF FAULTS

Various types of short-circuit faults that can occur in a power system are depicted in Figure 5.1. The frequency of occurrence decreases from part (*a*) to part (*f*). Although the balanced three-phase short circuit in Figure 5.1(*d*) is

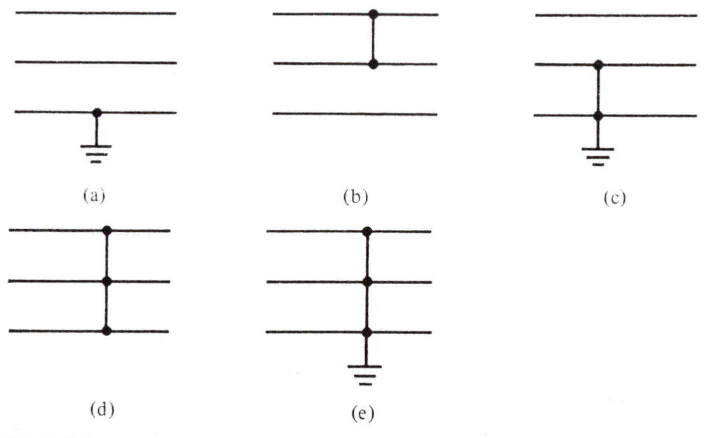

FIGURE 5.1
Types of faults on a power system.

relatively uncommon, it is often the most severe fault and therefore determines the rating of the line-protecting circuit breaker. A *fault study* includes the following:

1. Determination of the maximum and minimum three-phase short-circuit currents.

2. Determination of unsymmetrical fault currents, as in single line-to-ground, double line-to-ground and line-to-line faults.

3. Determination of the ratings of required circuit breakers.

4. Investigation of schemes of protective relaying.

5. Determination of voltage levels at strategic points during a fault.

The short-circuit faults depicted in Figure 5.1 are called *shunt faults*; open circuits, which may be caused by broken conductors, for instance, are categorized as *series faults*. The fault shown in Figure 5.1(*a*) is a line-to-ground fault and that of Figure 5.1(*b*) is a line-to-line fault. Faults shown in Figures 5.1(*c*) and (*d*) are respectively, double-line-to-ground fault and three-phase symmetrical short circuit.

Strictly speaking, for current and MVA calculations, the various components constituting the system must be represented by their respective equivalent circuits interconnected to form the system. However, under fault conditions, we are most often interested in the fault current during the first few cycles (sometimes called the subtransient period) as the fault must be cleared quickly. In such a case, transformers are represented by their (per phase) leakage reactances, transmission lines by their reactances and synchronous machines by constant voltage sources in series with their subtransient reactances.

5.2 SYMMETRICAL FAULTS

A balanced three-phase short circuit [Figure 5.1(*d*) or (*e*)] is an example of a symmetrical fault. Balanced three-phase fault calculations can be carried out on a per-phase basis, so that only single-phase equivalent circuits need be used in the analysis. Invariably, the circuit constants are expressed in per-unit terms, and all calculations are made on a per-unit basis. In short-circuit calculations, we often evaluate the short-circuit MVA (megavolt-amperes), which is equal to $\sqrt{3}V_l I_f$, where V_l is the nominal line voltage in kilovolts, and I_f is the fault current in kiloamperes.

An example of a three-phase symmetrical fault is a sudden short at the terminals of a synchronous generator. The symmetrical trace of such a short-circuited stator-current wave is shown in Figure 5.2. (The time-decaying dc component is neglected. See Section 5.12 for a discussion of this current.) The wave, whose envelope is shown in Figure 5.3, may be divided into three periods or time regimes: the *subtransient period*, lasting only for the first few cycles, during which the current decrement is very rapid; the *transient period*, covering a relatively longer time during which the current decrement is more moderate; and finally the *steady-state period*. The difference $\Delta i'$ (in Figure 5.3) between the transient envelope and the steady-state amplitude is plotted on a logarithmic scale as a function of time in Figure 5.4, along with the difference $\Delta i''$ between the subtransient envelope and an extrapolation of the transient envelope. Both plots closely approximate straight lines, illustrating the essentially exponential nature of the decrement. (See Section III.4 in Appendix III for a discussion of the physical causes behind these three time periods.)

The currents during these three regimes are limited primarily by various reactances of the synchronous machine (we neglect the armature resistance, which is relatively small). These currents and reactances are defined by the following equations, provided the alternator was operating at no load before the occurrence of a three-phase fault at its terminals:

$$|I| = \frac{Oa}{\sqrt{2}} = \frac{|E_g|}{X_d} \tag{5.1}$$

$$|i'| = \frac{Ob}{\sqrt{2}} = \frac{|E_g|}{X_d'} \tag{5.2}$$

$$|i''| = \frac{Oc}{\sqrt{2}} = \frac{|E_g|}{X_d''} \tag{5.3}$$

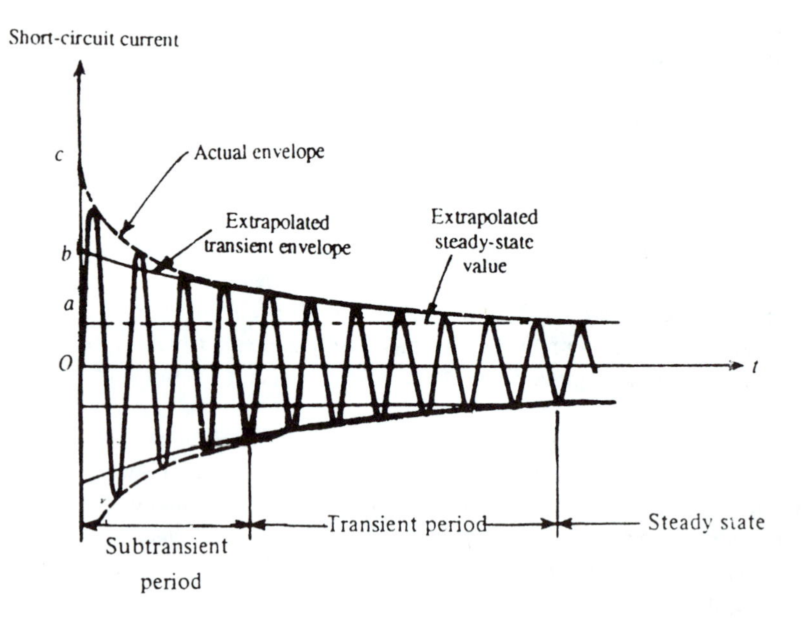

FIGURE 5.2
Plot of symmetrical three-phase short-circuit current.

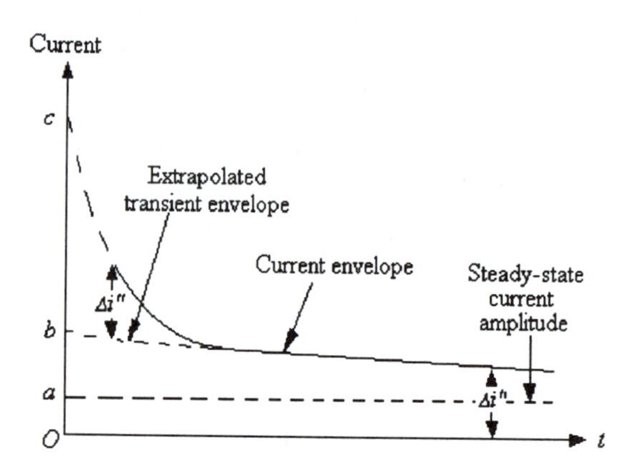

FIGURE 5.3
Envelope of the current waveform of Figure 5.2.

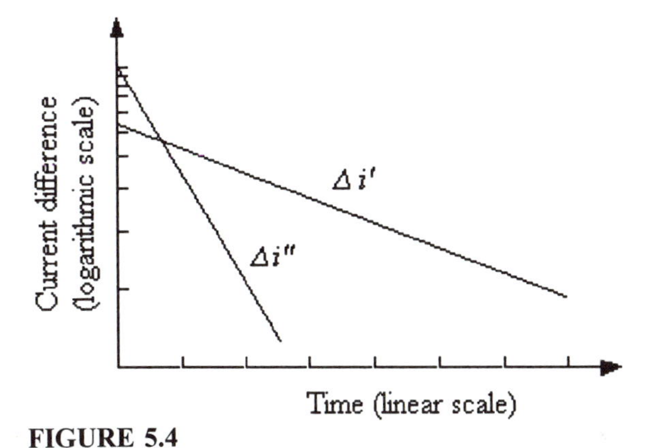

FIGURE 5.4

$\Delta i'$ and $\Delta i''$ plotted on a logarithmic scale.

where $|E_g|$ is the no-load voltage of the generator, the currents are rms currents, and O, a, b, and c are shown in Figure 5.2. The machine reactances X_d, X_d', and X_d'' are known as the *direct-axis synchronous reactance*, *direct-axis transient reactance*, and *direct-axis subtransient reactance*, respectively. The currents I, i', and i'' are known as the steady-state, transient, and subtransient currents. From (5.1) through (5.3) it follows that the fault currents in a synchronous generator can be calculated when the machine reactances are known.

Suppose now that a generator is loaded when a fault occurs. Figure 5.5(a) shows the corresponding equivalent circuit with the fault to occur at point P. Since we wish to represent the machine by a voltage source E_g'' in series with a reactance X_d'' immediately after the fault occurs (Figure 5.5(b)) we choose this same representation prior to the fault. Note that this is a departure from our normal steady-state model of a voltage source in series with the synchronous reactance.

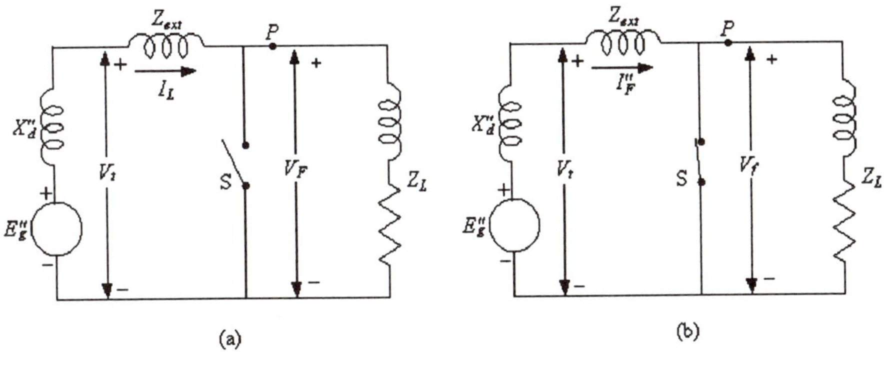

FIGURE 5.5

Fault on a loaded generator.

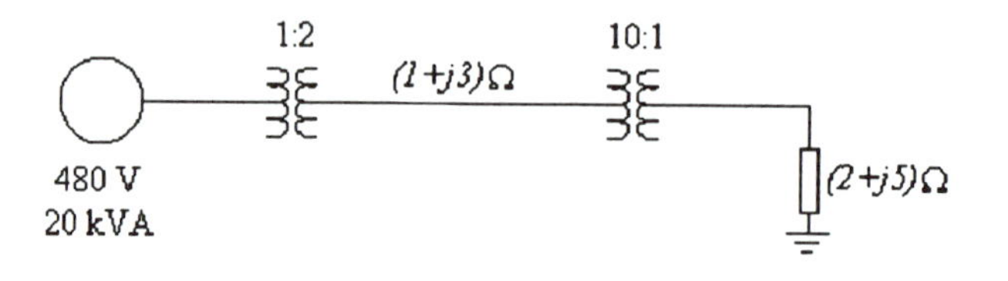

FIGURE 5.6
Example 5.1.

The current flowing before the fault occurs is I_L, the voltage at the fault is V_F, and the terminal voltage of the generator is V_t. When a three-phase fault occurs at P, the circuit shown in Figure 5.5*(b)* becomes the appropriate equivalent circuit (with switch S closed). Here the voltage E_g'' in series with X_d'' that supplies the steady-state current I_L when switch S is open, supplies the current to the short circuit through X_d'' and Z_{ext} when switch S is closed. If we can determine E_g'', we can find this current through X_d'', which will be I_F''. With switch S open, we have

$$E_g'' = V_t + jI_L X_d'' \tag{5.4}$$

which defines E_g'', the subtransient internal voltage. A similar argument for transient representation results in an expression for the transient internal voltage as

$$E_g' = V_t + jI_L X_d' \tag{5.5}$$

Clearly E_g'' and E_g' are dependent on the value of the load before the fault occurs.

Example 5.1 Figure 5.6 shows the one-line diagram for a three phase system in which a generator supplies a load through a step-up transformer, a transmission line, and a step-down transformer. Calculate the per-unit current if the terminal voltage of the generator is 450 V line-to-line. Determine E_g'' if X_d'' of the generator is 0.1 pu. The transformers are ideal and the turns ratios are line-to-line values.

Solution
Because the voltage (and current) levels change across the transformers, different base voltages are involved at different parts of the system. At the generator,

$$\text{kVA}_{\text{base}} = 20 \text{ kVA three phase}$$

(one kVA base is chosen for the entire problem)

$$\text{V}_{\text{base, gen}} = 480 \text{ V line-to-line}$$

$$\text{Z}_{\text{base, gen}} = \frac{(480)^2}{20{,}000} = 11.52\ \Omega \text{ from (3.9)}$$

Along the transmission line,

$$V_{\text{base, line}} = \frac{480}{0.5} = 960\text{V line-to-line}$$

$$Z_{\text{base, line}} = \frac{(960)^2}{20{,}000} = 46.08\ \Omega$$

$$Z_{\text{line}} = \frac{1 + j3}{46.08} = (0.022 + j0.065)\ \text{pu}$$

At the load,

$$V_{\text{base, load}} = \frac{960}{10} = 96\text{V line-to-line}$$

$$Z_{\text{base, load}} = \frac{(96)^2}{20{,}000} = 0.4608\ \Omega$$

$$Z_{\text{load}} = \frac{2 + j5}{0.4608} = (4.34 + j10.85)\ \text{pu}$$

The total impedance is then

$$Z_{\text{total}} = Z_{\text{line}} + Z_{\text{load}} = (0.022 + j0.065) + (4.34 + j10.85)$$

$$= 4.362 + j10.915 = 11.75\ \angle 68°\ \text{pu}$$

and

$$V_t = \frac{450}{480} = 0.9375\ \text{pu}$$

so that

$$I_L = \frac{0.9375\angle 0°}{11.75\angle 68°} = 0.0798\angle -68°\ \text{pu}$$

$$E_g'' = 0.9375 + (j0.1)(0.0798\angle -68°) = 0.945\ \angle 0.2°\ \text{pu}$$

Example 5.2 An interconnected generator-reactor system is shown in Figure 5.7*(a)*. The base values for the given percent reactances are the ratings of the individual pieces of equipment. A three-phase short-circuit occurs at point F. Determine the fault current if the busbar line-to-line voltage is 11 kV and the currents flowing prior to the fault are assumed to be negligible.

Solution

First, we arbitrarily choose 11 kV line-to-line as the base voltage and 50 MVA as the base MVA. The per-unit values for the system reactances, referred to this base are then

$$X_{G1} = \frac{50}{10}0.10 = 0.5\ \text{pu} \quad \text{from (3.6) or (3.7)}$$

$$X_{G2} = \frac{50}{20}0.15 = 0.375\ \text{pu}$$

$$X_{G3} = \frac{50}{20}\ 0.15 = 0.375\ \text{pu}$$

$$X_1 = \frac{50}{10} 0.05 = 0.25 \text{ pu}$$

$$X_2 = \frac{50}{8} 0.04 = 0.25 \text{ pu}$$

These reactances produce the per-phase reactance diagram of Figure 5.7(*b*), which is simplified to Figure 5.7(*c*).

Since prefault currents are negligible, the open circuit voltage in the Thevenin equivalent circuit at the fault point is 11 kV line-to-line or 1.0 pu. The Thevenin impedance in per unit is

$$\text{Per-unit impedance} = j\frac{0.5\,(0.2344 + 0.25)}{0.5 + (0.2344 + 0.25)} = j0.246$$

Then

$$I_F = \frac{1.0}{j0.246} = -j4.06 \text{ pu}$$

$$I_{\text{base}} = \frac{50{,}000{,}000}{\sqrt{3}\ (11{,}000)} = 2624.3 \text{ A}$$

and

$$|I_F| = (4.06)(2624.3) = 10{,}655 \text{ A}$$

Example 5.3 A three-phase short-circuit fault occurs at point F in the system shown in Figure 5.8(*a*). Calculate the fault current assuming that prefault currents are zero and that the generators are operating at rated voltage.

Solution
Let the base MVA be 30 MVA, and let 33 kV be the base line voltage at the high voltage side of the transformer. Then referred to these values, we have the following reactances and impedance:

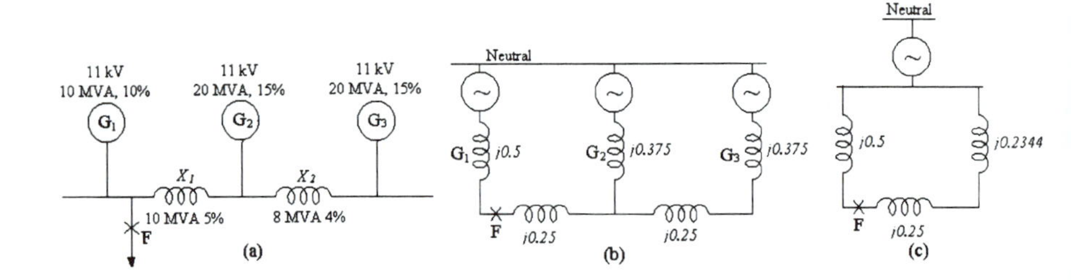

FIGURE 5.7
Example 5.2.

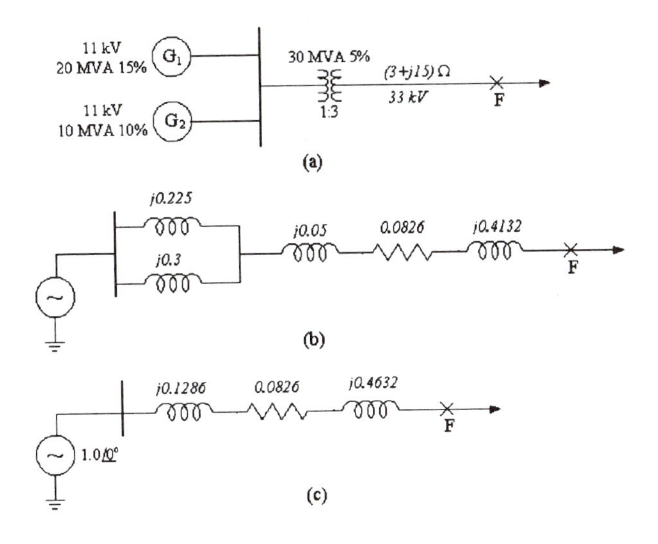

FIGURE 5.8
Example 5.3.

$$X_{G1} = \frac{30}{20} 0.15 = 0.225 \text{ pu}$$

$$X_{G2} = \frac{30}{10} 0.10 = 0.3 \text{ pu}$$

$$X_{trans} = \frac{30}{30} 0.05 = 0.05 \text{ pu}$$

$$Z_{line} = (3 + j15) \frac{30}{33^2}$$

$$= (0.0826 + j0.4132) \text{ pu}$$

These per-unit values are shown in Figure 5.8(b), which can be reduced to Figure 5.8(c). From that diagram we find that the total impedance from the generator neutral to the fault is

$$\text{Per-unit } Z_{total} = 0.0826 + j0.5918 = 0.5975 \angle 82° \text{ pu}$$

Then

$$I_F = \frac{1.0}{0.5975 \angle 82°} = 1.674 \angle -82° \text{ pu}$$

and

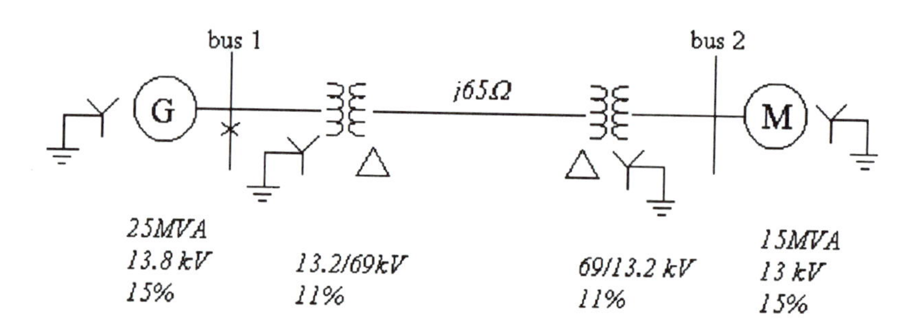

FIGURE 5.9
Example 5.4.

$$I_{\text{base}} = \frac{30{,}000{,}000}{\sqrt{3}\ (33{,}000)} = 524.8 \text{ A}$$

$$|I_F| = (1.674)(524.8) = 878.6 \text{ A}$$

Example 5.4 A three-phase short-circuit occurs at the generator bus (bus 1) of the system shown in Figure 5.9. Calculate the subtransient fault current using superposition. The generator is operating at its rated voltage. Neglect prefault current.

Solution
With the exception that the motor M_2 has been deleted, the given system is identical to that of Example 3.6 (of Chapter 3). So, we use the reactance diagram of Figure 3.15, excluding the second motor. We obtain the reactance diagram shown in Figure 5.10(*a*) where the switch SW simulates the short-circuit, and E_g'' and E_m'' are the machine prefault internal voltages. The circuit of Figure 5.10(*a*) may now be represented by that of Figure 5.10(*b*) where the voltages V_F in phase opposition replace the switch. To apply superposition, we replace the circuit of Figure 5.10(*b*) by the two circuits shown in Figure 5.10(*c*). If we now choose the value of V_F to be equal to the voltage at the fault point prior to the occurrence of the fault, then $V_F = E_m'' = E_g''$ (remember we are neglecting prefault currents) and $I_{F2}'' = 0$. As a result V_F may be open circuited as shown in Figure 5.10(*d*).

From the two circuits shown in Figure 5.10(*d*), we see that the fault current is simply $I_F'' = I_{F1}''$. (Note: We use double-primes as superscripts to reiterate the subtransient nature of currents and voltages).

The equivalent impedance between terminals *a* and *b* in the first network of Figure 5.10(*d*) is

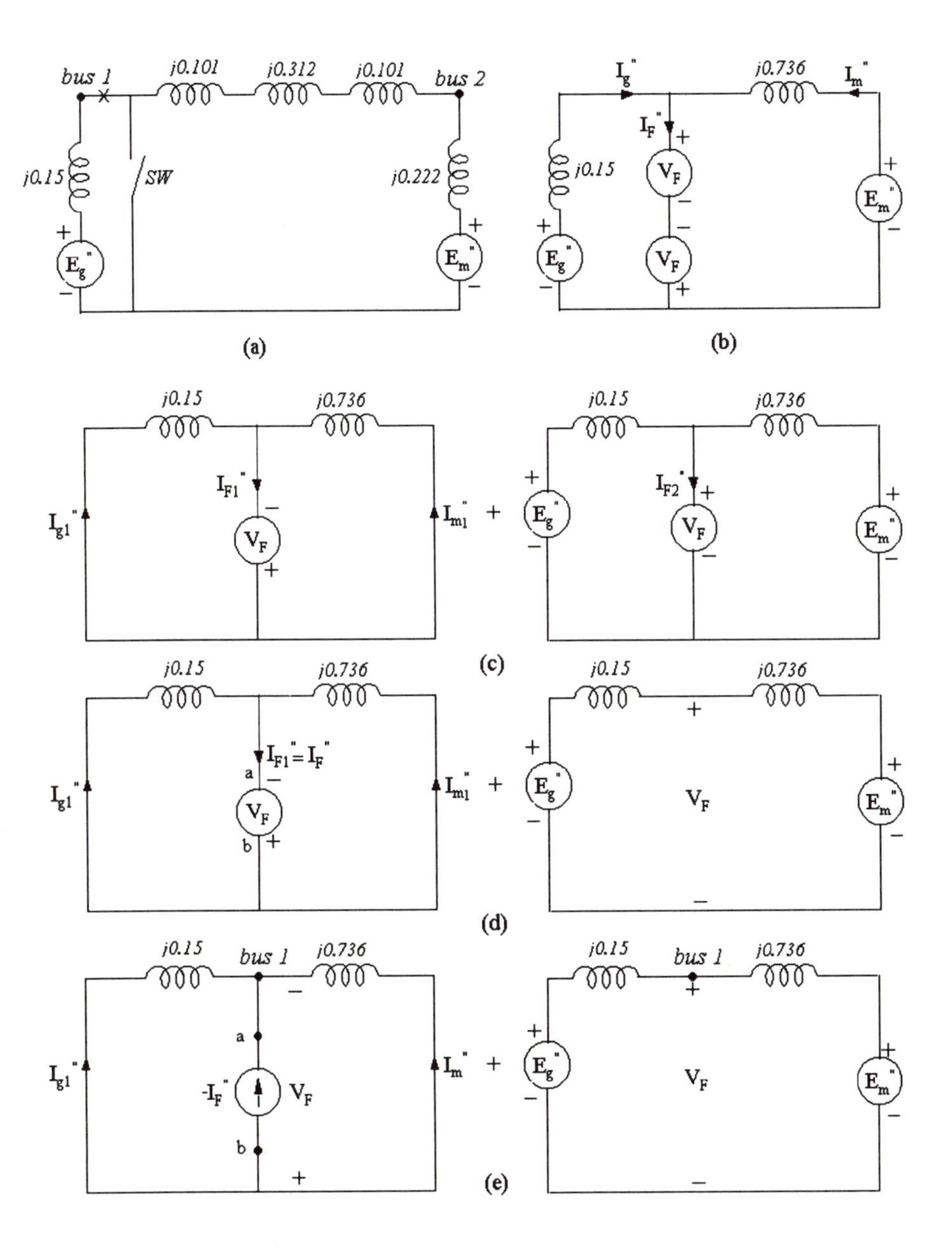

FIGURE 5.10
Example 5.4.

$$jX = \frac{j0.15 \times j0.736}{j0.15 + j0.736} = j0.1246 \text{ pu}$$

Since we are neglecting prefault currents, $E_g'' = E_m'' = V_F = 1.0 \angle 0°$ pu and

$$I_F'' = I_{F1}'' = \frac{V_F}{jX} = \frac{1\angle 0°}{j0.1246} = -j8.025 \text{ pu}$$

We may further note that if V_F in the first network of Figure 5.10(*d*) is replaced by a current source of value $-I_F''$ as shown in Figure 5.10(*e*), then $V_{ab} = (jX)\,(-I_f'') = -V_F$. In terms of the actual values, $(j0.1246)(j8.025) = -1.0$ pu $= -V_F$ or $V_F = 1.0$ pu and the networks of Figure 5.10(*e*) are equivalent to those of 5.10(*d*).

The procedure of the preceding example may be extended to calculate three-phase faults on an N-bus system via the concept of bus impedance matrix as presented in the next section.

5.3 BUS IMPEDANCE MATRIX

The impedance matrix is defined by

$$\boldsymbol{Z}_{\text{bus}} = \boldsymbol{Y}_{\text{bus}}^{-1} \tag{5.6}$$

Then (4.17) of Chapter 4 may be written as

$$\boldsymbol{V} = \boldsymbol{Z}_{\text{bus}}\boldsymbol{I} \tag{5.7}$$

where

$$\boldsymbol{Z}_{\text{bus}} = \begin{bmatrix} Z_{11} & Z_{12} & \cdots & Z_{1N} \\ Z_{21} & Z_{22} & \cdots & Z_{2N} \\ \cdots\cdots & \cdots\cdots & \cdots\cdots & \cdots\cdots \\ Z_{N1} & N_{N2} & \cdots & Z_{NN} \end{bmatrix} \tag{5.8}$$

and $\boldsymbol{Y}_{\text{bus}}$ has been defined by (4.18).

In order to calculate a three-phase short-circuit fault current for a fault at bus F due to voltage sources $E_1, E_2, \ldots, E_N$, we use superposition as applied in Example 5.4. Thus, corresponding to the first circuit of Figure 5.10(*e*), we have only one source at bus F. Remembering that $\boldsymbol{V}$ in (5.7) is the column matrix of each bus voltage measured with respect to reference and that $\boldsymbol{I}$ is the column matrix of currents injected into each bus, we may write (5.7) for the first network (denoted by the superscript 1) of Figure 5.10(*e*) as

$$\begin{bmatrix} V_1^{(1)} \\ V_2^{(1)} \\ \vdots \\ -V_F \\ \vdots \\ V_N^{(1)} \end{bmatrix} = \begin{bmatrix} Z_{11} & Z_{12} & \dots & Z_{1F} & \dots & Z_{1N} \\ Z_{22} & Z_{22} & . & . & . & . \\ \vdots & . & . & . & . & \vdots \\ Z_{F1} & . & \dots & Z_{FF} & \dots & Z_{FN} \\ \vdots & . & . & . & . & . \\ Z_{N1} & \dots & \dots & Z_{NF} & \dots & Z_{NN} \end{bmatrix} \begin{bmatrix} 0 \\ 0 \\ \vdots \\ -I_F'' \\ \vdots \\ 0 \end{bmatrix} \tag{5.9}$$

and the total fault current (remembering that $I_{F2}'' = 0$)

$$I_{F1}'' = I_F'' = \frac{V_F}{Z_{FF}} \tag{5.10}$$

where V_F is the voltage at bus F prior to the occurrence of the fault. Furthermore, combining (5.9) and (5.10), the voltage at any bus n in the first circuit $V_n^{(1)}$ is related to V_F by

$$V_n^{(1)} = Z_{nF}\,(-I_F'') = \frac{-Z_{nF}}{Z_{FF}}\,V_F \tag{5.11}$$

Recall from Example 5.4 that the second circuit of Figure 5.10(e) corresponds to prefault conditions and we will neglect prefault currents. Thus, the prefault voltage at each bus n (of the second circuit) is V_F; that is

$$V_n^{(2)} = V_F \tag{5.12}$$

By superposition, from (5.11) and (5.12) we have

$$V_n = \left(1 - \frac{Z_{nF}}{Z_{FF}}\right) V_F \tag{5.13}$$

We now illustrate the application of the above procedure to solve the problem of Example 5.4.

Example 5.5 Repeat Example 5.4 using the Z_{bus} concept.

Solution

In order to find $\boldsymbol{Z}_{\text{bus}}$, we first find $\boldsymbol{Y}_{\text{bus}}$. To obtain $\boldsymbol{Y}_{\text{bus}}$ we redraw the circuit of Figure 5.10(a) in terms of admittances as shown in Figure 5.11, from which we have

$$\boldsymbol{Y}_{\text{bus}} = \begin{bmatrix} -j8.612 & j1.945 \\ j1.945 & -j6.45 \end{bmatrix} \text{pu}$$

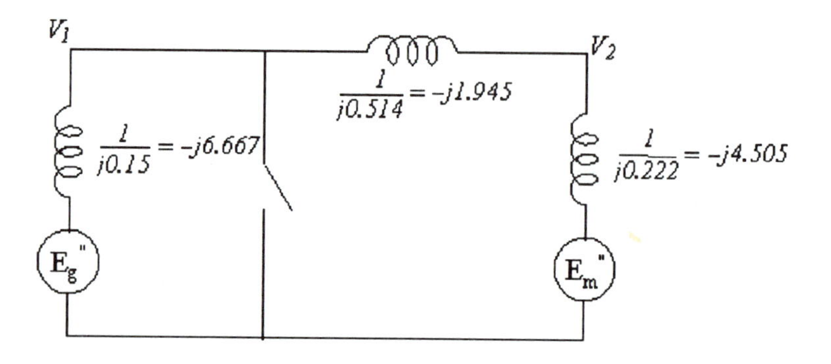

FIGURE 5.11
Example 5.5.

and

$$Z_{bus} = Y_{bus}^{-1} = \begin{bmatrix} j0.1246 & j0.0376 \\ j0.0376 & j0.1664 \end{bmatrix} \text{ pu}$$

Thus, from (5.10), we obtain for $F = 1$

$$I_F'' = \frac{V_F}{Z_{11}} = \frac{1.0\angle 0°}{j0.1246} = -j8.025 \text{ pu}$$

which (obviously) agrees with the result of Example 5.4.

Example 5.6 For the system of Example 5.4, calculate the fault current through the transmission line.

Solution
In order to find the current through the transmission line, with a three-phase short-circuit at bus 1 (Figure 5.11), we must determine the voltages at buses 1 and 2. Thus, from (5.13) we have:

$$V_1 = \left(1 - \frac{Z_{11}}{Z_{11}}\right) V_F = 0$$

$$V_2 = \left(1 - \frac{Z_{21}}{Z_{11}}\right) V_F = \left(1 - \frac{j0.0376}{j0.1246}\right) 1\angle 0° = 0.698\angle 0° \text{ pu}$$

From Figure 5.10(a), we may write

$$I_{12} = \frac{0 - 0.698\angle 0}{j0.514} = j1.36 \text{ pu}$$

5.4 UNSYMMETRICAL FAULTS AND SYMMETRICAL COMPONENTS

Unsymmetrical faults such as line-to-line and line-to-ground faults (which occur more frequently than three-phase short circuits) cannot be analyzed on a per-phase basis, as we were able to do for symmetrical faults. For unsymmetrical faults, the method of symmetrical components is used. This method is based on the fact that a set of three-phase unbalanced phasors can be resolved into three sets of *symmetrical components*, which are termed the *positive-sequence*, *negative-sequence*, and *zero-sequence components*. The phasors of the set of positive-sequence components have equal magnitude and a counterclockwise rotation (or phase sequence *abc*); the negative-sequence components have equal magnitudes and counterclockwise rotation but have the reverse phase sequence *acb*; and the equal magnitude zero-sequence components are all in phase with each other. These sequence components are represented geometrically in Figure 5.12. The positive-sequence components are designated with the subscript 1, and the subscripts 2 and 0 are used for negative- and zero-sequence components, respectively.

Thus, the unbalanced system of Figure 5.13 can be resolved into symmetrical components as shown in Figure 5.12. In particular, we have

$$V_a = V_{a0} + V_{a1} + V_{a2} \tag{5.14}$$

$$V_b = V_{b0} + V_{b1} + V_{b2} \tag{5.15}$$

$$V_c = V_{c0} + V_{c1} + V_{c2} \tag{5.16}$$

We now introduce an operator a. The a-operator, when operating on a phasor, causes a counterclockwise rotation of the phasor by 120° (just as the j operator produces a 90° rotation), such that

$$a = 1\ \angle 120° = 1 \times e^{j120} = -0.5 + j0.866$$

$$a^2 = 1\ \angle 240° = -0.5 - j0.866 = a^*$$

$$a^3 = 1\ \angle 360° = 1\ \angle 0°$$

$$1 + a + a^2 = 0$$

A graphical representation of the properties of the a-operator is given in Figure 5.14. Using these properties, we may write the components of a given sequence in terms of any chosen component. From Figures 5.12 and 5.14, we have

$$V_{b1} = a^2 V_{a1}$$

$$V_{c1} = a V_{a1}$$

$$V_{b2} = a V_{a2}$$

$$V_{c2} = a^2 V_{a2}$$

$$V_{a0} = V_{b0} = V_{c0}$$

Consequently, (5.14) to (5.16) become, in terms of components of phase a,

$$V_a = V_{a0} + V_{a1} + V_{a2} \tag{5.17}$$

$$V_b = V_{a0} + a^2 V_{a1} + a V_{a2} \tag{5.18}$$

$$V_c = V_{a0} + a V_{a1} + a^2 V_{a2} \tag{5.19}$$

or in matrix notation $\boldsymbol{V} = \boldsymbol{A}\boldsymbol{V}'$ where

$$\boldsymbol{V}' = \begin{bmatrix} V_{a0} \\ V_{a1} \\ V_{a2} \end{bmatrix} \quad \boldsymbol{V} = \begin{bmatrix} V_a \\ V_b \\ V_c \end{bmatrix} \quad \boldsymbol{A} = \begin{bmatrix} 1 & 1 & 1 \\ 1 & a^2 & a \\ 1 & a & a^2 \end{bmatrix}$$

Solving for the sequence components from (5.17) through (5.19) yields

$$V_{a0} = \frac{1}{3}(V_a + V_b + V_c) \tag{5.20}$$

$$V_{a1} = \frac{1}{3}(V_a + a V_b + a^2 V_c) \tag{5.21}$$

$$V_{a2} = \frac{1}{3}(V_a + a^2 V_b + a V_c) \tag{5.22}$$

or in matrix notation $\boldsymbol{V}' = \boldsymbol{A}^{-1}\boldsymbol{V}$ where

$$A^{-1} = \begin{bmatrix} 1 & 1 & 1 \\ 1 & a & a^2 \\ 1 & a^2 & a \end{bmatrix}$$

Equations similar to (5.20) to (5.22) hold for currents as well.

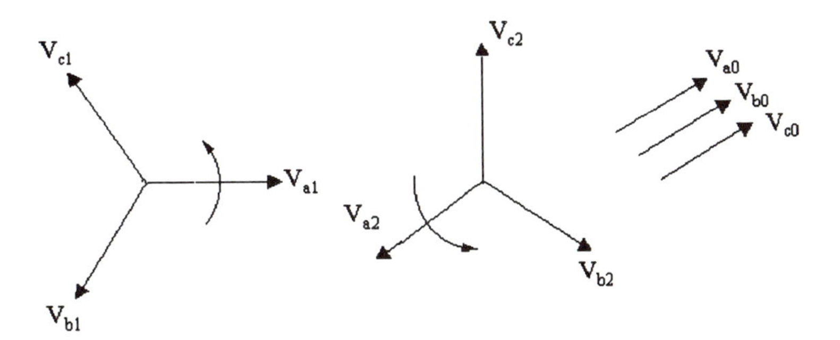

FIGURE 5.12
Symmetrical components of a three-phase system.

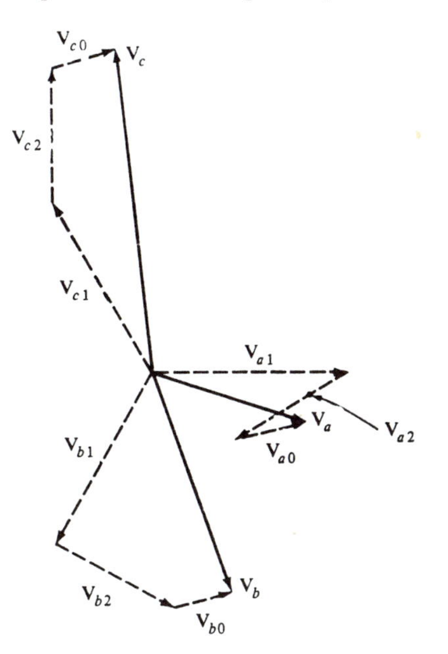

FIGURE 5.13
Graphical addition of the symmetrical components to obtain the three-phase unbalanced system.

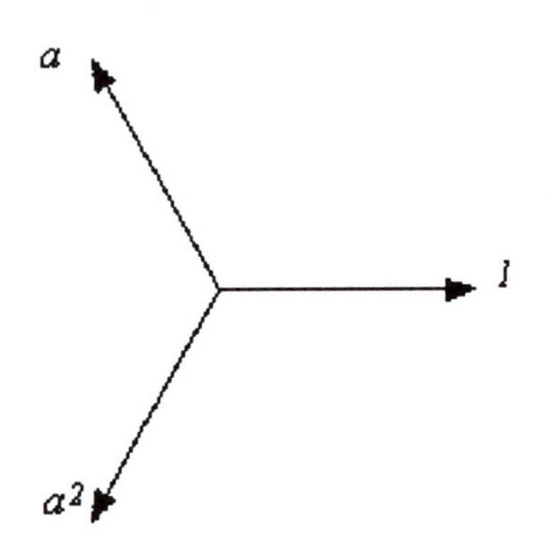

FIGURE 5.14
Properties of the a-operator.

A quantity (current, voltage, impedance, power) that is given in terms of symmetrical components is sometimes called the *sequence quantity*, as in "sequence current." We now illustrate the use of the concept of symmetrical components by the following examples.

Example 5.7 The phase currents in a wye-connected, unbalanced load with its neutral connected to ground are $I_a = (44 - j33)$, $I_b = -(32 + j24)$, and $I_c = (-40 + j25)$A. Determine the sequence currents.

Solution
From (5.20) through (5.22), adapted for currents, we obtain

$$I_{a0} = \frac{1}{3}[(44 - j33) - (32 + j24) + (-40 + j25)]$$

$$= -9.33 - j10.67 = 14.17 \angle -131.2^\circ \text{ A}$$

$$I_{a1} = \frac{1}{3}[(44 - j33) - (-0.5 + j0.866)(32 + j24)$$

$$+ (-0.5 - j0.866)(-40 + j25)]$$

$$= 40.81 - j8.86 = 41.76 \angle -12.2^\circ \text{ A}$$

$$I_{a2} = \frac{1}{3}[(44 - j33) - (-0.5 - j0.866)(32 + j24)$$

$$+ (-0.5 - j0.866)(-40 + j25)]$$

$$= 12.52 - j13.48 = 18.37 \angle -47^\circ \text{ A}$$

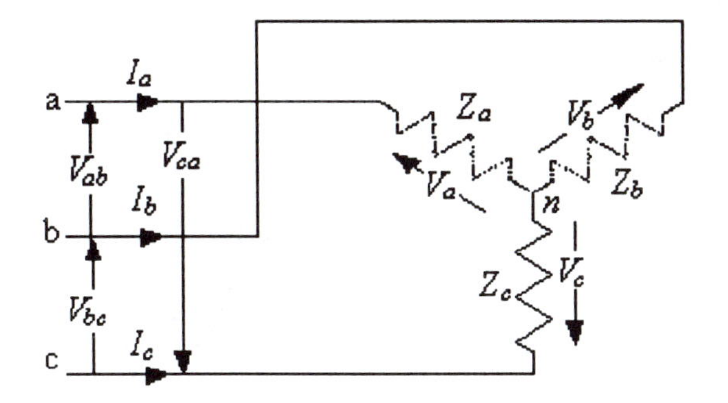

FIGURE 5.15
Example 5.8.

Example 5.8 A three-phase, wye-connected load is connected across a three-phase, balanced supply system. Obtain a set of equations relating the symmetrical components of the line and phase voltages.

Solution
The symmetrical system, the assumed directions of voltages, and the nomenclature are shown in Figure 5.15, from which we have

$$V_{ab} = V_a - V_b \qquad V_{bc} = V_b - V_c \qquad V_{ca} = V_c - V_a$$

Because $V_{ab} + V_{bc} + V_{ca} = 0$, we get

$$V_{ab0} = V_{bc0} = V_{ca0} = 0$$

We choose V_{ab} as the reference phasor. For the positive-sequence component, we have

$$\begin{aligned}
V_{ab1} &= \frac{1}{3}(V_{ab} + aV_{bc} + a^2V_{ca}) \\
&= \frac{1}{3}[(V_a - V_b) + a(V_b - V_c) + a^2(V_c - V_a)] \\
&= \frac{1}{3}[(V_a + aV_b + a^2V_c) - (a^2V_a + V_b + aV_c)] \\
&= \frac{1}{3}[(V_a + aV_b + a^2V_c) - a^2(V_a + aV_b + a^2V_c)] \\
&= \frac{1}{3}[(1 - a^2)(V_a + aV_b + a^2V_c)] = (1 - a^2)V_{a1} \\
&= \sqrt{3}\,V_{a1}e^{j30°} = (\sqrt{3}\,\angle 30°)\,V_{a1}
\end{aligned} \tag{5.23}$$

Similarly, for the negative-sequence component, we obtain

$$V_{ab2} = \frac{1}{3}(V_{ab} + a^2 V_{bc} + a V_{ca})$$

$$= \frac{1}{3}[(V_a - V_b) + a^2(V_b - V_c) + a(V_c - V_a)]$$

$$= \frac{1}{3}[(V_a + a^2 V_b + a V_c) - (a V_a + V_b + a^2 V_c)]$$

$$= \frac{1}{3}[(V_a + a^2 V_b + a V_c) - a(V_a + a^2 V_b + a V_c)]$$

$$= \frac{1}{3}[(1 - a)(V_a + a^2 V_b + a V_c)] = (1 - a) V_{a2}$$

$$= \sqrt{3}\, V_{a2} e^{-j30°} = (\sqrt{3}\ \angle -30°)\, V_{a2} \tag{5.24}$$

In (5.23) and (5.24), V_{a1} and V_{a2} are, respectively, the positive- and negative-sequence components of the phase voltage V_a.

Proceeding as in the derivation of (5.23) and (5.24), but now choosing V_{bc} as the reference phasor, we can show that

$$V_{bc1} = -j\sqrt{3}\, V_{a1} \tag{5.25}$$

$$V_{bc2} = j\sqrt{3}\, V_{a2} \tag{5.26}$$

5.5 RELATIONSHIP BETWEEN SEQUENCE VOLTAGES AND SEQUENCE CURRENTS

In Figure 5.15, consider a general situation where the impedance Z_a has three components, an impedance Z_{a0} to zero-sequence current, an impedance Z_{a1} to positive-sequence current, and an impedance Z_{a2} to negative-sequence current. Assuming similar quantities for phases b and c, we then have

$$V_a = Z_{a0}I_{a0} + Z_{a1}I_{a1} + Z_{a2}I_{a2} = Z_{a0}I_{a0} + Z_{a1}I_{a1} + Z_{a2}I_{a2} \tag{5.27}$$

$$V_b = Z_{b0}I_{b0} + Z_{b1}I_{b1} + Z_{b2}I_{b2} = Z_{b0}I_{a0} + a^2 Z_{b1}I_{a1} + aZ_{b2}I_{a2} \tag{5.28}$$

$$V_c = Z_{c0}I_{c0} + Z_{c1}I_{c1} + Z_{c2}I_{c2} = Z_{c0}I_{a0} + aZ_{c1}I_{a1} + a^2 Z_{c2}I_{a2} \tag{5.29}$$

or upon substitution of sequence quantities for V_a, V_b and V_c

$$V_{a0} + V_{a1} + V_{a2} = Z_{a0}I_{a0} + Z_{a1}I_{a1} + Z_{a2}I_{a2} \tag{5.30}$$

$$V_{a0} + a^2V_{a1} + aV_{a2} = Z_{b0}I_{a0} + (a^2Z_{b1})I_{a1} + (aZ_{b2})I_{a2} \tag{5.31}$$

$$V_{a0} + aV_{a1} + a^2V_{a2} = Z_{c0}I_{a0} + (aZ_{c1})I_{a1} + (a^2Z_{c2})I_{a2} \tag{5.32}$$

Solving (5.30) - (5.32) for V_{a0}, V_{a1} and V_{a2} gives

$$V_{a0} = z_{00}I_{a0} + z_{21}I_{a1} + z_{12}I_{a2} \tag{5.33}$$

$$V_{a1} = z_{10}I_{a0} + z_{01}I_{a1} + z_{22}I_{a2} \tag{5.34}$$

$$V_{a2} = z_{20}I_{a0} + z_{11}I_{a1} + z_{02}I_{a2} \tag{5.35}$$

where

$$z_{00} = \frac{Z_{a0} + Z_{b0} + Z_{c0}}{3} \qquad z_{21} = \frac{Z_{a1} + a^2Z_{b1} + aZ_{c1}}{3} \qquad z_{12} = \frac{z_{a2} + aZ_{b2} + a^2Z_{c2}}{3}$$

$$z_{10} = \frac{Z_{a0} + aZ_{b0} + a^2Z_{c0}}{3} \qquad z_{01} = \frac{Z_{a1} + Z_{b1} + Z_{c1}}{3} \qquad z_{22} = \frac{Z_{a2} + a^2Z_{b2} + aZ_{c2}}{3}$$

$$z_{20} = \frac{Z_{a0} + a^2Z_{b0} + aZ_{c0}}{3} \qquad z_{11} = \frac{Z_{a1} + aZ_{b1} + a^2Z_{c1}}{3} \qquad z_{02} = \frac{Z_{a2} + Z_{b2} + Z_{c2}}{3}$$

Whereas (5.33) - (5.34) are chiefly of academic interest, there are two cases of practical importance.

1. Unbalanced static network

In this situation, impedances are the same to all sequences of current but are unbalanced. Thus, Z_a, Z_b and Z_c may have different values but $Z_a = Z_{a0} = Z_{a1} = Z_{a2}$; $Z_b = Z_{b0} = Z_{b1} = Z_{b2}$ and $Z_c = Z_{c0} = Z_{c1} = Z_{c2}$. In this case,

$$z_{00} = z_{01} = z_{02} = \frac{Z_a + Z_b + Z_c}{3} = z_0 \tag{5.36}$$

$$z_{10} = z_{11} = z_{12} = \frac{Z_a + aZ_b + a^2Z_c}{3} = z_1 \tag{5.37}$$

$$z_{20} = z_{21} = z_{22} = \frac{Z_a + a^2Z_b + aZ_c}{3} = z_2 \tag{5.38}$$

2. Balanced nonstatic network

This special situation corresponds to three-phase rotating machines which will generally be balanced impedances which have a different impedance to different sequences of current. Thus $Z_{a0} = Z_{b0} = Z_{c0} = Z_0$; $Z_{a1} = Z_{b1} = Z_{c1} = Z_1$ and $Z_{a2} = Z_{b2} = Z_{c2} = Z_2$. In this case, $z_{00} = Z_0$, $z_{01} = Z_1$, $z_{02} = Z_2$ and all the remaining terms are zero (remember $1 + a + a^2 = 0$). Of course, if we combine these special cases to obtain a balanced static network, then $z_{00} = z_{01} = z_{02} = Z$ where Z is the impedance in each phase and we have returned to the familiar balanced network.

Example 5.9 The line voltages across a three-phase, wye-connected load such as that shown in Figure 5.15 are unbalanced such that $V_{ab} = 220\angle 131.7°$ V, $V_{bc} = 252\angle 0°$ V, and $V_{ca} = 195\angle -122.6°$ V. Determine the sequence phase voltages. Then find the voltages across the load, and calculate the line currents for the following load conditions:

1. a 10 Ω resistance in each phase
2. a rotating machine have a zero-sequence impedance of $j5$ Ω, a positive-sequence impedance of $j2$ Ω and a negative-sequence impedance of $j1$ Ω.
3. a 7 Ω resistance in phase a and a 4 Ω resistance in phases b and c.

Solution

Since the line voltages are given, we first choose V_{bc} as the reference phasor and determine the sequence components of the line voltages. Using (5.21) and (5.22) for line voltages yields

$$V_{bc1} = \frac{1}{3}(V_{bc} + aV_{ca} + a^2 V_{ab})$$

$$= \frac{1}{3}(252\ \angle 0° + 1\ \angle 120° \times 195\ \angle -122.6° + 1\angle -120° \times 220\ \angle 131.7°)$$

$$= 221 + j11.9 \text{ V}$$

$$V_{bc2} = \frac{1}{3}(V_{bc} + a^2 V_{ca} + aV_{ab})$$

$$= \frac{1}{3}(252\ \angle 0° + 1\ \angle -120° \times 195\ \angle -122.6° + 1\ \angle 120° \times 220\ \angle 131.7°)$$

$$= 31 - j11.9 \text{ V}$$

From (5.20) we have

$$V_{bc0} = \frac{1}{3}(V_{bc} + V_{ca} + V_{ab}) = \frac{1}{3}(252\angle 0° + 195\ \angle -122.6° + 220\ \angle 131.7°)$$

$$= 0 \text{ V [Note that } V_{bc} + V_{ca} + V_{ab} \text{ must always equal zero.]}$$

The sequence components of the phase voltages are from (5.25) and (5.26)

$$V_{a1} = \frac{V_{bc1}}{-j\sqrt{3}} = \frac{221 + j11.9}{-j\sqrt{3}} = (-6.9 + j127.5)\ \text{V}$$

$$V_{a2} = \frac{V_{bc2}}{j\sqrt{3}} = \frac{31 - j11.9}{j\sqrt{3}} = (-6.9 - j17.9)\ \text{V}$$

For the 10Ω balanced load, $z_{00} = z_{01} = z_{02} = 10\Omega$ and the other zs are all equal to zero. Then from (5.33) through (5.35)

$$V_{a0} = 10\ I_{a0}$$

$$V_{a1} = 10\ I_{a1}$$

$$V_{a2} = 10\ I_{a2}$$

Now since applying Kirchhoff's current law at the junction point of the wye connection gives $I_a + I_b + I_c = 0$ and $3I_{a0} = I_a + I_b + I_c$ from (5.20), it is apparent that $V_{a0} = 10\ I_{a0}$ must also be zero.

Hence, from (5.17) through (5.19) we obtain

$$V_a = -6.9 + j127.5 - 6.9 - j17.9 = (-13.8 + j109.6)\ \text{V}$$

$$V_b = a^2 V_{a1} + a V_{a2} = (132.8 - j54.8)\ \text{V}$$

$$V_c = aV_{a1} + a^2 V_{a2} = (-119 - j54.8)\ \text{V}$$

The line currents are given by

$$I_a = \frac{V_a}{R} = \frac{1}{10}(-13.8 + j109.6) = -1.38 + j10.96 = 11.05\ \angle 97.2°\ \text{A}$$

$$I_b = \frac{V_b}{R} = \frac{1}{10}(132.8 - j54.8) = 13.28 - j5.48 = 14.37\ \angle -22.4°\ \text{A}$$

and

$$I_c = \frac{V_c}{R} = \frac{1}{10}(119 - j54.8) = -11.9 - j5.48 = 13.1\ \angle -155°\ \text{A}$$

Note that $I_a + I_b + I_c = 0$ as expected.

Another method would be to calculate

$$I_{a1} = \frac{V_{a1}}{10} = \frac{-6.9 + j127.5}{10} = -0.69 + j12.75 \text{ A}$$

$$I_{a2} = \frac{V_{a2}}{10} = \frac{-6.9 - j17.9}{10} = -0.69 - j1.79 \text{ A}$$

$$I_{a0} = 0$$

and

$$I_a = I_{a0} + I_{a1} + I_{a2} = -1.38 + j10.96 \text{ A}$$

$$I_b = I_{a0} + a^2 I_{a1} + aI_{a2} = 13.28 - j5.48 \text{ A}$$

$$I_c = I_{a0} + aI_{a1} + a^2 I_{a2} = -11.9 - j5.48 \text{ A}$$

For the rotating machine, $z_{00} = j5\Omega = Z_0$, $z_{01} = j2\Omega = Z_1$ and $z_{02} = j1\Omega = Z_2$ so that

$$I_{a1} = \frac{V_{a1}}{Z_1} = \frac{-6.9 + j127.5}{j2} = 63.75 + j3.45 = 63.84 \angle 3.1° \text{ A}$$

$$I_{a2} = \frac{V_{a2}}{Z_2} = \frac{-6.9 - j17.9}{j1} = -17.9 + j6.9 = 19.2 \angle 158.9° \text{ A}$$

$$I_{a0} = \frac{V_{a0}}{Z_0} = (I_a + I_b + I_c)/3 = 0$$

and

$$I_a = I_{a0} + I_{a1} + I_{a2} = 45.8 + j10.35 \text{ A}$$

$$I_b = I_{a0} + a^2 I_{a1} + aI_{a2} = -25.9 - j75.9 \text{ A}$$

$$I_c + I_{a0} + aI_{a1} + a^2 I_{a2} = -19.9 + j65.55 \text{ A}$$

Note that it is not possible to first determine V_a, V_b and V_c and then divide by the impedance to get I_a, I_b and I_c as it was with the static balanced load.

Finally, we consider the case where $Z_a = 7\ \Omega$ and $Z_b = Z_c = 4\ \Omega$. Now from (5.36) - (5.38),

$$z_0 = \frac{Z_a + Z_b + Z_c}{3} = \frac{7 + 4 + 4}{3} = 5\ \Omega$$

$$z_1 = \frac{Z_a + aZ_b + a^2 Z_c}{3} = \frac{7 - 4}{3} = 1\ \Omega$$

$$z_2 = \frac{Z_a + a^2 Z_b + aZ_c}{3} = \frac{7 - 4}{3} = 1\ \Omega$$

and

$$V_{a0} = z_0 I_{a0} + z_2 I_{a1} + z_1 I_{a2} = 5I_{a0} + I_{a1} + I_{a2}$$

$$V_{a1} = z_1 I_{a0} + z_0 I_{a1} + z_2 I_{a2} = I_{a0} + 5I_{a1} + I_{a2}$$

$$V_{a2} = z_2 I_{a0} + z_1 I_{a1} + z_0 I_{a2} = I_{a0} + I_{a1} + 5I_{a2}$$

Note that now there is interaction between the sequences and V_{a0} is no longer necessarily equal to zero even though I_{a0} must still equal zero. Using the last two equations

$$-6.9 + j127.5 = 5I_{a1} + I_{a2}$$

$$-6.9 - j17.9 = I_{a1} + 5I_{a2}$$

and solving gives

$$I_{a1} = -1.15 + j27.31 = 27.33 \angle 92.4°$$

$$I_{a2} = -1.15 - j9.04 = 9.11 \angle -97.25°$$

and $I_{a0} = 0$

so that

$$I_a = I_{a0} + I_{a1} + I_{a2} = -2.3 + j18.27 \text{ A}$$

$$I_b = I_{a0} + a^2 J_{a1} + aI_{a2} = 32.6 - j9.14 \text{ A}$$

$$I_c = I_{a0} + aI_{a1} + a^2 I_{a2} = -30.3 - j9.13 \text{ A}$$

Also

$$V_{a0} = I_{a1} + I_{a2} = -2.3 + j18.27 \text{ V}$$

and

$$V_a = V_{a0} + V_{a1} + V_{a2} = (-2.3 + j18.7) + (-6.9 + j127.5) + (-6.9 - j17.9)$$

$$= -16.1 + j127.9 \text{ V}$$

$$V_b = V_{a0} + a^2 V_{a1} + aV_{a2} = 130.4 - j36.56 \text{ V}$$

$$V_c = V_{a0} + aV_{a1} + a^2 V_{a2} = -121.2 - j36.52 \text{ V}$$

or

$$V_a = Z_a I_a = 7(-2.3 + j18.27) = -16.1 + j127.9 \text{ V}$$

$$V_b = Z_b I_b = 4(32.6 - j9.14) = 130.4 - j36.56 \text{ V}$$

$$V_c = Z_c I_c = 4(-30.3 - j9.13) = -121.2 - j36.52 \text{ V}$$

5.6 SEQUENCE POWER

To obtain the power in a three-phase system in terms of symmetrical components, we rewrite (5.17) through (5.19) in matrix notation as follows:

$$V = AV' \tag{5.39}$$

where

$$\mathbf{V} = \begin{bmatrix} V_a \\ V_b \\ V_c \end{bmatrix} \quad \mathbf{V}' = \begin{bmatrix} V_{a0} \\ V_{a1} \\ V_{a2} \end{bmatrix} \quad \mathbf{A} = \begin{bmatrix} 1 & 1 & 1 \\ 1 & a^2 & a \\ 1 & a & a^2 \end{bmatrix}$$

Similarly, for the currents we have

$$\mathbf{I} = \mathbf{AI}' \tag{5.40}$$

where

$$\mathbf{I} = \begin{bmatrix} I_a \\ I_b \\ I_c \end{bmatrix} \quad \text{and} \quad \mathbf{I}' = \begin{bmatrix} I_{a0} \\ I_{a1} \\ I_{a2} \end{bmatrix}$$

The complex power $\boldsymbol{S}$ may now be written as

$$\boldsymbol{S} = \mathbf{V}^T\mathbf{I}^* \tag{5.41}$$

where $\mathbf{V}^T$ is the transpose of $\mathbf{V}$, and $\mathbf{I}^*$ is the complex conjugate of $\mathbf{I}$. From (5.39) and (5.40), we have (remembering that $(\mathbf{AB})^T = (\mathbf{B}^T\mathbf{A}^T)$)

$$\mathbf{V}^T = (\mathbf{V}')^T\mathbf{A}^T \tag{5.42}$$

and

$$\mathbf{I}^* = \mathbf{A}^*(\mathbf{I}')^* \tag{5.43}$$

Consequently, (5.41) through (5.43) yield

$$\boldsymbol{S} = (\mathbf{V}')^T\mathbf{A}^T\mathbf{A}^*(\mathbf{I}')^* \tag{5.44}$$

Now, since

$$\mathbf{A}^T\mathbf{A}^* = \begin{bmatrix} 1 & 1 & 1 \\ 1 & a^2 & a \\ 1 & a & a^2 \end{bmatrix}\begin{bmatrix} 1 & 1 & 1 \\ 1 & a & a^2 \\ 1 & a^2 & a \end{bmatrix} = \begin{bmatrix} 3 & 0 & 0 \\ 0 & 3 & 0 \\ 0 & 0 & 3 \end{bmatrix}$$

(5.44) becomes

$$S = 3(V_{a0}I_{a0}^* + V_{a1}I_{a1}^* + V_{a2}I_{a2}^*) = V_aI_a^* + V_bI_b^* + V_cI_c^* \qquad (5.45)$$

Example 5.10 For Example 5.9 part 1, compute the total real and reactive powers using (1) the phase quantities and (2) the sequence quantities. Compare the results.

$$S = V_aI_a^* + V_bI_b^* + V_cI_c^* = (-13.8 + j109.6)(-1.38 - j10.96)$$

$$+ (132.8 - j54.8)(13.28 + j5.48) + (-119.0 - j54.8)(-11.9 + j5.48)$$

$$= P + jQ = 5 \times 10^3 + j0$$

and

$$P = 5 \text{ kW} \qquad Q = 0$$

Alternatively,

$$S = 3V_{a0}I_{a0}^* + 3V_{a1}I_{a1}^* + 3V_{a2}I_{a2}^* = 3(0)(0)$$

$$+ 3(-6.9 + j127.5)(-0.69 - j12.75) + 3(-6.9 - j17.9)(-0.69 + j1.79)$$

$$S = P + jQ \approx 5 \times 10^3 + j0$$

and

$$P = 5 \text{ kW}, \quad Q = 0$$

5.7 SEQUENCE IMPEDANCES AND SEQUENCE NETWORKS

We will now consider the situation where we have a power system which is balanced everywhere except at the point where the unbalanced fault occurs. We will however allow for the situation that rotating machines may have different impedances to different sequences of current (the balanced nonstatic network of Section 5.5). In this case, we may define sequence impedances corresponding to the sequence currents. An impedance through which only positive-sequence currents flow is called the *positive-sequence impedance*. Similarly, when only negative-sequence currents flow, the impedance is known as the *negative-sequence impedance*; and when zero-sequence currents alone are present, the impedance is called the *zero-sequence impedance.*

Unsymmetrical (or unbalanced) fault calculations are facilitated by the use of the concepts of sequence voltages, currents, and impedances. Because a voltage of a specific sequence produces a current of the same sequence only, the various sequence networks representing an unbalanced condition have no mutual coupling. This feature of sequence networks simplifies the calculations considerably. We now proceed to develop sequence networks for certain commonly encountered situations.

Sequence Network for a Generator on No-Load

A three-phase synchronous generator, grounded through an impedance Z_n, is shown in Figure 5.16(*a*). The generator is not supplying any load, but because of a fault at the generator terminals, currents I_a, I_b, and I_c flow through phases *a*, *b*, and *c*, respectively. We now develop and draw the *sequence networks* for the generator for this condition. Let the generator no-load voltages in the three phases be E_a, E_b, and E_c. These generator voltages are balanced and may be assumed to be of positive sequence only. As a result $E_{a1} = E_a$ and $E_{a2} = E_{a0} = 0$ from (5.20) - (5.22). The coils shown in Figure 5.16(*a*) represent the internal impedance of the generator. There is one important difference between this internal impedance and other internal impedances with which we are most familiar. While this internal impedance is the same in all three phases (balanced), it may have different values for positive, negative and zero-sequence currents. We will call these impedances Z_1, Z_2 and Z_0, respectively. The currents

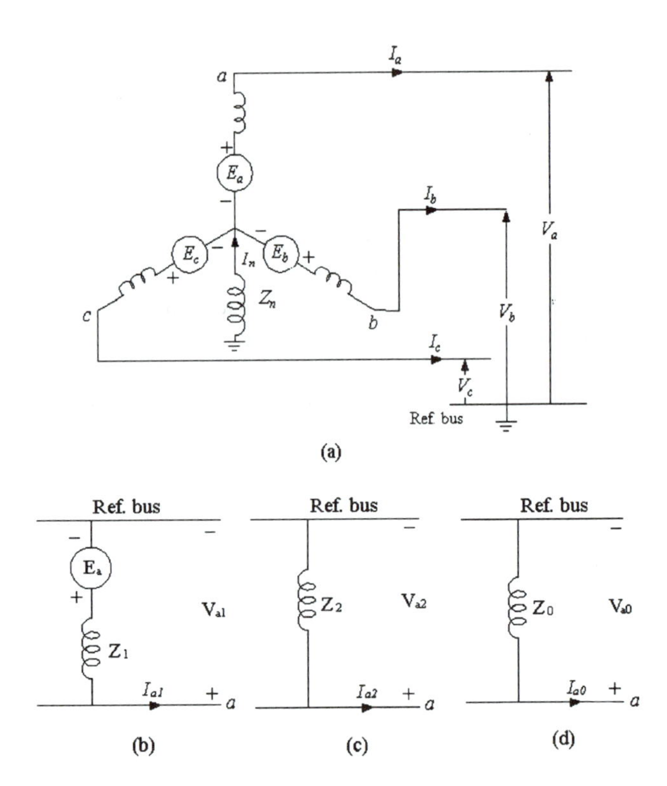

FIGURE 5.16
(*a*) An unloaded synchronous generator, with neutral grounded through Z_n; (*b*)-(*d*) sequence networks.

I_a, I_b, and I_c are represented by their symmetrical components as given by (5.20) - (5.22). For the positive-sequence network, we have

$$V_{a1} = E_a - I_{a1}Z_1 \tag{5.46}$$

where $I_{a1}Z_1$ is the positive-sequence voltage drop in the positive-sequence impedance Z_1 of the generator and V_{a1} is the positive-sequence component of the phase voltages V_a, V_b and V_c. If Z_2 is the negative-sequence impedance of the generator, the negative-sequence voltage at the terminal of phase a is simply

$$V_{a2} = -I_{a2}Z_2 \tag{5.47}$$

since there is no negative-sequence generated voltage. The generator zero-sequence currents flow through Z_n as well as through Z_{g0}, the generator zero-sequence impedance. The total zero-sequence current through Z_n is $I_{a0} + I_{b0} + I_{c0} = 3I_{a0}$, but the current through Z_{g0} is I_{a0} . Hence

$$V_{a0} = -I_{a0}Z_{g0} - 3I_{a0}Z_n = -I_{a0}(Z_{g0} + 3Z_n) = -I_{a0}Z_0 \tag{5.48}$$

where $Z_0 = Z_{g0} + 3Z_n$. Equations (5.46), (5.47) and (5.48) are represented by Figure 5.16*(b)*, *(c)* and *(d)*, respectively.

Sequence Network for a Generator with a Line-to-Ground Fault

We now let a line-to-ground short-circuit fault occur on phase a of the generator of Figure 5.16(a), which was operating without a load, and derive a sequence-network representation of this condition. The constraints corresponding to the fault are $I_b = I_c = 0$ (lines remain open-circuited) and $V_a = 0$ (line-to-ground short-circuit). Consequently, the symmetrical components of the current in phase a are

$$I_{a0} = \frac{1}{3}(I_a + I_b + I_c) = \frac{1}{3}I_a$$

$$I_{a1} = \frac{1}{3}(I_a + aI_b + a^2I_c) = \frac{1}{3}I_a$$

$$I_{a2} = \frac{1}{3}(I_a + a^2I_b + aI_c) = \frac{1}{3}I_a$$

so that

$$I_{a0} = I_{a1} = I_{a2} = \frac{1}{3}I_a \tag{5.49}$$

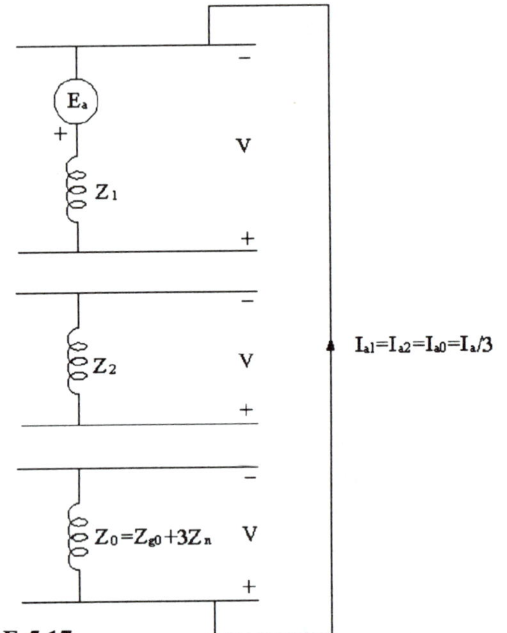

FIGURE 5.17
Sequence network for a generator with line-to-ground fault.

In addition,

$$V_a = V_{a1} + V_{a2} + V_{a3} = 0$$

Hence the sequence networks must be connected in series, as shown in Figure 5.17. The sequence voltages appear as marked in the figure.

We may use this network to determine the fault current as follows. From Figure 5.17 we have:

$$V_{a0} + V_{a1} + V_{a2} = E_a - I_{a1}Z_1 - I_{a1}Z_2 - I_{a1}Z_0$$

But since

$$V_a = V_{a0} + V_{a1} + V_{a2} = 0$$

we have

$$I_{a1} = \frac{E_a}{Z_1 + Z_2 + Z_0} = \frac{1}{3}I_a$$

and

$$I_a = \frac{3E_a}{Z_1 + Z_2 + Z_0} \tag{5.50}$$

where

$$Z_0 = Z_{g0} + 3Z_n$$

Example 5.11 The positive-, negative-, and zero-sequence reactances of a 20-MVA, 13.2-kV synchronous generator are 0.3 pu, 0.2 pu, and 0.1 pu, respectively. The generator is solidly grounded and is not loaded. A line-to-ground fault occurs on phase a. Neglecting all the resistances, determine the fault current.

Solution
The sequence network corresponding to this fault is shown in Figure 5.17. For a solidly grounded generator $Z_n = 0$ and $Z_0 = Z_{g0} + 3Z_n = j0.1 + 3(0) = j0.1$ pu. Since $E_a = 1\angle 0°$ and the total impedance is $j0.3 + j0.2 + j0.1 = j0.6$ pu, from (5.50) we have

$$I_{a1} = \frac{1\ \angle 0°}{j0.6} = 1.67\ \angle -90° = -j1.67 \text{ pu}$$

and

$$I_a = 3I_{a1} = 5\ \angle -90° \text{ pu}$$

Choosing the rated values as base quantities, we have

$$\text{Base current} = \frac{20{,}000}{\sqrt{3} \times 13.2} = 874.8 \text{ A}$$

$$\text{Fault current} = |I_a| = 5 \text{ pu} = 5 \times 874.8 = 4374 \text{ A}$$

Example 5.12 Determine the terminal voltages for the generator of Example 5.11.

Solution
From Figure 5.17 and (5.46) - (5.48), we have the per-unit sequence voltages

$$V_{al} = 1 - (-j1.67)\ (j0.3) = 0.5\angle 0° \text{ pu}$$

$$V_{a2} = -\ (-j1.67)\ (j0.2) = -\ 0.333\ \angle 0° \text{ pu}$$

$$V_{a0} = -\ (-j1.67)\ (j0.1) = -\ 0.167\angle 0° \text{ pu}$$

Then

$$V_a = 0.5 - 0.333 - 0.167 = 0 \text{ pu}$$

To determine V_b and V_c, we determine their sequence components. Thus, we have

$$V_{b1} = a^2 V_{a1} = (-0.5 - j0.866)\,0.5 = -0.25 - j0.433 \text{ pu}$$

$$V_{b2} = a V_{a2} = (-0.5 + j0.866)\,(-0.333) = 0.167 - j0.288 \text{ pu}$$

$$V_{b0} = V_{a0} = -0.167 \text{ pu}$$

Hence,

$$V_b = -1.67 - 0.25 - j0.433 + 0.167 - j0.288 = -0.25 - j0.721 \text{ pu} = 0.7631 \angle -109.1^\circ$$

In a similar fashion, we find

$$V_{c0} = V_{a0} = -0.167 \text{ pu}$$

$$V_{c1} = aV_{a1} = (-0.5 + j0.866)\,0.5 = -0.25 + j0.433 \text{ pu}$$

$$V_{c2} = a^2 V_{a2} = (-0.5 - j0.866)(-0.333) = 0.167 + j0.288 \text{ pu}$$

and

$$V_c = -1.67 - 0.25 + j0.433 + 0.167 + j0.288 = -0.25 + j0.721 \text{ pu} = 0.7631 \angle 109.1^\circ$$

Since the line-to-neutral base voltage is $13.2/\sqrt{3}$ kV, we have

$$|V_b| = |V_c| = 0.7631 \left(\frac{13.2}{\sqrt{3}} \right) = 5.82 \text{ kV}$$

The line-to-line voltages are

$$V_{ab} = V_a - V_b = -V_b = 0.25 + j0.721 = 0.7631 \angle 70.87^\circ \text{ pu}$$

$$V_{bc} = V_b - V_c = -j1.442 = 1.442 \angle -90^\circ \text{ pu}$$

$$V_{ca} = V_c - V_a = V_c = -0.25 + j0.721 = 0.7631 \angle 109.1^\circ \text{ pu}$$

and

$$|V_{ab}| = |V_{ca}| = 0.7631 \left(\frac{13.2}{\sqrt{3}} \right) = 5.82 \text{ kV}$$

$$|V_{bc}| = 1.442 \left(\frac{13}{2\sqrt{3}} \right) = 10.99 \text{ kV}$$

Note that the line-to-neutral voltage base is used here because these line-to-line voltages were obtained from the line-to-neutral voltages.

Sequence Network for a Generator with a Line-to-Line Fault

Proceeding in a similar fashion as above, we may obtain the sequence network of a solidly grounded unloaded generator having a short circuit between phases b and c as shown in Figure 5.18(a).

From Figure 5.18(a), the current and voltage constraints are

$$I_a = 0 \qquad \text{(line open)} \tag{5.51}$$

$$I_b + I_c = 0 \qquad \text{(KCL)} \tag{5.52}$$

$$V_b = V_c \text{ (line-to-line short circuit)} \tag{5.53}$$

Substituting (5.51) and (5.52) in (5.20) through (5.22) expressed as currents, we obtain

$$I_{a0} = \frac{1}{3}(0 + 0) = 0 \tag{5.54}$$

$$I_{a1} = \frac{1}{3}(0 + aI_b - a^2 I_b) = \frac{1}{3}(a - a^2) I_b \tag{5.55}$$

$$I_{a2} = \frac{1}{3}(0 + a^2 I_b - aI_b) = \frac{1}{3}(a^2 - a) I_b \tag{5.56}$$

From (5.55) and (5.56) we observe that

$$I_{a1} = -I_{a2} \tag{5.57}$$

whereas (5.54) shows that the zero-sequence current is absent.

Now, from (5.53), (5.18) and (5.19)

$$V_{a0} + a^2 V_{a1} + a V_{a2} = V_{a0} + a V_{a1} + a^2 V_{a2}$$

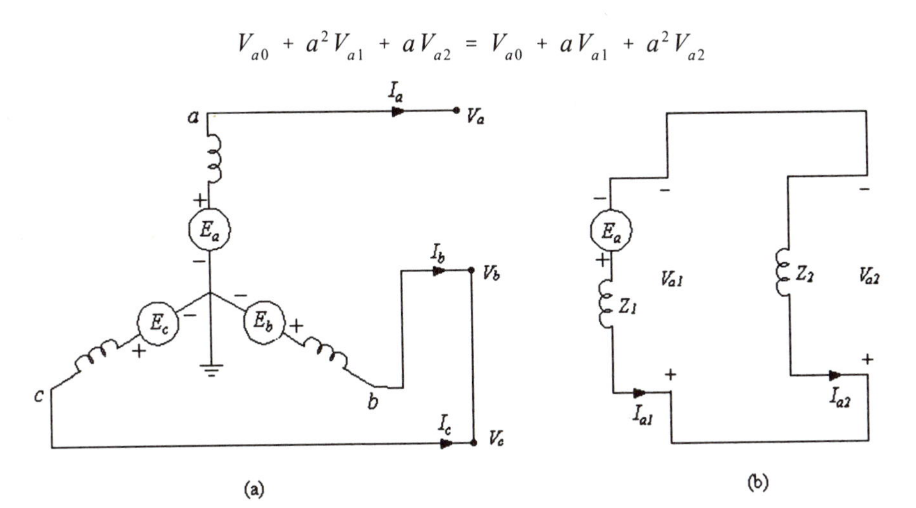

FIGURE 5.18
(*a*) A generator with line-to-line fault; (*b*) Sequence network.

Upon substituting values for a and a^2, we obtain

$$V_{a1} = V_{a2} \tag{5.58}$$

In terms of sequence impedances, (5.58) may be written as

$$E_a - I_{a1}Z_1 = -I_{a2}Z_2 \tag{5.59}$$

Combining (5.57) and (5.59) and solving for I_{a1} yields

$$I_{a1} = \frac{E_a}{Z_1 + Z_2} \tag{5.60}$$

Equations (5.57) through (5.60) may be represented by the sequence network in Figure 5.18(*b*).

Example 5.13 A line-to-line fault occurs at the terminals of the unloaded generator of Example 5.11. Calculate the fault current.

Solution
For this fault condition, the sequence network is that shown in Figure 5.18(*b*). Since we are assuming that the generator is producing rated voltage and that it is unloaded prior to the fault, $E_a = 1\angle 0°$ pu. If the generator were loaded prior to the fault, then an equation similar to (5.4) or (5.5) would be used to determine the internal voltage E_a. Then, from (5.54), (5.57) and (5.60), we obtain

$$I_{a0} = 0$$

$$I_{a1} = -I_{a2} = \frac{1\ \angle 0°}{j0.3 + j0.2} = 2\ \angle -90° = -j2.0 \text{ pu}$$

Hence, the fault current is given by

$$I_b = I_{b0} + I_{b1} + I_{b2} = 0 + a^2 I_{a1} + aI_{a2}$$

$$= (-0.5 - j0.866)(-j2.0) + (-0.5 + j0.866)(j2.0)$$

$$= (-j1.732)(-j2.0) = (-j\sqrt{3})\ I_{a1} = -3.464\angle 0° \text{ pu}$$

As calculated in Example 5.11, the base current is 874.8A. Hence,

$$\text{Fault current} = |I_b| = 874.8 \times 3.464 = 3030 \text{ A}$$

Example 5.14 Calculate the line-to-line voltages for the generator of Example 5.13 (which has a line-to-line fault).

Solution

To determine the line voltages, we must first determine their sequence components. From Figure 5.18*(b)* and Example 5.13, we have

$$V_{a1} = E_a - I_{a1}Z_1 = 1\ \angle 0^\circ - (-j2)\ (j0.3) = 0.4\ \angle 0^\circ$$

$$V_{a2} = -I_{a2}Z_2 = -(j2)\ (j0.2) = 0.4\ \angle 0^\circ = V_{a1}$$

$$V_{a0} = -\ Z_0 I_{a0} = -Z_0 \times 0 = 0$$

$$V_a = V_{a0} + V_{a1} + V_{a2} = 0 + 0.4 + 0.4 = 0.8\ \angle 0^\circ \text{ pu}$$

$$V_b = a^2 V_{a1} + a V_{a2} = (-0.5 - j0.866)\ (0.4\ \angle 0^\circ) + (-0.5 + j0.866)\ (0.4\ \angle 0^\circ)$$

$$= -0.4\ \angle 0^\circ \text{ pu}$$

$$V_c = V_b = -0.4 \text{ pu}$$

The line voltages are then

$$V_{ab} = V_a - V_b = 0.8 - (-0.4) = 1.2 \text{ pu}$$

$$V_{ac} = V_a - V_c = 1.2 \text{ pu}$$

$$V_{bc} = V_b - V_c = 0 \text{ pu}$$

and

$$|V_{ab}| = |V_{ac}| = 1.2 \left(\frac{13.2}{\sqrt{3}} \right) = 9.145 \text{ kV}$$

Sequence Network for a Generator with a Double Line-To-Ground Fault

The final case we wish to consider is the development of the sequence network for a solidly grounded synchronous generator having a double line-to-ground fault as shown in Figure 5.19(*a*).

For this case, the current and voltage constraints are

$$I_a = 0 \tag{5.61}$$

$$V_b = V_c = 0 \tag{5.62}$$

Proceeding as above, we use (5.62) and (5.20) to (5.22) to find, for the sequence components of the voltages,

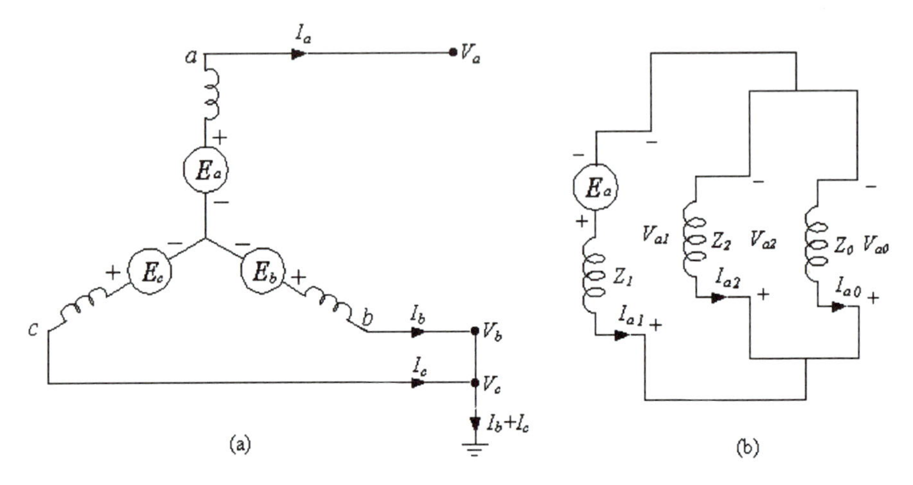

FIGURE 5.19
(*a*) A generator with double line-to-ground fault; (*b*) Sequence network.

$$V_{a0} = V_{a1} = V_{a2} = \frac{1}{3} V_a \tag{5.63}$$

Consequently, the sequence network equations become

$$E_a - I_{a1}Z_1 = -I_{a2}Z_2 = -I_{a0}Z_0 \tag{5.64}$$

Solving for I_{a0} and I_{a2} from (5.64), we obtain

$$I_{a0} = -\frac{E_a - I_{a1}Z_1}{Z_0} \quad \text{and} \quad I_{a2} = -\frac{E_a - I_{a1}Z_1}{Z_2} \tag{5.65}$$

From (5.17), (5.61) and (5.65), we then obtain

$$I_a = -\frac{E_a - I_{a1}Z_1}{Z_0} + I_{a1} - \frac{E_a - I_{a1}Z_1}{Z_2} = 0 \tag{5.66}$$

from which

$$I_{a1} = \frac{E_a}{Z_1 + \dfrac{Z_0 Z_2}{Z_0 + Z_2}} \tag{5.67}$$

The denominator of (5.67) shows that Z_0 and Z_2 are connected in parallel, and this parallel combination is connected in series with Z_1. Hence, the sequence network representing (5.67) is that shown in Figure 5.19*(b)*.

Example 5.15 The generator of Example 5.11 is initially unloaded and operating at rated voltage. A double line-to-ground fault occurs at the generator terminals.

a. Calculate the fault current and the line voltages.
b. Calculate the fault current if the generator is connected to a $j0.2$ pu balanced load when the fault occurs. Neglect prefault currents and assume $Z_n = j0.05$ pu. The load is wye connected with a solidly grounded neutral.

Solution

a. The sequence network for this case is shown in Figure 5.19(*b*). We let $E_a = 1\angle 0°$ pu. Then, from (5.67) we get

$$I_{a1} = \frac{1 \underline{/0°}}{j0.3 + \dfrac{(j0.2)\,(j0.1)}{j0.2 + j0.1}} = -j2.73 \text{ pu}$$

Also, from Figure 5.19(*b*),

$$V_{a1} = E_a - I_{a1}Z_1 = 1\angle 0° - (-j2.73)\,(j0.3) = 0.181 \text{ pu}$$

and, from (5.63) , $V_{a2} = V_{a0} = 0.181$ pu. Therefore,

$$I_{a2} = -\frac{V_{a2}}{Z_2} = \frac{-0.181}{j0.2} = j0.905 \text{ pu}$$

$$I_{a0} = -\frac{V_{a0}}{Z_0} = \frac{-0.181}{j0.1} = j1.81 \text{ pu}$$

From Figure 5.19(*a*), the fault current is $I_b + I_c$. Expressing this sum in terms of sequence components, we have

$$\begin{aligned} I_b + I_c &= (I_{a0} + a^2 I_{a1} + aI_{a2}) + (I_{a0} + aI_{a1} + a^2 I_{a2}) \\ &= 2I_{a0} + (a + a^2)\,(I_{a1} + I_{a2}) \qquad (5.68) \\ &= 2I_{a0} - (I_{a1} + I_{a2}) \end{aligned}$$

Since $I_a = 0$ from (5.61), we may write

$$I_a = I_{a0} + I_{a1} + I_{a2} = 0$$

or

$$-(I_{a1} + I_{a2}) = I_{a0} \qquad (5.69)$$

Now (5.68) and (5.69) yield

$$\text{Fault current} = I_b + I_c = 3I_{a0}$$

Now

$$I_b = I_{a0} + a^2 I_{a1} + a I_{a2} = 4.16 \angle 139.1° \text{ pu}$$

$$I_c = I_{a0} + a I_{a1} + a^2 I_{a2} = 4.16 \angle 40.85° \text{ pu}$$

and

$$I_b + I_c \approx 5.43 \angle 90° = j5.43 \text{ pu}$$

or

$$I_b + I_c = 3I_{a0} = 3(j1.81) = j5.43 \text{ pu}$$

and

$$|\text{Fault current}| = 5.43 \times 874.8 = 4750 \text{ A}$$

To calculate the line voltages, we use (5.62) and (5.63). They yield

$$V_a = 3V_{a1} = 3 \times 0.181 = 0.543 \text{ pu}$$

and

$$V_b = V_c = 0$$

Hence,

$$|V_{ab}| = |V_a - V_b| = 0.543 \times \frac{13.2}{\sqrt{3}} = 4.14 \text{ kV}$$

$$|V_{bc}| = |V_b - V_c| = 0$$

$$|V_{ca}| = |V_c - V_a| = 0.543 \frac{13.2}{\sqrt{3}} = 4.14 \text{ kV}$$

b. Since prefault currents are neglected, E_a is still equal to 1.0 $\angle 0°$ pu. However, we now have a $j0.2$ pu load connected between the fault point and reference such that the sequence network must be modified as shown below.

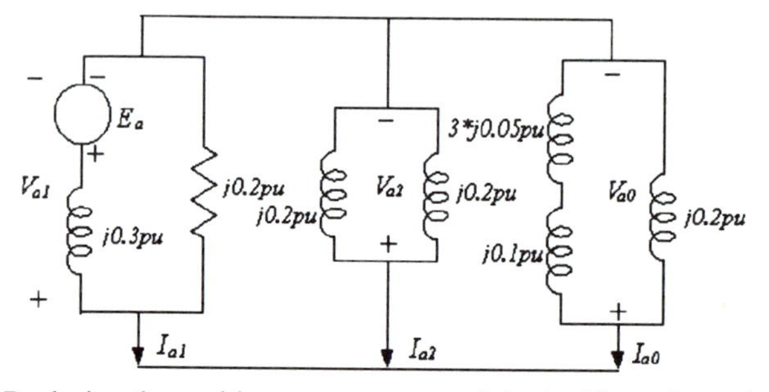

Replacing the positive-sequence network by its Thevenin equivalent, we have

$$E_{th} = E_a = 1.0\angle 0° \text{ (no prefault current)}$$

$$E_{th} = Z_1 = \frac{(j0.3)(j0.2)}{j0.3 + j0.2} = j0.12 \text{ pu}$$

and then

$$Z_2 = \frac{(j0.2)(j0.2)}{j0.2 + j0.2} = j0.1 \text{ pu}$$

$$Z_0 = \frac{(j0.1 + j0.15)(j0.2)}{j0.1 + j0.15 + j0.2} = j0.111 \text{ pu}$$

We may now use (5.67) such that

$$I_{a1} = \frac{1.0 \angle 0^\circ}{j0.12 + \dfrac{(j0.111)(j0.1)}{j0.111 + j0.1}} = -j5.8 \text{ pu}$$

and by current division

$$I_{a0} = -\frac{j0.1}{j0.1 + j0.111}(-j5.8) = j2.75 \text{ pu}$$

$$\text{Fault current} = 3I_{a0} = 3(j2.75) = j8.25 \text{ pu}$$

We have now seen that a fault in a power system can lead to abnormal currents and voltages. For example, during a short circuit, the currents may become excessively large and the voltages may go to zero. The system must be protected against such occurrences, and steps must be taken to remove a fault as quickly as possible. In Sections 5.9 through 5.13 we examine briefly some of the means for doing so. First, we will consider the components required for protection, and then we will survey some approaches for the protection of generators, transformers, and transmission lines. Before this discussion, however, we will consider general procedures for fault current calculations and the use of digital computers to perform such analyses.

5.8 GENERAL PROCEDURES FOR FAULT ANALYSIS AND COMPUTER TECHNIQUES

In Section 5.2 we discussed calculation of fault currents when a three-phase fault occurs on a power system. The presentation in Section 5.3 extended this analysis through development of a general procedure, using the Z_{bus} matrix, for the determination of postfault currents and voltages when a three-phase short circuit is applied at any bus in a power system. This procedure may be applied to a power system of arbitrary size although it is evident that the formulation of the required Z_{bus} matrix is not a simple task for practical systems. In particular, a system having N buses where Z_{bus} is formed by the inversion of Y_{bus} will require the inversion of an Nth order matrix having complex elements. It seems clear that even for a small system with perhaps 10-20 buses, this task must be delegated to a digital computer. While there are procedures for the formulation of Z_{bus} without having to invert Y_{bus} (see Section 5.14 for a discussion of methods for algorithmic formation of Z_{bus}), these are also computationally intensive such that a digital computer is a necessary tool for practical systems.

In Section 5.7, we have shown how the method of symmetrical components may be combined with the formation of sequence networks for unbalanced fault situations such that postfault currents and voltages may be determined for situations such as single-line-to-ground, line-to-line and double-line-to-ground short circuits on a generator. Although it is beyond the scope of this text, it may be shown that this approach may be extended to power systems of arbitrary size, allowing the determination of postfault currents and voltages when an unbalanced fault appears at any system bus. In addition, methods similar to the Z_{bus} procedure for balanced faults have also been developed for unbalanced fault situations. While this latter scheme involves separate formation of a Z_{bus} matrix for each of the positive-, negative-, and zero-sequence networks, the general formation and calculation algorithms are similar. Thus, it has been possible to create software for digital computers which will allow for the determination of postfault voltages and currents when a balanced or unbalanced fault appears at any system bus. An example of such an application is given in the following.

Example 5.16. For the generator shown in Figure 5.20(*a*), calculate the fault currents for balanced, single-line-to-ground, line-to-line, and double-line-to-ground faults. The data for this generator are the same as in Examples 5.11 - 5.15. Then repeat the process for three-phase and single-line-to-ground faults on the power system shown in Figure 5.20(*b*). The data for this system are given in Table 5.1.

Solution
This problem was solved using SKM Systems Analysis Power* Tools for Windows software and a digital computer. From Figure 5.20(*a*), it may be seen that the fault currents printed on the single-line diagram for the generator of Examples 5.11 - 5.15 agree with those calculated in these examples. These currents are labeled as the initial symmetrical RMS currents which correspond to the subtransient fault

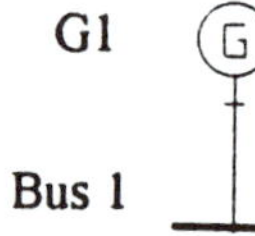

FIGURE 5.20(a)
Generator for Example 5.16.

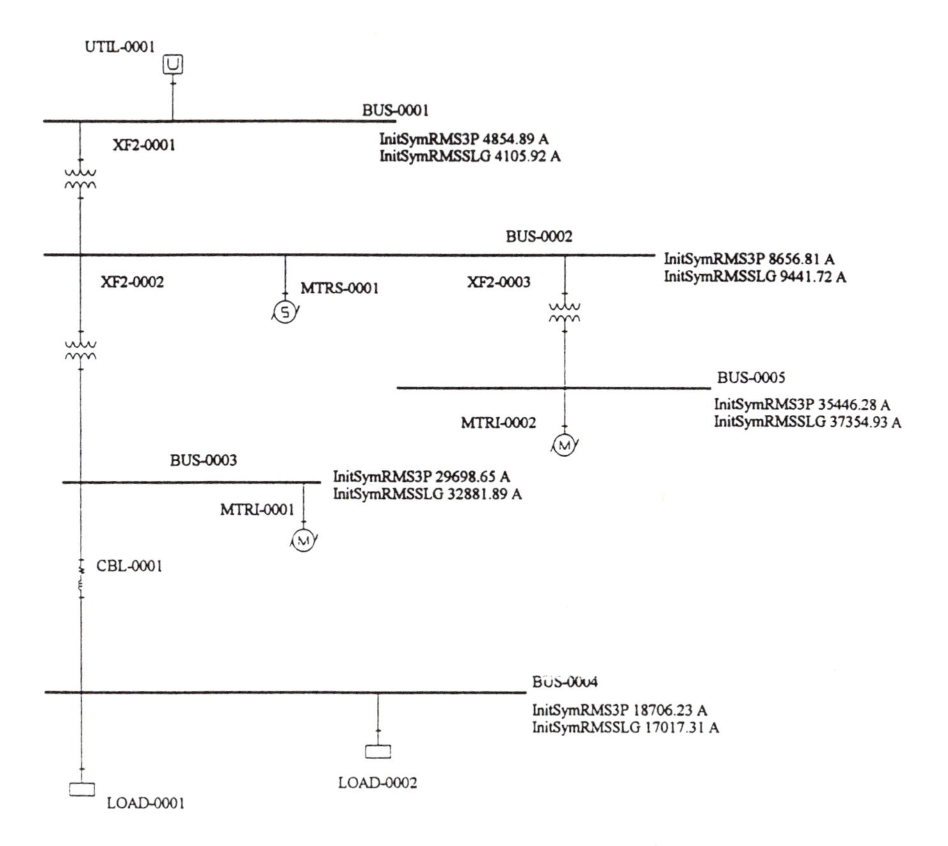

FIGURE 5.20(b)
Power system for Example 5.16.

TABLE 5.1 Portions of computer output for example 5.16.

FEEDER INPUT DATA

FEEDER FROM NAME	FEEDER TO NAME	QTY /PH	VOLTS L-L	LENGTH		FEEDER DESCRIPTION SIZE	TYPE	DUCT	INSUL
Bus 3	Bus 4	1	480	100.00	FT	250	C	N	THWN
+/- Impedance:	0.0541 + J	0.0396	OHMS/k	Length	2.35 + J	1.72	PU		
Z0 Impedance:	0.0860 + J	0.1007	OHMS/k	Length	3.73 + J	4.37	PU		

MOTOR LOAD DATA

BUS NAME	LOAD NAME	VOLT	SIZE	# UNIT	TYPE	EFF	PF	LAG/LEAD
Bus 2	Sync Motor	4160.	2000*1.0	HP	KVA	1.00	0.85	LAG
Bus 3	Ind Motor 2	480.0	500.*1.0	HP	KVA	0.96	0.85	LAG
Bus 5	Ind Motor 1	480.0	500.*3.0	HP	KVA	0.96	0.85	LAG

ENERGY AUDIT LOADS

BUS NAME	LOAD NAME	VOLTS	SIZE	LOADTYPE	PF	LAG/LEAD
Bus 4	Load 1	480.00	200.00*1.00 kVA	KVA	0.85	LAG
Bus 4	Load 2	480.00	9.70*1.00 kVA	KVA	0.90	LAG

Table 5.1 (*Cont'd*)

```
                              GENERATION  DATA
---------------------------------------------------------------------------
BUS NAME        GENERATION                      VOLT    SIZE   InitKW  MaxKVAR   TYPE
---------------------------------------------------------------------------
Bus 1           Utility                  1 pu                                 SB
                Three Phase   Contribution:  100.00 MVA  X/R :   15.00
                Line to Earth Contribution:   30.00 MVA  X/R :   15.00
   Pos  sequence impedance (100 MVA base) 0.0665 + J  0.9978 PU
   Zero sequence impedance (100 MVA base) 0.0887 + J    1.33 PU


                             TRANSFORMER  INPUT  DATA
===========================================================================
PRIMARY RECORD        VOLTS   * SECONDARY RECORD     VOLTS   FULL-LOAD    NOMINAL
   NO NAME             L-L          NO NAME            L-L      KVA          KVA
===========================================================================
Bus 1             D  13800.0   Bus 2             YG   4160.00 5750.00      5000.00
Pos. Seq. Z%:      1.000 + J   6.93 0.200 + j   1.39 PU
Zero Seq. Z%:      1.000 + J   6.93 0.200 + j   1.39 PU
Taps  Pri. 0.000 %  Sec. 0.000 %  Phase Shift (Pri. Leading Sec.):  0.000 Deg.

Bus 2             D  4160.00   Bus 3             YG    480.00 2300.00      2000.00
Pos. Seq. Z%:      1.000 + J   5.66 0.500 + j   2.83 PU
Zero Seq. Z%:      1.000 + J   5.66 0.500 + j   2.83 PU
Taps  Pri. 0.000 %  Sec. 0.000 %  Phase Shift (Pri. Leading Sec.):  0.000 Deg.

Bus 2             D  4160.00   Bus 5             YG    480.00 2300.00      2000.00
Pos. Seq. Z%:      1.000 + J   5.66 0.500 + j   2.83 PU
Zero Seq. Z%:      1.000 + J   5.66 0.500 + j   2.83 PU
Taps  Pri. 0.000 %  Sec. 0.000 %  Phase Shift (Pri. Leading Sec.):  0.000 Deg.
```

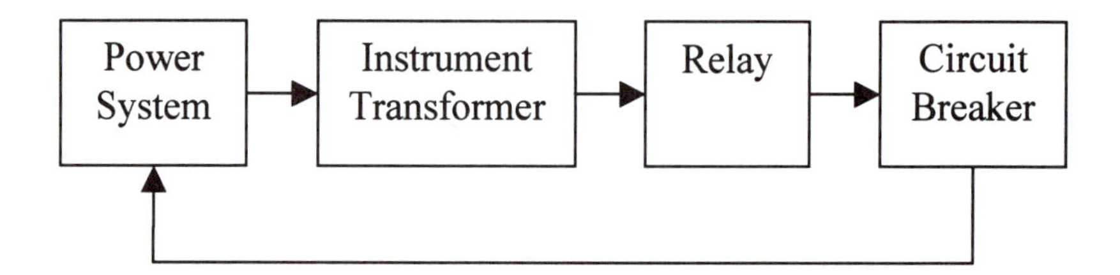

FIGURE 5.21
Power system protection components.

current (no dc component is included as discussed later in Sec. 5.12). Note that the fault current listed for the double line-to-ground fault is $|I_b| = |I_c| = (4.16)(874.8) \approx 3644.1$ A. The minor differences in the listed fault currents are due to the finite *X*/*R* ratios that are used for the generator impedances in the program.

Results for the power system of Table 5.1 are shown on the single-line diagram of Figure 5.20(*b*). The fault current for three-phase and single-line-to-ground faults at each system bus are printed next to the faulted bus. Information such as the postfault bus voltages or the branch currents is also available and may be printed on the diagram or obtained by printing an output report. Thus, the data required by a practicing engineer for design of protection for this power system are readily available.

5.9 COMPONENTS OF A PROTECTION SYSTEM

The preceding sections of this chapter have addressed calculation of the voltages and currents that will exist on a power system upon occurrence of a system fault. The purpose of such an analysis is to obtain the data that is necessary to design and specify a protection system. Components used for power system protection consist of instrument transformers, relays and circuit breakers, and the purpose and applications of each of these devices will be surveyed in the following discussions. The basic function of each device is shown schematically in Figure 5.21. The purpose of the instrument transformer is to sense a measurement parameter (normally voltage or current) and send this information to a relay. The relay then processes this information to determine if an abnormal situation exists and acts to achieve coordination with other protection points in the system. If it is determined that action is required at its own point of protection, the relay sends a trip signal to the circuit breaker which must then act to open the power system circuit at that point.

5.10 INSTRUMENT TRANSFORMERS

Instrument transformers are used to sense abnormal current and voltage levels and transmit input signals to the relays of a protection system. These transformers take the form of current and voltage (or potential) transformers. In contrast to power transformers, the power ratings of instrument transformers are rather low, perhaps 25 to 500 VA, depending on the load or *burden* on the transformer.

A *current transformer* (CT) is symbolically represented as in Figure 5.22. The primary generally consists of a transmission line or other power system conductor and the secondary winding consists of a multiturn coil. In practice, instrument transformers have phase angle and ratio errors. Standard CT transformation ratios range from 50:5 to 1200:5.

Voltage transformers (VTs) for application at or below 12 kV (primary voltage) generally have a 57-V secondary winding. For higher-voltage applications, a coupling capacitor configuration of the type shown in Figure 5.23 is used. The values of C_1 and C_2 are chosen so that only a few kilovolts appear across C_2 when A is at the bus voltage, and the tapped voltage is then reduced to the relay operating voltage.

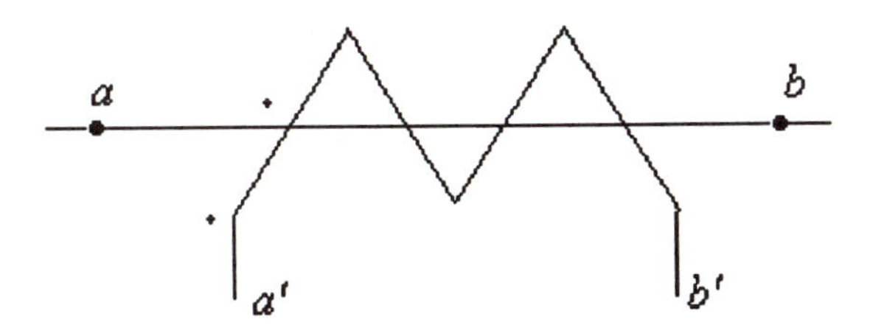

FIGURE 5.22
Representation of a CT.

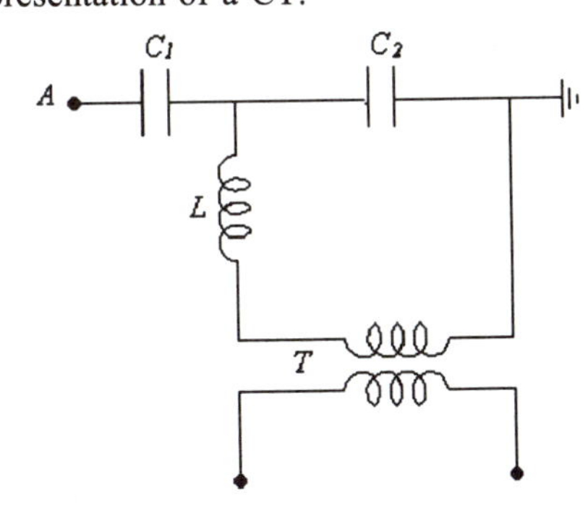

FIGURE 5.23
A coupling-capacitor voltage transformer.

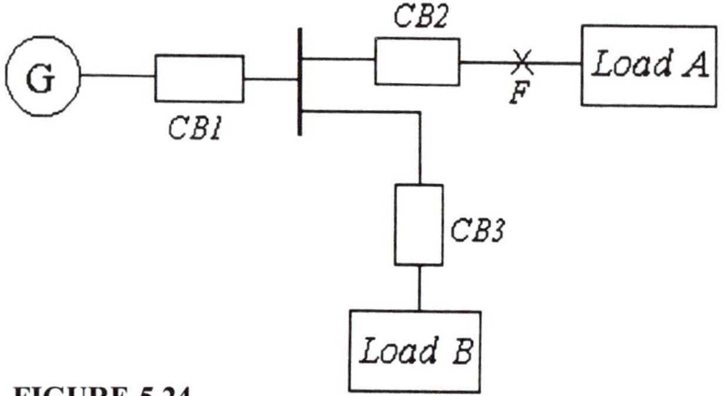

FIGURE 5.24
Time delay overcurrent protection.

5.11 RELAYS

In power system terminology, a relay is used as a somewhat generic term to describe a device that processes voltage and/or current input data from an instrument transformer and provides an appropriate trip signal to a circuit breaker. Some major types include overcurrent (or magnitude) relays, directional relays, distance (or impedance or ratio) relays, differential relays, pilot relays and digital (or microprocessor-based) relays.

Overcurrent or *magnitude relays* respond to a current input. They operate when the current on the secondary side of the CT that feeds the relay exceeds the pickup current of the relay. The relay may then send an instantaneous trip signal to its circuit breaker or there may be an intentionally introduced time delay before such a signal is sent. The reason for such delays (remember that there is a conflicting requirement that we would like to interrupt fault currents as soon as possible) is to achieve coordination in the protective system. This concept may be visualized by referring to Figure 5.24. In this simple system, the generator G supplies power to loads A and B. There are three points of protection denoted by the circuit breakers CB1, CB2, and CB3, each of which has its own CT and overcurrent relay. When a fault occurs at point F, it is desirable for CB2 to operate to clear the fault and for CB1 and CB3 to remain closed. Thus, load B would continue to operate even though load A may be disconnected for some time.

For this fault situation, it should be clear that CB3 will not trip because its CT and relay do not see the fault current and the relay's pickup current will not be exceeded. However, the relays at CB1 and CB2 will both see the fault current and each pickup current may be exceeded. It is therefore desirable to have CB2 clear the fault before the relay at CB1 sends a trip signal to CB1. Thus, a time delay may be introduced by the overcurrent relay at CB1, allowing CB2 to clear the fault and load B to continue its operation. Such time delays

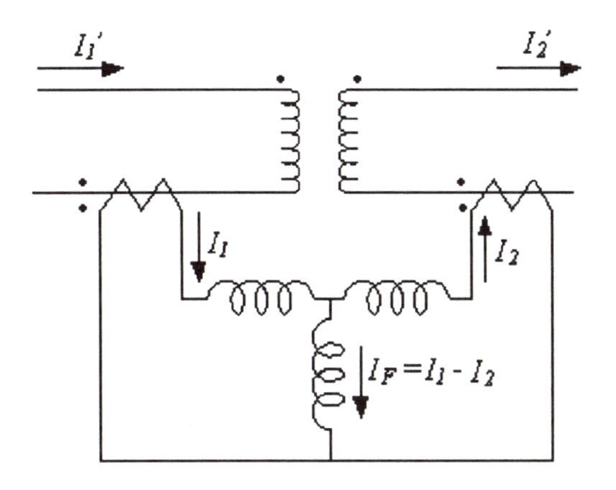

FIGURE 5.25
Differential protection of a transformer.

must be carefully coordinated with the fault current magnitude, however, as large fault currents must be cleared quickly.

A *directional relay* operates for fault currents in only one direction. Thus, such a relay responds to faults either to the left or to the right of its location. Its operation depends upon the phase angle of the fault current with respect to a reference voltage. A 180° range of possible current phase angles for positive currents flowing in the trip direction are then defined with respect to the reference voltage. For currents of fault magnitude that have a phase angle within the defined range, the relay will then trip the circuit breaker. Currents in the opposite direction will have a 180° phase shift and will be outside of the defined range and a trip signal is not sent to the circuit breaker.

A *distance relay* (also called an impedance or ratio relay) responds to the ratio of the voltage and current at the protection point. Since the ratio of the voltage divided by the current represents an impedance, and the amount of impedance between the protection point and the fault depends upon the length of transmission line (distance) between these two points, it seems clear how the various names are derived. The major point is that this type of relay may be adjusted to respond only to short circuit faults that occur within a set distance from the point of protection. Other parts of the line may be protected by distance relays at other points of protection so that coordination and backup may be achieved.

The operation of a *differential relay* may be understood by referring to Figure 5.25 for a transformer winding. The CT turns ratios are chosen such that under normal conditions we have $I_1 - I_2 = 0$. Under fault conditions, $I_1 - I_2 = I_F$, where I_F is the fault current referred to the secondary of the CTs. If I_F is greater than the relay pickup current, the relay will then trip the appropriate circuit breakers.

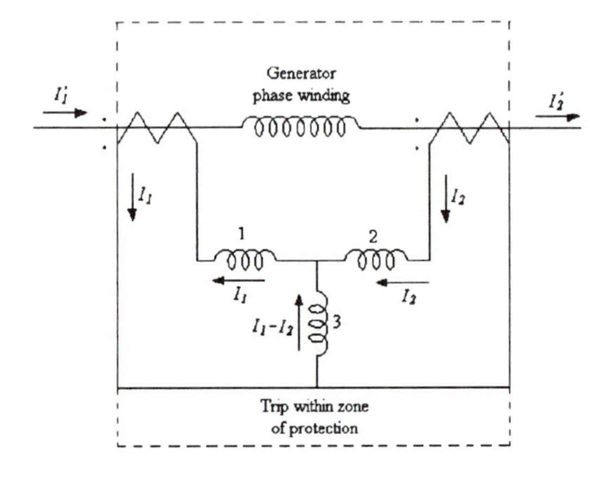

FIGURE 5.26
Differential protection of a generator.

In the differential protection scheme for a generator shown in Figure 5.26, the CTs have the same turns ratio. We further note that the zone of protection of a differential relay is small; that is, the boundary points of the zone are close to each other. A *pilot relay* provides a means of transmitting fault signals from a remote zone boundary to relays at the terminals of a long transmission line. Pilot protection is therefore simply a differential protection scheme where information must be transmitted between measurement points.

Digital relays are devices that receive current and/or voltage signals and analyze these signals in some way in order to make a trip decision or provide coordination. An example might be a relay that calculates the negative and zero-sequence current components in order to make a trip decision.

5.12 CIRCUIT BREAKERS

Circuit breakers perform the difficult and important job of interrupting the fault current. Cooling of the arc that will be established when trying to interrupt large currents in an inductive circuit is an integral part of this process. It must be done in such a way that the arc will not be reestablished after a current zero and a discussion of such techniques is beyond the scope of this text. It is sufficient to state, however, that one result of these considerations is that circuit breaker interrupting speeds are normally rated in cycles (i.e., 2, 3, 8, etc.). In addition to the time required to clear a fault, circuit breakers must be rated to interrupt any possible fault current which may occur at the point of protection (interrupting current). Finally, they must be able to carry the fault current prior to interruption (withstand current).

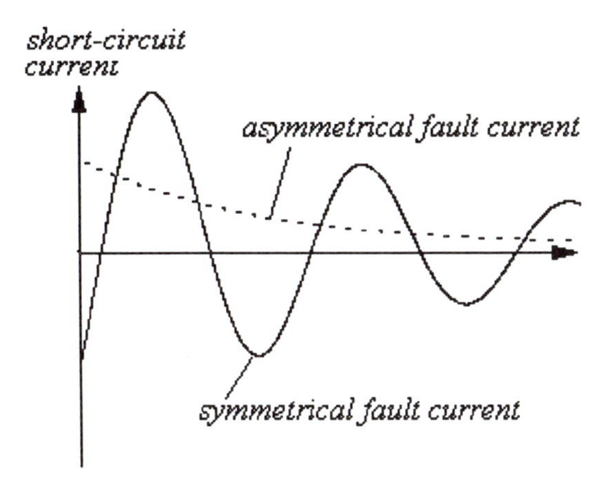

FIGURE 5.27
Symmetrical and asymmetrical fault currents.

The necessary withstand and interrupting currents are available from a subtransient fault analysis as discussed earlier in this chapter. However, there is one additional complication. The plot of symmetrical short-circuit current given in Figure 5.2 is for a very special case. It is for the case where the fault occurs at exactly the time where the system voltages will cause the ac fault current to begin at zero (i.e., $i_F = 0$ at $t = 0+$ when the fault occurs at $t = 0$ in an inductive circuit). This is, of course, not often the case as the fault is a random occurrence. Suppose, for example, that the fault were to occur at a time when the phase of the short-circuit current is such that this ac current would be a negative maximum as shown in Figure 5.27. In order for the total fault current to begin at zero as it must in an inductive circuit, an exponentially decaying dc current must also be present as indicated. The initial value of this dc component is random and ranges from zero to ± the maximum value of the ac component. The existence of such a dc component may be easily verified by the analysis of a series network containing a voltage source, resistance, inductance and a short circuit that occurs at $t = 0$. In power systems terminology, the subtransient ac component of the fault current is often called the initial symmetrical current (not to be confused with symmetrical components) and the dc component is often called the asymmetrical current. The withstand and interrupting ratings of a circuit breaker must include the sum of both currents, however, and a multiplication factor is often utilized to account for the random asymmetrical component.

5.13 PROTECTION OF LINES, TRANSFORMERS, AND GENERATORS

A radial transmission line like that of Figure 5.28 might be protected with time-overcurrent relays. These relays can be set to provide primary protection for one line and remote backup protection for a neighboring line. For instance, the relay at bus 1 will protect the line from bus 1 to bus 2 and act as a backup for the line between buses 2 and 3. It should, however, be adjusted to provide an adequate time delay, such that the relay at bus 2 operates first for a fault on line 2.

To protect lines fed from both ends (Figure 5.29) or loop systems (Figure 5.30), directional relays with coordinated time settings may be used. (In the figures, the arrows show the direction protected by each relay.) The relays associated with circuit breakers 1, 3, and 5 may be coordinated, as may relays associated with breakers 2, 4, and 6. Overcurrent relays may be used and made directional by adding a directional relay at each location, and then arranging the outputs of the directional relay and the overcurrent relay so that their breaker will not operate unless both relays provide a trip signal. Transmission lines belonging to a complex interconnected system are often protected by *impedance relays*, which respond to the impedance between their own location and the location of a fault.

Transformers and generators are protected against certain types of faults by differential relays. Figures 5.25 and 5.26, respectively, show arrangements of differential relays to protect against faults in a transformer or a generator. Notice the similarity between the scheme for the protection of a generator shown in Figure 5.26 and that for the transformer given in Figure 5.25. The differential current ($I_1 - I_2$) is a measure of the fault current. If the relay is set for a sufficiently low pickup current, the differential relay would trip in case of an internal fault. At the same time, external faults are ignored.

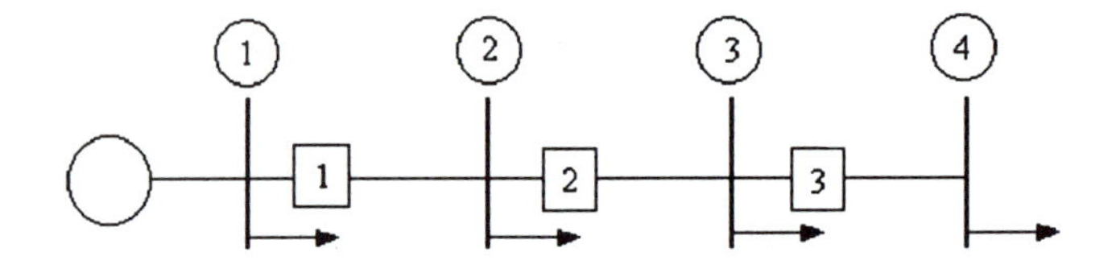

FIGURE 5.28
Radial transmission system.

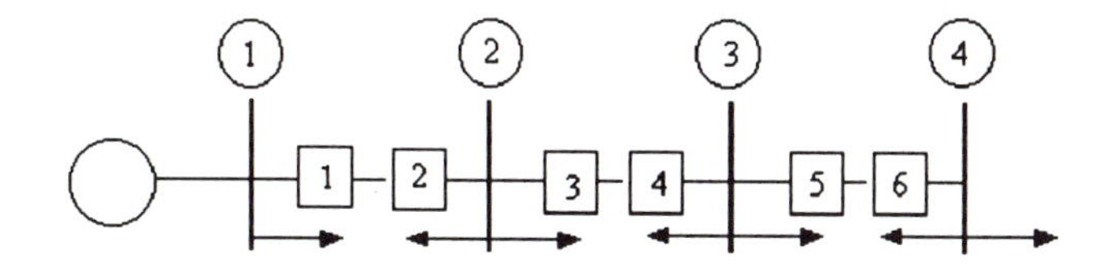

FIGURE 5.29
Line protection from two ends.

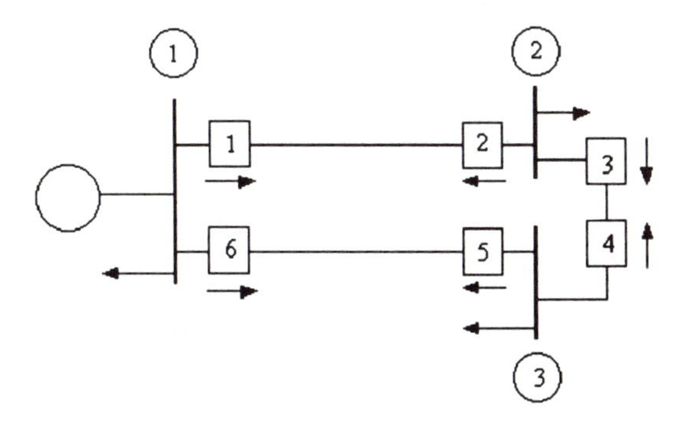

FIGURE 5.30
Protection of a loop system.

5.14 BUS IMPEDANCE MATRIX FORMATION

Examples 5.5 and 5.6 show the usefulness of the $\boldsymbol{Z}_{\text{bus}}$ concept for a two-bus system. Whereas the two-bus system merely illustrates the procedure using long-hand calculations, computer algorithms may be developed for an N-bus system where manual calculations are not feasible.

An N-bus system is shown in Figure 5.31,* where all the (synchronous machine) voltage sources have been replaced by an equivalent voltage source V_F. A short-circuit fault on bus N is represented by the switch SW closed. In terms of $\boldsymbol{Z}_{\text{bus}}$ for the circuit of Figure 5.31, we have

$$\boldsymbol{Z}_{\text{bus}}\boldsymbol{I} = \boldsymbol{V}_F - \boldsymbol{E} \tag{5.70}$$

indicating that we can solve for the fault currents $\boldsymbol{I}$ if we know $\boldsymbol{Z}_{\text{bus}}$. Recall that

*This representation has been called the *rake* equivalent in the literature.

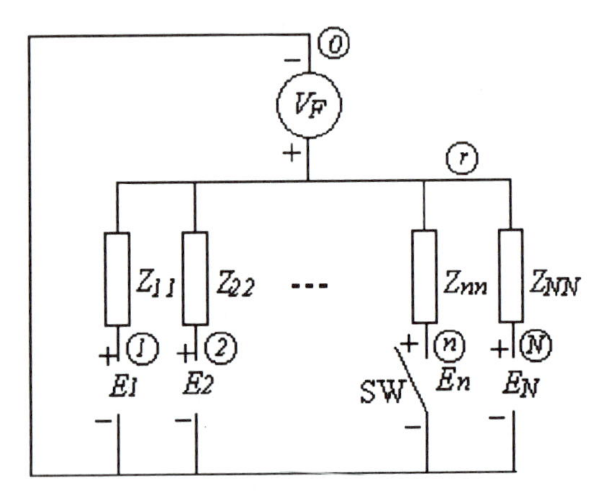

FIGURE 5.31
An equivalent circuit for N-bus system.

(5.6) was used to find the $\boldsymbol{Z}_{\text{bus}}$ in Section 5.3. However, constructing $\boldsymbol{Y}_{\text{bus}}$ and then inverting it to find $\boldsymbol{Z}_{\text{bus}}$ is not an efficient method for a system having a large number of buses. In such a case, $\boldsymbol{Z}_{\text{bus}}$ is formed in a step-by-step fashion by adding one branch at a time. This approach may be used to construct a new $\boldsymbol{Z}_{\text{bus}}$ or to modify a given $\boldsymbol{Z}_{\text{bus}}$.

The addition of a branch having an impedance Z_b to a system having an original known bus impedance matrix $\boldsymbol{Z}_{\text{bus}}$ may take one of the following forms:

1. Inserting Z_b from the reference bus r to a new bus p (Figure 5.32).

2. Inserting Z_b from an old bus k to a new bus p (Figure 5.33).

3. Inserting Z_b from the reference bus r to an old bus k (Figure 5.34).

4. Inserting Z_b between two old buses k and m (Figure 5.35).

Referring to Figures 5.32 through 5.35, we may obtain the new bus impedance matrix that results from the addition of Z_b as follows.

1. For Figure 5.32, the new bus p is connected to the reference bus r through an impedance Z_b, without altering any of the original buses. Thus, the original bus voltages do not change. Since there is no mutual coupling between the new bus and the original buses, we obtain

$$
Z_{bus(new)} = \left[\begin{array}{cccc:c} & & & & 0 \\ & Z_{bus(old)} & & & 0 \\ & N \times N & & & \cdots \\ \hdashline 0 & 0 & \cdots & 0 & Z_b \end{array}\right] \tag{5.71}
$$

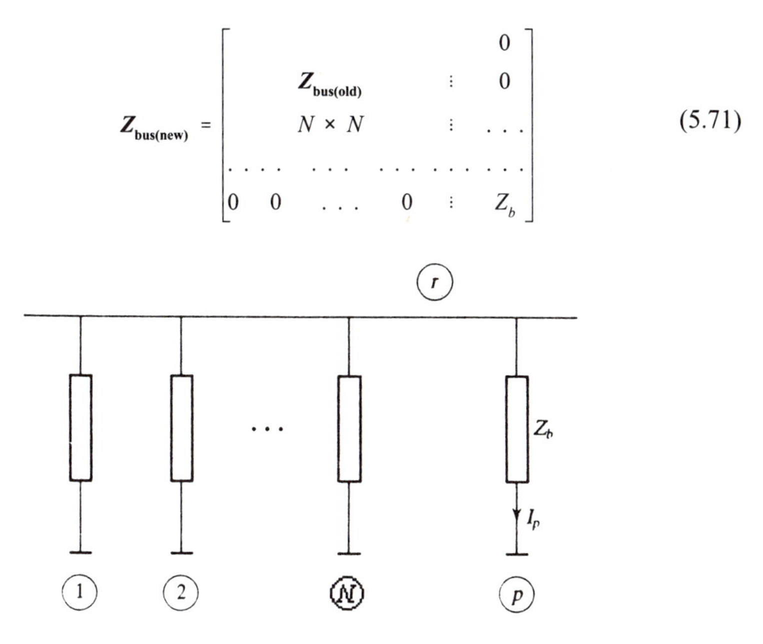

FIGURE 5.32
Type 1 modification.

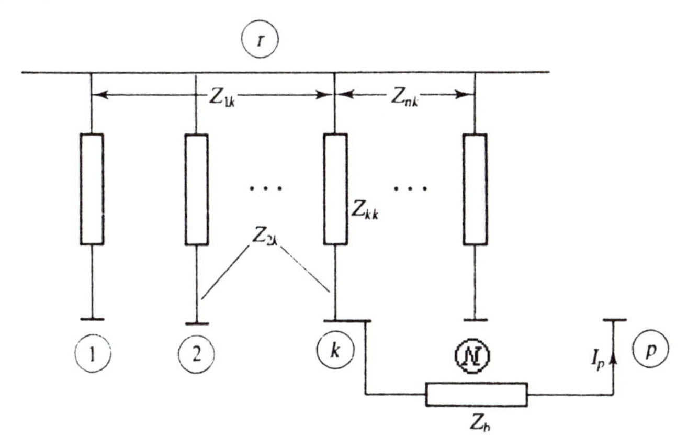

FIGURE 5.33
Type 2 modification.

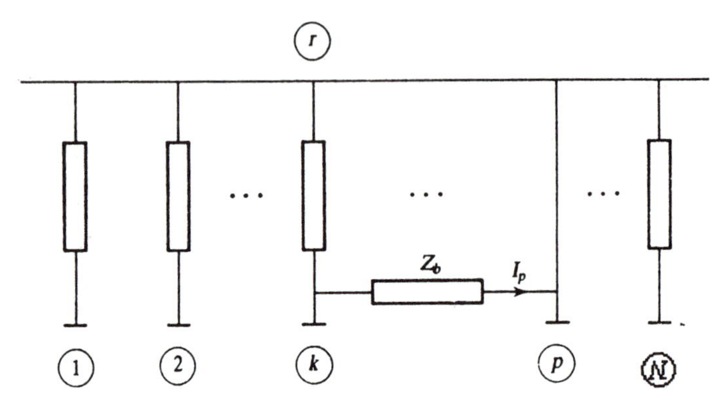

FIGURE 5.34
Type 3 modification.

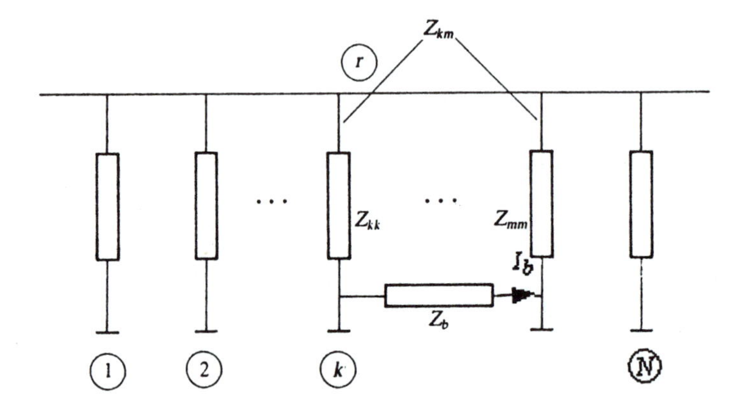

FIGURE 5.35
Type 4 modification.

2. Connecting the new bus p to an existing bus k (Figure 5.33) will make the current entering the original network at bus k the sum of currents I_k and I_p, where I_p is the current entering bus p. Consequently, the voltage at bus k becomes

$$V_{k(\text{new})} = V_{k(\text{old})} + I_p Z_{kk} \tag{5.72}$$

With reference to Figure 5.31, it is to be noted that $V_k = V_F - E_k$.
And

$$V_p = V_{k(\text{old})} + I_p Z_{kk} + I_p Z_b \tag{5.73}$$

Furthermore,

$$V_{k(\text{old})} = I_1 Z_{k1} + I_2 Z_{k2} + \ldots + I_k Z_{kn} \tag{5.74}$$

Combining (5.73) and (5.74) yields

$$V_p = I_1 Z_{k1} + I_2 Z_{k2} + \ldots + I_k Z_{kn} + I_p(Z_{kk} + Z_b) \tag{5.75}$$

The new voltage-current relationship becomes

$$\begin{bmatrix} V_1 \\ V_2 \\ \vdots \\ V_N \\ \cdots \\ V_P \end{bmatrix} = \begin{bmatrix} & & & & \vdots & Z_{1k} \\ & & & & \vdots & Z_{2k} \\ & & \boldsymbol{Z}_{\text{bus(old)}} & & \vdots & \vdots \\ & & & & \vdots & Z_{Nk} \\ \cdots & \cdots & \cdots & \cdots & \vdots & \cdots \\ Z_{k1} & Z_{k2} & \cdots & Z_{Nk} & \vdots & Z_{kk} + Z_b \end{bmatrix} \begin{bmatrix} I_1 \\ I_2 \\ \vdots \\ I_N \\ \cdots \\ I_p \end{bmatrix} \tag{5.76}$$

Thus, we have

$$\boldsymbol{Z}_{\text{bus(new)}} = \begin{bmatrix} & & & & \vdots & Z_{1k} \\ & & & & \vdots & Z_{2k} \\ & & \boldsymbol{Z}_{\text{bus(old)}} & & \vdots & \cdots \\ & & & & \vdots & Z_{Nk} \\ \cdots & \cdots & \cdots & \cdots & \cdots & \cdots \\ Z_{1k} & Z_{2k} & \cdots & Z_{Nk} & \vdots & Z_{kk} + Z_b \end{bmatrix} \tag{5.77}$$

3. For the modification shown in Figure 5.34, we have the same volt-amp equation as (5.76) except that now bus p is shorted to bus r making $V_p = 0$. So, we may eliminate I_p from (5.76) by expanding it in terms of submatrices such that

$$\begin{bmatrix} V_1 \\ V_2 \\ \vdots \\ V_N \end{bmatrix} = \begin{bmatrix} \boldsymbol{Z}_{\text{bus(old)}} \end{bmatrix} \begin{bmatrix} I_1 \\ I_2 \\ \vdots \\ I_N \end{bmatrix} + \begin{bmatrix} Z_{1k} \\ Z_{2k} \\ \vdots \\ Z_{Nk} \end{bmatrix} I_p \tag{5.78}$$

And

$$[Z_{1k}\ Z_{2k}\ .\ .\ .Z_{Nk}] \begin{bmatrix} I_1 \\ I_2 \\ \vdots \\ I_N \end{bmatrix} + (Z_{kk} + Z_b)\ I_p = 0 \tag{5.79}$$

We may now eliminate I_p from (5.78) and (5.79) to obtain

$$\begin{bmatrix} V_1 \\ V_2 \\ \vdots \\ V_N \end{bmatrix} = \begin{bmatrix} \boldsymbol{Z}_{\text{bus(old)}} \end{bmatrix} \begin{bmatrix} I_1 \\ I_2 \\ \vdots \\ I_N \end{bmatrix}$$

$$\frac{1}{Z_{kk} + Z_b} \begin{bmatrix} Z_{1k}^2 & Z_{1k}Z_{2k} & \dots & Z_{1k}Z_{NK} \\ Z_{2k}Z_{1k} & Z_{2k}^2 & \dots & Z_{2k}Z_{Nk} \\ \dots & \dots & \dots & \dots \\ Z_{Nk}Z_{1k} & Z_{Nk}Z_{2k} & \dots & Z_{Nk}^2 \end{bmatrix} \begin{bmatrix} I_1 \\ I_2 \\ \vdots \\ I_N \end{bmatrix} \tag{5.80}$$

Or, the new $\boldsymbol{Z}_{\text{bus}}$ is given by

$$\boldsymbol{Z}_{\text{bus(new)}} = \boldsymbol{Z}_{\text{bus(old)}} - \frac{1}{Z_{kk} + Z_b} \begin{bmatrix} Z_{1k}^2 & Z_{1k}Z_{2k} & \dots & Z_{1k}Z_{Nk} \\ Z_{2k}Z_{1k} & Z_{2k}^2 & \dots & Z_{2k}Z_{Nk} \\ \dots & \dots & \dots & \dots \\ Z_{Nk}Z_{1k} & Z_{Nk}Z_{2k} & \dots & Z_{Nk}^2 \end{bmatrix} \tag{5.81}$$

The term on the right-hand side is often written in a modified form such that

$$\boldsymbol{Z}_{\text{bus(new)}} = \boldsymbol{Z}_{\text{bus(old)}} - \frac{1}{Z_{kk} + Z_b} \begin{bmatrix} Z_{1k} \\ Z_{2k} \\ \vdots \\ Z_{Nk} \end{bmatrix} [Z_{1k}Z_{2k}\ .\ .\ .\ Z_{Nk}] \tag{5.82}$$

4. Finally, when a system is modified by connecting an impedance Z_b between two existing buses k and m, as shown in Figure 5.35, we may write for

nodes k and m:

$$V_k = Z_{k1}I_1 + \ldots + Z_{kk}I_k + Z_{km}I_m + \ldots + (Z_{kk} - Z_{km})\, I_b \tag{5.83}$$

$$V_m = Z_{ml}I_1 + \ldots + Z_{mk}I_k + Z_{mm}I_m + \ldots + (Z_{mk} - Z_{mm})I_b \tag{5.84}$$

The impedance drop across Z_b is

$$I_b Z_b = V_m - V_k \tag{5.85}$$

Combining (5.83) through (5.85) yields

$$(Z_{kl} - Z_{ml})I_1 + \ldots + (Z_{kk} - Z_{mk})\, I_k + (Z_{km} - Z_{mm})\, I_m + \ldots + Z_{bb}I_b \tag{5.86}$$

where

$$Z_{bb} = Z_b + Z_{kk} + Z_{mm} - 2Z_{km} \tag{5.87}$$

The volt-ampere equations for the modified system in matrix form become

$$\begin{bmatrix} V_1 \\ V_2 \\ \vdots \\ V_N \\ \cdots \\ 0 \end{bmatrix} = \left[\begin{array}{ccc|c} & & & Z_{1k} - Z_{1m} \\ & & & \vdots \\ & \boldsymbol{Z}_{\text{bus(old)}} & & Z_{kk} - Z_{km} \\ & & & \vdots \\ & & & Z_{Nk} - Z_{Nm} \\ \cdots & \cdots & \cdots & \cdots \\ (Z_{k1} - Z_{m1}) & \ldots & (Z_{mk} - Z_{mm}) \ \ldots & Z_{bb} \end{array}\right] \begin{bmatrix} I_1 \\ \vdots \\ I_k \\ \vdots \\ I_N \\ \cdots \\ I_b \end{bmatrix} \tag{5.88}$$

As was done for type 3 modification, we may eliminate I_b from (5.88), and modified $\boldsymbol{Z}_{\text{bus}}$ takes the following form

$$\boldsymbol{Z}_{\text{bus(new)}} = \boldsymbol{Z}_{\text{bus(old)}} - \frac{1}{Z_b + Z_{kk} + Z_{mm} - 2Z_{km}}$$

$$\begin{bmatrix} (Z_{1k} - Z_{1m})^2 & (Z_{1k} - Z_{1m})(Z_{2k} - Z_{2m}) & \ldots & (Z_{1k} - Z_{1m})(Z_{Nk} - Z_{Nm}) \\ (Z_{2k} - Z_{2m})(Z_{1k} - Z_{1m}) & (Z_{2k} - Z_{2m})^2 & \ldots & (Z_{2k} - Z_{2m})(Z_{Nk} - Z_{Nm}) \\ \cdots & \cdots & \cdots & \cdots \\ (Z_{Nk} - Z_{Nm})(Z_{1k} - Z_{1m}) & (Z_{Nk} - Z_{Nm})(Z_{2k} - Z_{2m}) & \ldots & (Z_{Nk} - Z_{Nm})^2 \end{bmatrix} \tag{5.89}$$

This equation may also be written in the form given as (5.82); that is

$$
\mathbf{Z}_{\text{bus(new)}} = \mathbf{Z}_{\text{bus(old)}} - \frac{1}{Z_b}\begin{bmatrix} Z_{1k} - Z_{1m} \\ \vdots \\ Z_{kk} - Z_{km} \\ \vdots \\ Z_{Nk} - Z_{Nm} \end{bmatrix} [(Z_{1k} - Z_{1m}) \ldots (Z_{kk} - Z_{Km}) \ldots (Z_{Nk} - Z_{Nm})] \tag{5.90}
$$

We now illustrate the application of the bus modification procedure to obtain the $\mathbf{Z}_{\text{bus}}$ of a given system.

Example 5.17 Using step-by-step method (or $\mathbf{Z}_{\text{bus}}$ modification approach presented above) obtain the $\mathbf{Z}_{\text{bus}}$ for the system shown in Figure 5.9.

Solution
We choose the reference bus r to which we add the generator impedance $j0.15$ from a new bus 1 which is type 1 addition as shown in Figure 5.36(a). For this step, $\mathbf{Z}_{\text{bus}} = j[0.15]$ pu. Next, from bus 1, we add an impedance $j(0.101 + 0.312 + 0.101) = j0.514$ from bus 1 to bus 2, which is a type 2 addition as shown in Figure 5.36(b).

In accordance with (5.77), the new $\mathbf{Z}_{\text{bus}}$ becomes

$$
\mathbf{Z}_{\text{bus}} = j\begin{bmatrix} 0.15 & 0.15 \\ 0.15 & 0.514 + 0.15 \end{bmatrix} = j\begin{bmatrix} 0.15 & 0.15 \\ 0.15 & 0.664 \end{bmatrix} \text{pu}
$$

Thus, we obtain the representation shown in Figure 5.36(c). Finally, we add the motor impedance, $j0.222$ pu, from bus r to bus 2, which is a type 3 modification as shown in Figure 5.36(d). Thus, using (5.82), we obtain

$$
\mathbf{Z}_{\text{bus}} = j\begin{bmatrix} 0.15 & 0.15 \\ 0.15 & 0.664 \end{bmatrix} - \frac{j}{0.664 + 0.222}\begin{bmatrix} 0.15 \\ 0.664 \end{bmatrix}[0.15 \quad 0.664]
$$

$$
= j\begin{bmatrix} 0.15 - \dfrac{0.15^2}{0.886} & \vdots & 0.15 - \dfrac{0.664 \times 0.15}{0.886} \\ \cdots\cdots\cdots & \vdots & \cdots\cdots\cdots \\ 0.15 - \dfrac{0.664 \times 0.15}{0.886} & \vdots & 0.664 - \dfrac{0.664^2}{0.886} \end{bmatrix}
$$

$$
= j\begin{bmatrix} 0.1246 & 0.0376 \\ 0.0376 & 0.1664 \end{bmatrix} \text{pu}
$$

which agrees with the result of Example 5.5, as expected.

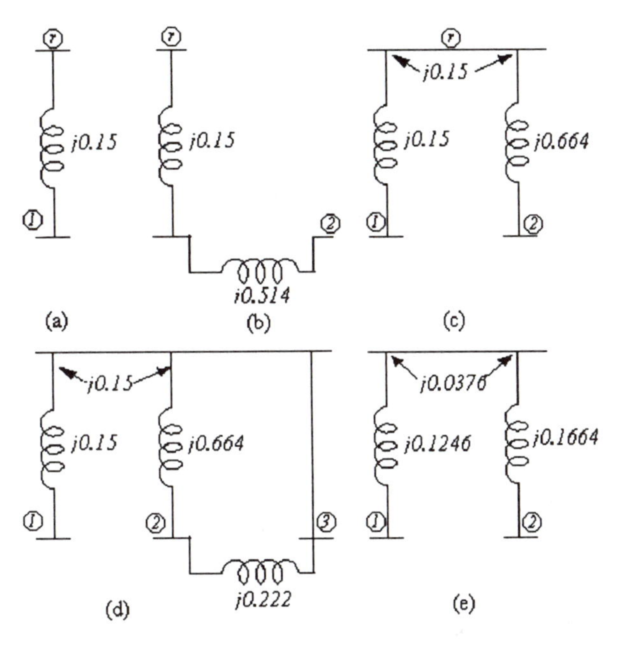

FIGURE 5.36
Step-by-step formation of $\boldsymbol{Z}_{\text{bus}}$.

PROBLEMS

5.1. The per-unit reactances of a synchronous generator are $X_d = 1.0$, $X_d' = 0.35$ and $X_d'' = 0.25$. The generator is supplying 1.0 per-unit current at rated voltage to a 0.8 power factor lagging load. Calculate the internal voltages behind the direct axis synchronous, transient, and subtransient reactances.

5.2. The three-phase system shown in Figure 5.37 is initially unloaded. Calculate the subtransient fault current that results when a three-phase short circuit occurs at F, given that the transformer voltage on the high-voltage side is 66 kV prior to the fault. Assume a 75 MVA 69 kV base at the transformer secondary.

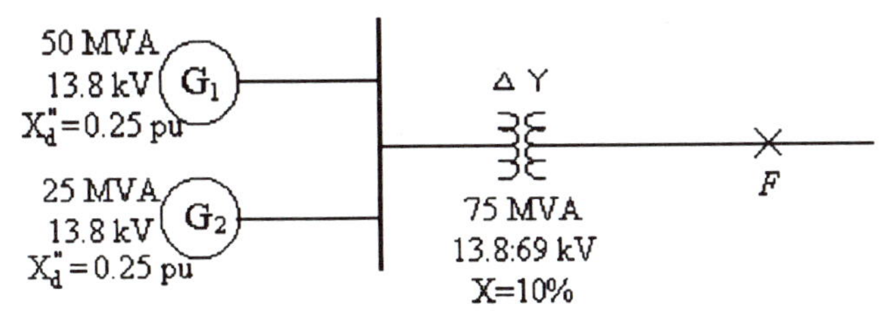

FIGURE 5.37
Problem 5.2.

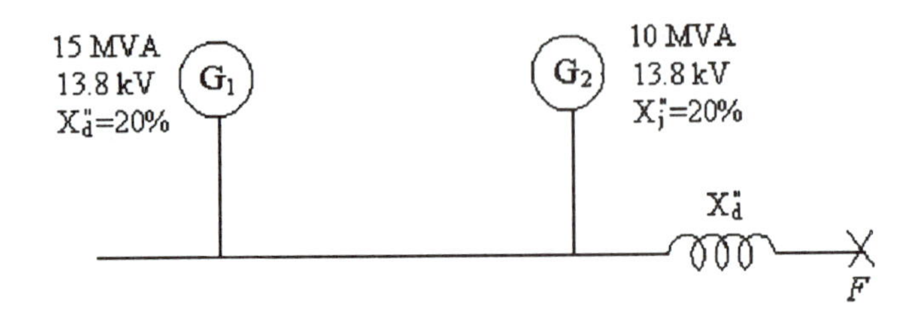

FIGURE 5.38
Problem 5.3

5.3. A portion of a power system is shown in Figure 5.38, which also shows the ratings of the generators and their respective percent reactances. A symmetrical short circuit appears on a feeder F. Find the value of the reactance X (in percent) such that the subtransient short-circuit current will not exceed 1.0 pu. Use a 50 MVA 13.8 kV base and neglect prefault currents. The voltage at F prior to the fault is 13.8 kV.

5.4. Three generators, each rated at 10 MVA 4160V and having a subtransient reactance of 10% are connected to a common bus and supply a load through a 30 MVA 4160:66kV step-up transformer. The transformer has a reactance of 7%. Determine the 3-phase short circuit subtransient fault current in per unit for a fault on (*a*) the high voltage side and (*b*) the low voltage side of the transformer. Neglect prefault currents and assume rated voltage exists prior to the fault. Use a 10 MVA base.

5.5. For the system of Figure 5.39, calculate the subtransient fault current for a three-phase short circuit at A or B. Neglect prefault currents and assume that the voltage at B is 1.0 pu prior to the fault. Use a 30 kVA base.

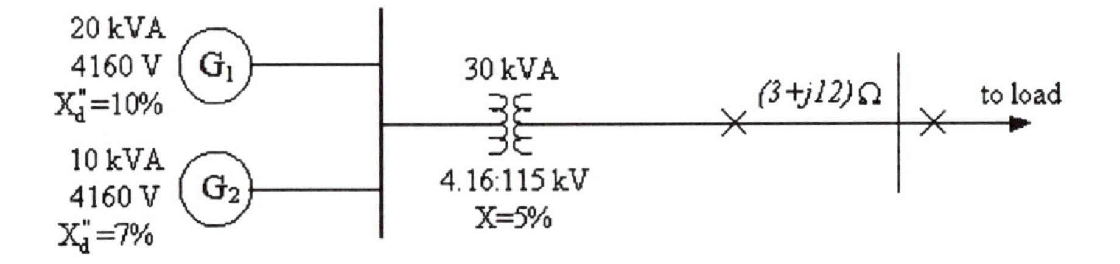

FIGURE 5.39
Problem 5.5.

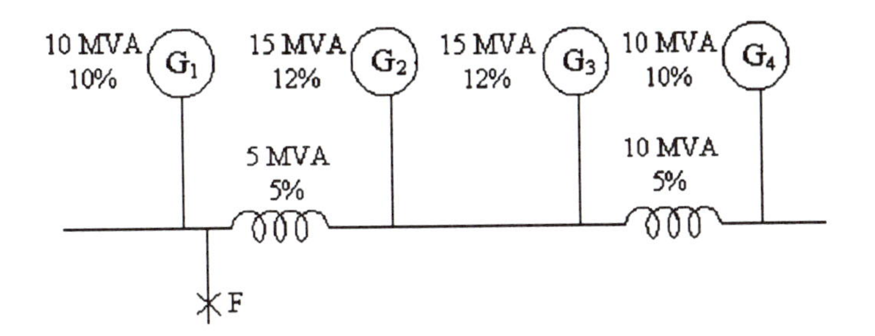

FIGURE 5.40
Problem 5.6.

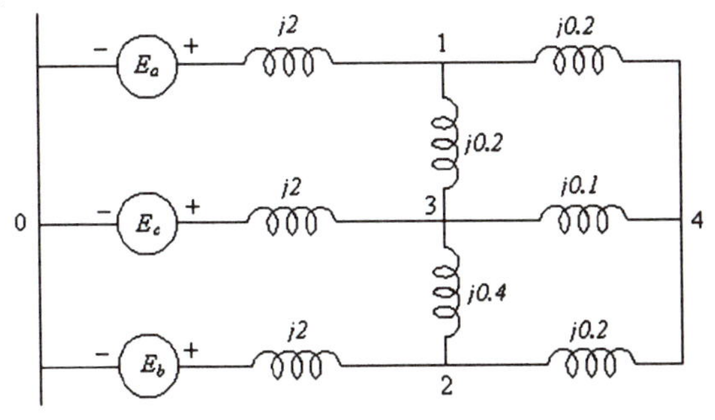

FIGURE 5.41
Problem 5.7.

5.6. A three-phase short circuit occurs at F in Figure 5.40. Calculate the subtransient fault current in per unit assuming a 15 MVA base. All voltages are 1.1 pu prior to the fault and the generator reactances are subtransient values. All rated voltages are 1.0 pu.

5.7. Determine the bus admittance matrix for the network shown in Figure 5.41. All values shown are per-unit impedances.

5.8. The bus impedance matrix for a 5-bus system is

$$Z_{bus} = j\begin{bmatrix} 0.112 & 0.071 & 0.034 & 0.05 & 0.08 \\ 0.071 & 0.228 & 0.055 & 0.079 & 0.1 \\ 0.034 & 0.055 & 0.073 & 0.065 & 0.05 \\ 0.05 & 0.079 & 0.065 & 0.094 & 0.071 \\ 0.08 & 0.1 & 0.05 & 0.071 & 0.118 \end{bmatrix} \text{pu}$$

Determine the fault current and the bus voltages per unit for a three-phase short circuit at bus 4. The prefault voltage is 1.0 pu and prefault currents are negligible.

5.9. Repeat Problem 5.8 for the fault located at bus 2.

5.10. Repeat Problem 5.8 for a short circuit at bus 3 and a prefault voltage of 1.1 pu.

5.11. If the admittance between buses 4 and 5 is $-j9.5$ pu, determine I_{45} in per unit for the situation given in Problem 5.8.

5.12. The line currents in a 4-wire system (see Figure 2.6) are $I_{a'a} = I_a = 300 + j400$ A, $I_{b'b} = I_b = 200 + j200$ A and $I_{c'c} = I_c = 300 + j400$ A. Determine the positive, negative and zero-sequence components of these line currents.

5.13. What is the current in the fourth wire of the system in Problem 5.12? How is this current related to I_{a0}?

5.14. Derive equations (5.25) and (5.26).

5.15. The line currents to a delta-connected load are $I_a = 4.95\ \angle 0°$, $I_b = 7\angle 135°$, and $I_c = 4.95\angle -90°$A. Calculate the sequence components of the line currents and determine the positive and negative-sequence components of I_{ab}.

5.16. Show that equation (5.33) may be obtained by adding (5.30) - (5.32).

5.17. An unbalanced, wye-connected load consisting of the phase resistances $R_a = 60\ \Omega$, $R_b = 40\ \Omega$, and $R_c = 80\ \Omega$ is connected to a 440-V, three-phase, balanced supply. Calculate the line currents by the method of symmetrical components.

5.18. A wye-connected induction machine with an ungrounded neutral has a positive-sequence impedance of $j2\ \Omega$, a negative-sequence impedance of $j1\ \Omega$ and a zero-sequence impedance of $j0.05\ \Omega$. Determine the line currents if the machine is connected to a 240 V line-to-line balanced voltage source.

5.19. Repeat Problem 5.18 if the machine is supplied from a wye-connected source having voltages $V_a = 100\ \angle 0°$, $V_b = 100\ \angle 90°$ and $V_c = 100\ \angle -90°$ V.

5.20. A three-phase, delta-connected load draws 100 A of line current with an open-circuit fault in line c. Determine the sequence components of the line currents.

5.21. The positive-, negative-, and zero-sequence reactances of a 15-MVA, 11-kV, three-phase, wye-connected generator are 11 percent, 8 percent, and 3 percent, respectively. The neutral of the generator is grounded, and the generator is excited to the rated voltage on open circuit. A line-to-ground short circuit occurs on phase a of the generator. Calculate the phase voltages and currents.

5.22. A line-to-line short circuit occurs between phases b and c of the generator of Problem 5.21 while phase a remains open-circuited. Determine the phase voltages and currents.

5.23. The sequence reactances of a three-phase alternator are $X_1 = X_2 = 0.15$ pu and $X_0 = 0.05$ pu, based on the machine rating of 30 MVA and 13.2 kV. The alternator is producing rated voltage and is connected to an open-circuited transmission line having $X_1 = X_2 = 0.1$ pu and $X_0 = 0.4$ pu. A line-to-line short circuit occurs on the line at the receiving end. Draw the sequence networks for the faulted system, and calculate the line currents.

5.24. A double line-to-ground short circuit occurs at F in the unloaded system shown in Figure 5.42. Draw the sequence networks for the system, and calculate the line current I_b. Assume that the generator is producing 90% of its rated voltage prior to the fault.

5.25. Repeat Example 5.15(b) for a single-line to ground short-circuit fault.

5.26. Repeat Example 5.15(b) for a line-to-line short-circuit fault.

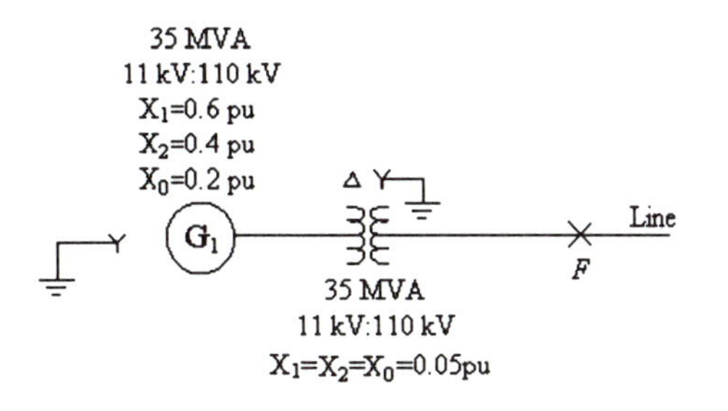

FIGURE 5.42
Problem 5.24.

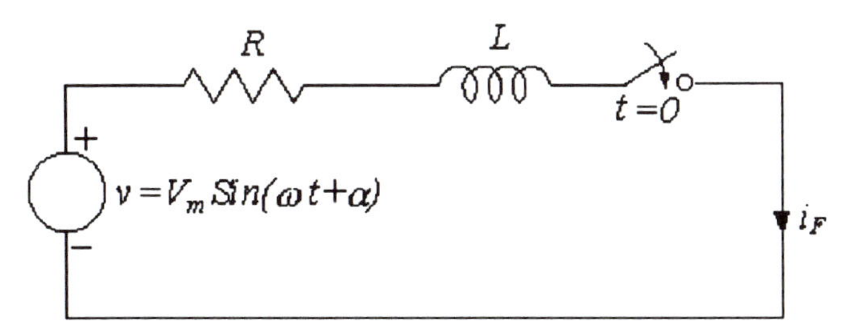

FIGURE 5.43
Problem 5.28.

5.27. Using the results of Problem 5.19, determine the complex power supplied by the source using sequence quantities and then repeat this process using phase values. Compare your answers.

5.28. For the circuit shown in Figure 5.43, the governing differential equation is

$$Ri_F + L\frac{di_F}{dt} = V_m \sin(\omega t + \alpha).$$

Solving this equation gives

$$i_F = \frac{V_m}{|Z|}\left[\sin(\omega t + \alpha - \theta) - e^{-\frac{R}{L}t}\sin(\alpha - \theta)\right]$$

where

$$|Z| = \sqrt{R^2 + (\omega L)^2} \qquad \alpha = \tan^{-1}\frac{\omega L}{R}$$

Explain how the components of i_F are related to the symmetrical and asymmetrical currents discussed in Section 5.12. Why doesn't the ac portion of this solution exhibit the decaying amplitude shown in Figures 5.2 and 5.27?

5.29. A two-bus system has $\boldsymbol{Z}_{bus} = \begin{bmatrix} j0.11565 & j0.04580 \\ j0.04580 & j0.13893 \end{bmatrix}$ pu. If an impedance $Z_b = j0.4$ pu is connected between buses 1 and 2, what is the new $\boldsymbol{Z}_{bus}$?

CHAPTER 6

POWER SYSTEM STABILITY

Power system stability implies that the system remains in a state of equilibrium under normal operating conditions, and returns to equilibrium subsequent to the occurrence of a disturbance. The state of equilibrium, or stability, of a power system commonly alludes to maintaining synchronous operation of the system. In this chapter, we will focus on this aspect of stability whereby a loss of synchronism will mean to render the system unstable. Three types of stability are of concern: steady-state, dynamic, and transient stability.

Steady-state stability relates to the response of a synchronous machine to a gradually increasing load.

Dynamic stability involves the response to small disturbances that occur on the system, producing oscillations. If these oscillations are of successively smaller amplitudes, the system is considered dynamically stable. If the oscillations grow in amplitude, the system is dynamically unstable. The source of this type of instability is usually an interaction between control systems. The system's response to the disturbance may not become apparent for some 10 to 30 s.

Transient stability involves the response to large disturbances, which may cause rather large changes in rotor speeds, power angles, and power transfers. The system's response to such a disturbance is usually evident within a few seconds.

Invariably, stability studies of power systems are carried out on a digital computer. In the following, we present special cases to illustrate certain principles and basic concepts.

6.1 INERTIA CONSTANT AND THE SWING EQUATION

The angular momentum and inertia constant play an important role in determining the transient stability of a synchronous machine. The *per-unit inertia constant H* in MJ/MVA is defined as the kinetic energy stored in the rotating parts of the machine *at synchronous speed* per-unit megavoltampere (MVA) rating of the machine. Thus, if G is the MVA rating of the machine, then

$$GH = \frac{1}{2} J\omega_m^2 \times 10^{-6} \tag{6.1}$$

where J is the polar moment of inertia of all rotating parts in kg-m^2 and ω_m is the angular synchronous velocity in mechanical radians per second. If M is the corresponding angular momentum, then

$$M = J\omega_m \tag{6.2}$$

From (6.1) and (6.2) we have

$$GH = \frac{1}{2} M\omega_m$$

or

$$M = \frac{2GH}{\omega_m} \quad \text{MJ} \cdot \text{s/mechanical radian} \tag{6.3}$$

Consider a synchronous generator developing an electromagnetic torque T_e (and a corresponding electromagnetic power P_e) while operating at the synchronous speed ω_m. If the input torque provided by the prime mover at the generator shaft is T_i, then under steady-state conditions (with no disturbance)

$$T_e = T_i$$

where we have neglected any retarding torque due to rotational losses. Thus we have

$$T_e \omega_m = T_i \omega_m$$

and

$$T_i \omega_m - T_e \omega_m = P_i - P_e = 0 \tag{6.4}$$

If a departure from steady state occurs, such as a change in load or a fault, the "power in" P_i no longer equals the "power out", which is P_e if the armature resistance is neglected. This point is explained further in the discussion of power angle characteristics in Appendix III. Thus, the left side of (6.4) is not zero and an accelerating torque comes into play. If P_a is the corresponding accelerating (or decelerating) power, then

$$P_a = P_i - P_e = M \frac{d^2 \theta_m}{dt^2} + D \frac{d\theta_m}{dt} \tag{6.5}$$

where M has been defined in (6.3), P_a is in megawatts, D is a damping coefficient, and θ_m is the mechanical angular position of the rotor. Further, in the steady state,

$$\frac{d\theta_m}{dt} = \omega_m = \text{constant}$$

so

$$\theta_m = \omega_m t + \delta_m \tag{6.6}$$

where δ_m is the *power angle* of the synchronous machine in mechanical radians. Neglecting damping and substituting (6.6) in (6.5) yields

$$M \frac{d^2 \delta_m}{dt^2} = P_i - P_e = P_a \tag{6.7a}$$

which is known as the *swing equation*. Note that we are now allowing for a change in the angle δ_m, which may be visualized as the angle between the rotor magnetic axis and a synchronously rotating reference axis. It should be noted that the rotor magnetic axis will always rotate with the rotor at mechanical speed while the synchronously rotating reference axis is always rotating at synchronous speed. Thus any deviations of rotor mechanical speed from the synchronous value will result in a decrease or increase in the power angle δ_m.

If we combine (6.3) and (6.7a) and divide by G, we obtain the per-unit swing equation as

$$\frac{2H}{\omega_m} \frac{d^2 \delta_m}{dt^2} = P_i - P_e = P_a \quad \text{per unit} \tag{6.7b}$$

where δ_m is in mechanical radians.

If we now convert both δ_m and ω_m to electrical radians (multiply each by the number of pole pairs of the machine), then

$$\frac{2H}{\omega}\frac{d^2\delta}{dt^2} = \frac{H}{\pi f}\frac{d^2\delta}{dt^2} = P_i - P_e \quad \text{per unit} \tag{6.7c}$$

where δ is in electrical radians and $\omega = 2\pi f$ is the radian frequency in electrical radians/s.

If it is desired to work in electrical degrees, then δ and ω in (6.7*c*) may be multiplied by $180/\pi$ degrees/radian and we obtain

$$\frac{H}{180f}\frac{d^2\delta}{dt^2} = P_i - P_e \quad \text{per unit} \tag{6.8}$$

where δ is in electrical degrees. The power angles in (6.7*c*) and (6.8) correspond to that given in (III.8) of Appendix III.

We may rewrite the swing equation as a set of two first-order differential equations as follows:

$$\frac{d\delta}{dt} = \dot{\delta} \tag{6.9a}$$

$$\frac{H}{180f}\frac{d\dot{\delta}}{dt} = P_i - P_e \tag{6.9b}$$

These various forms of the swing equation are used according to the nature of the problem at hand. Having formulated the governing equation for δ, to investigate the stability of the machine, we solve for δ as a function of time. A plot of $\delta(t)$ is known as the swing curve, a study of which often shows if the machine will remain in synchronism after a disturbance. The swing equation contains information regarding the machine dynamics and stability. However, it is important to realize that we made two basic assumptions in deriving it: (1) In (6.2) we took M to be constant, although, strictly speaking, this is not so; (2) the damping term proportional to $d\delta/dt$ has been neglected.

Referring to Appendix III and neglecting armature resistance, we may write the electromagnetic power in terms of the power angle equation as

$$P_e = \frac{|E_g|\,|V_t|}{X_d}\sin\delta = P_{\max}\sin\delta \tag{6.10}$$

where $|E_g|$ is the internal voltage of the machine, X_d is its reactance and $|V_t|$ is the terminal voltage.

6.2 EVALUATION OF THE *H* CONSTANT

We have seen in earlier chapters that in a power system, different components may have different ratings. Thus, many synchronous machines of different ratings may be operating in parallel. In such a case, the *H* constants of various machines are expressed on a common base.

An inertia constant *H* based on a machine's own MVA rating *G* may be converted to a value H_{syst} relative to the system base S_{syst} with the formula

$$H_{syst} = H\frac{G}{S_{syst}} \quad . \tag{6.11}$$

A convenient system base value is 100 MVA.

The moment of inertia of a synchronous machine is given by WR^2 lb-ft^2, where W is the weight of the rotating part of the machine in pounds, and R is its radius of gyration in feet. Machinery manufacturers generally supply the value of WR^2 for their machines. To express WR^2 in terms of H, we proceed as follows.

Let the machine rating be G in MVA. The kinetic energy of rotation of the rotor at synchronous speed is given by

$$\text{KE} = \frac{1}{2}J\omega_m^2 = \frac{1}{2}\frac{(WR^2)}{23.7}\left[\frac{2\pi}{60}n_s\right]^2 \tag{6.12}$$

where n is the rotor mechanical speed in revolutions per minute, and 23.7 is the conversion factor which will change WR^2 in lb-ft^2 to kg-m^2. The inertia constant is then

$$H = \frac{KE \times 10^{-6}}{G}\ \frac{\text{MJ}}{\text{MVA}} \tag{6.13}$$

where KE is computed from (6.12).

The range of inertia constants for certain machines are given in Table 6.1

Example 6.1 The inertia constant *H* for a 60-Hz, 100-MVA hydroelectric generator is 4.0 MJ/MVA. How much kinetic energy is stored in the rotor at synchronous speed? If the input to the generator is suddenly increased by 20 MVA, what acceleration is imparted to the rotor?

Solution

The energy stored in the rotor at synchronous speed is given by (6.1) and is

$$GH = 100 \times 4 = 400 \text{ MJ}$$

Table 6.1 Typical inertia constants of synchronous machines

Type of machine	Inertia constant H MJ/MVA
Turbine generator:	
Condensing, 1800 r/min	9-6
3600 r/min	7-4
Noncondensing, 3600 r min	4-3
Waterwheel generator:	
Slow-speed, < 200 r/min	2-3
High-speed, > 200 r/min	2-4
Synchronous Condenser	
Large	1.25
Small	1.0
Synchronous motor with load varies from 1.0 to 5.0 and higher for heavy flywheels	2.0

*Where range is given, the first value applies to machines of smaller MVA rating. [From Westinghouse Electrical Transmission and Distribution Reference Book, 1964, p. 486.]

The rotor acceleration $d^2\delta/dt^2$ is given by (6.8) with $P_a = P_i - P_e = 20$ MVA of accelerating power.

$$\frac{H}{180f} = \frac{4}{180 \times 60} = \frac{1}{2700}$$

and (6.8) becomes

$$\frac{1}{2700}\frac{d^2\delta}{dt^2} = \frac{20}{100}$$

so $d^2\delta/dt^2 = 2700 \times 0.2 = 540$ electrical degrees/s.

Example 6.2 A 500-MVA synchronous machine has $H_1 = 4.6$ MJ/MVA, and a 1500-MVA machine has $H_2 = 3.0$ MJ/MVA. The two machines operate in parallel in a power station. What is the equivalent H constant for the two, relative to a 100-MVA base?

Solution

The total kinetic energy of the two machines is

$$\text{KE} = 4.6 \times 500 + 3 \times 1500 = 6800 \text{ MJ}$$

Thus, the equivalent H relative to a 100-MVA base is

$$H = \frac{6800}{100} = 68 \text{ MJ/MVA}$$

or

$$H_1 = 4.6\left(\frac{500}{100}\right) = 23 \text{ MJ/MVA on a 100 MVA base}$$

$$H_2 = 3\left(\frac{1500}{100}\right) = 45 \text{ MJ/MVA on a 100 MVA base}$$

$$H = H_1 + H_2 = 68 \text{ MJ/MVA}$$

Example 6.3 A 100-MVA, two-pole, 60-Hz generator has a moment of inertia of 50×10^3 kg•m². What is the energy stored in the rotor at the rated speed? What is the corresponding angular momentum? Determine the inertia constant H.

Solution

$$n_s = \frac{120f}{\text{poles}} = \frac{120(60)}{2} = 3600 \text{ rpm}$$

The stored energy is

$$\text{KE (stored)} = \frac{1}{2}J\omega_m^2 = \frac{1}{2}(50 \times 10^3)\left(\frac{2\pi \times 3600}{60}\right)^2 = 3553 \text{ MJ}$$

Then

$$H = \frac{\text{KE (stored)}}{\text{MVA}} = \frac{3553}{100} = 35.53 \text{ MJ/MVA}$$

$$M = J\omega_m = (50 \times 10^3)\left(\frac{2\pi \times 3600}{60}\right) = 18.8 \frac{\text{MJ} \cdot \text{s}}{\text{mech rad}}$$

Example 6.4 The input to the generator of Example 6.3 is suddenly increased by 25 MW. Determine the rotor acceleration.

Solution
From (6.8)

$$\frac{35.53}{180 \times 60}\frac{d^2\delta}{dt^2} = \frac{25}{100}$$

Thus,

$$\frac{d^2\delta}{dt^2} = 76 \text{ electrical degrees/s}^2$$

Example 6.5 Assuming the acceleration calculated in Example 6.4 remains constant for twelve cycles, calculate the change in the rotor speed in rpm that occurs during those twelve cycles.

Solution
Twelve cycles are equivalent to 12/60 = 0.2s. Now

$$\frac{d^2\delta}{dt^2} = 76°/s^2 = \frac{\pi}{180}(76) = 1.326 \text{ rad/s}^2$$

and

$$\frac{d\delta}{dt} = 1.326t + \left.\frac{d\delta}{dt}\right|_{t=0} = 1.326t \text{ rad/s}$$

After twelve cycles

$$\text{rotor speed} = (1.326)(0.2)\left(\frac{60}{2\pi}\right) + 3600 = 3602.5 \text{ rev/min}$$

$$\text{rotor speed change} = 3602.5 - 3600 = 2.5 \text{ rev/min.}$$

6.3 POWER FLOW UNDER STEADY STATE

We briefly considered power flow in a short lossless transmission line in Chapter 4, Section 4.1. One result of this discussion was an expression for the real power flow/phase at the sending and receiving ends of a line which was given as (4.5) and (4.7). These equations will be repeated here for convenience as

$$P_S = P_R = \frac{|V_S|\,|V_R|}{X} \sin\delta \text{ W/phase} \tag{6.14}$$

where $|V_S|$ is the rms sending end voltage, $|V_R|$ is the rms receiving end voltage, X is the line reactance and δ is the power angle.

In a similar manner, we have developed the equation for steady-state power delivered by a lossless synchronous machine in Appendix III. This expression is also repeated here as

$$P_e = P_d = \frac{|E_g|\,|V_t|}{X_d} \sin\delta \text{ W/phase} \tag{6.15}$$

where $|E_g|$ is the rms internal voltage, $|V_t|$ is the rms terminal voltage, X_d is the direct axis reactance (or the synchronous reactance in a round rotor machine) and δ is the electrical power angle.

With this information, we consider the following examples.

Example 6.6 The sending-end and receiving-end voltages of a three-phase transmission line at a 100-MW load are equal at 115kV. The per-phase line impedance is $j7\Omega$. Calculate the maximum steady-state power that can be transmitted over the line.

Solution
Since $|V_R| = |V_S| = 115{,}000/\sqrt{3} = 66{,}400$, we have, from (6.14)

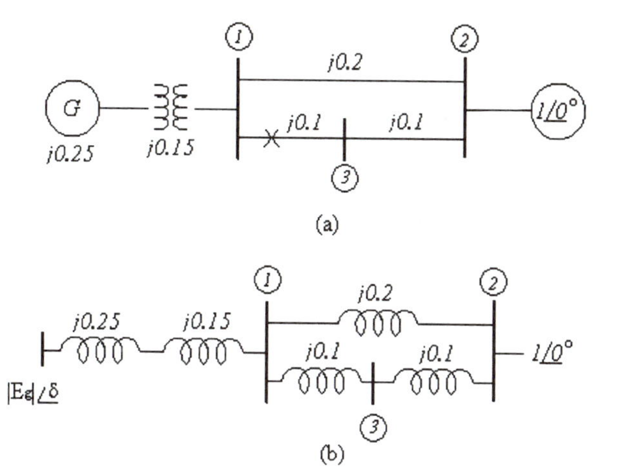

FIGURE 6.1
Example 6.7.

$$P_{R(\max)} = P_{S(\max)} = \frac{(66.4)^2}{7} = 629.9 \text{ MW/phase}$$

$$= 1889.7 \text{ MW total}$$

Example 6.7 A 1-line diagram of a system is shown in Figure 6.1(*a*). The corresponding reactance diagram is given in Figure 6.1(*b*). All values are in per unit on a common base. The power delivered into bus 2 (an infinite bus having a voltage of 1.0 pu) is 1.0 per unit at 0.90 power factor lagging. Obtain the power angle equation, and the swing equation for the system. Neglect all losses.

Solution
The equivalent reactance as seen by the generator is given by

$$X_e = 0.25 + 0.15 + 0.2 \text{ [in parallel with } (0.1 + 0.1)] = 0.5 \text{ per unit}$$

Current into bus 2 is

$$I = \frac{1.0}{1.0 \times 0.9} = 1.11 \angle -\cos^{-1} 0.9 = 1.11 \angle -25.84° \quad \text{per unit}$$

The voltage E_g is then given by

$$|E_g| \angle \delta = |V_2| \angle 0° + jX_e |I| \angle -25.84°$$

$$= 1.0 \angle 0° + (0.5 \angle 90°)(1.11 \angle -25.84)$$

$$= 1.339 \angle 21.93° \qquad \text{per unit}$$

Thus, the power-angle equation, from (6.15), becomes

$$P_e = \frac{1.339}{0.5} \sin\delta = 2.678 \sin\delta$$

Finally, from (6.8), the swing equation may be written as

$$\frac{H}{180f} \frac{d^2\delta}{dt^2} = 1.0 - 2.678 \sin\delta \qquad \text{per unit}$$

where $P_i = 1.0$, is the per-unit mechanical power input to the generator.

As a verification of these results, under steady-state, we must have $P_i = P_e = 1.0$; that is $2.678 \sin \delta = 2.678 \sin 21.93° = 1.0$.

6.4 EQUAL-AREA CRITERION

In the preceding discussions we have indicated that a solution to the swing equation, for the variation of the power angle with time $\delta(t)$, leads to the determination of the stability of a single machine operating as part of a large power system. However, it is not always necessary to solve the swing equation to investigate the system stability. Rather, in certain cases, a direct approach may be taken. Such an approach is based on the equal-area criterion.

Consider δ in the swing equation (6.8), which describes the motion, or *swing*, of the rotor. As is shown in Figure 6.2, in an unstable system, δ increases indefinitely with time and the machine loses synchronism. In a stable system, δ undergoes oscillations, which eventually die out due to damping. From the figure it is clear that, for a system to be stable, it must be that $d\delta/dt = 0$ at some instant. This criterion (that $d\delta/dt$ be zero) can be obtained simply from (6.8). Furthermore, if we assume that M is constant and that damping is negligible and we ignore the control system, then we have, from (6.8),

$$2 \frac{d\delta}{dt} \frac{d^2\delta}{dt^2} = \frac{2P_a}{M} \frac{d\delta}{dt}$$

which, upon integration with respect to time, gives

$$\left(\frac{d\delta}{dt}\right)^2 = \frac{2}{M} \int_{\delta_0}^{\delta} P_a \, d\delta \tag{6.16}$$

where $P_a = P_i - P_e$ = accelerating power. From (6.16), it follows that

$$\frac{d\delta}{dt} = \sqrt{\frac{2}{M} \int_{\delta_0}^{\delta} P_a \, d\delta} \tag{6.17}$$

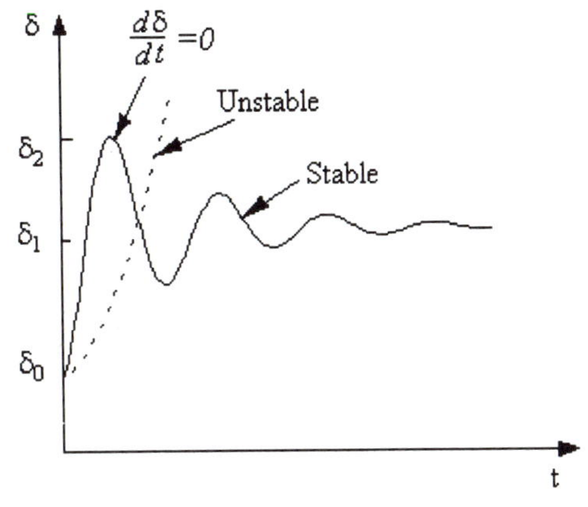

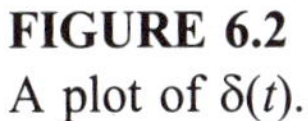
FIGURE 6.2
A plot of $\delta(t)$.

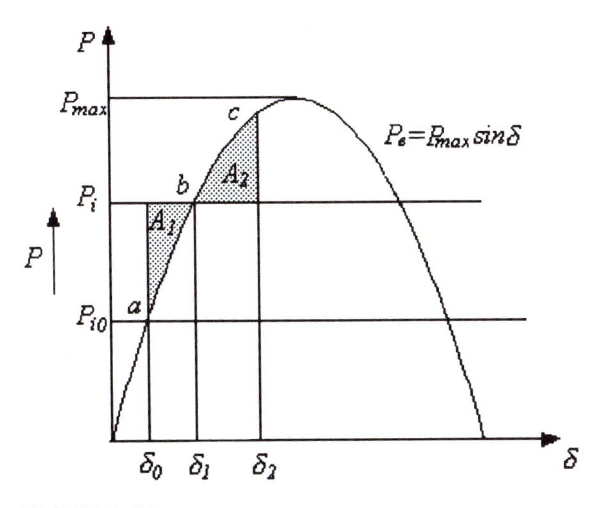

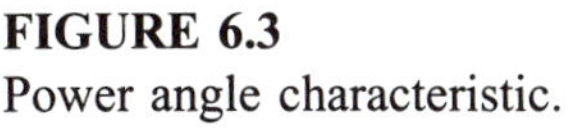
FIGURE 6.3
Power angle characteristic.

where δ_0 is the initial power angle before the rotor begins to swing because of a disturbance. The stability criterion $d\delta/dt = 0$ (at some moment) implies that

$$\int_{\delta_0}^{\delta} P_a \, d\delta = 0 \tag{6.18}$$

This condition requires that, for stability, the area under the graph of accelerating power P_a versus δ must be zero for some value of δ; that is, the positive (or accelerating) area under the graph must be equal to the negative (or decelerating) area. This criterion is therefore known as the *equal-area criterion* for stability. Referring to Figure 6.3, point *a* corresponding to the δ_0 is the initial steady-state operating point. At this point, the input power to the machine, P_{io}, equals the developed power P_{eo}. When a sudden increase in shaft input power occurs to P_i, the accelerating power, P_a, becomes positive and the rotor moves towards point *b*. We assume that the machine is connected to a large power system so that $|V_t|$ does no change. We also assume that X_d does not change and that a constant field current maintains $|E_g|$ constant. Consequently, the rotor accelerates and the power angle begins to increase. At the point *b*, $P_i = P_e$ and $\delta = \delta_1$. But $\dot{\delta}$ is still positive and δ overshoots *b*, the final steady-state operating point. Now P_a is negative and δ ultimately reaches a maximum value δ_2, or point *c* and then swings back toward *b*. Even though neglected in our formulation of (6.18), there is some inherent damping in the system. Therefore, the rotor settles to the point *b*, which is the ultimate steady-state stable operating point, as shown in Figure 6.2. In accordance with (6.18) the equal-area criterion requires that, for stability,

$$\text{Area } A_1 = \text{area } A_2$$

or

$$\int_{\delta_0}^{\delta_1} (P_i - P_{max} \sin\delta)\, d\delta = \int_{\delta_1}^{\delta_2} (P_{max} \sin\delta - P_i)\, d\delta \tag{6.19}$$

or, after the integrations are performed,

$$P_i(\delta_1 - \delta_0) + P_{max}(\cos\delta_1 - \cos\delta_0) = P_i(\delta_1 - \delta_2) + P_{max}(\cos\delta_1 - \cos\delta_2) \tag{6.20}$$

But,

$$P_i = P_{max} \sin\delta_1$$

which, when substituted in (6.20), yields

$$\begin{aligned} &P_{max} \sin\delta_1(\delta_1 - \delta_0) + P_{max}(\cos\delta_1 - \cos\delta_0) \\ &= P_{max} \sin \delta_1(\delta_1 - \delta_2) + P_{max}(\cos\delta_1 - \cos \delta_2) \end{aligned} \tag{6.21}$$

Upon simplification, (6.21) becomes

$$(\delta_2 - \delta_0) \sin\delta_1 + \cos\delta_2 - \cos\delta_0 = 0 \tag{6.22}$$

We now apply the equal-area criterion to the following examples.

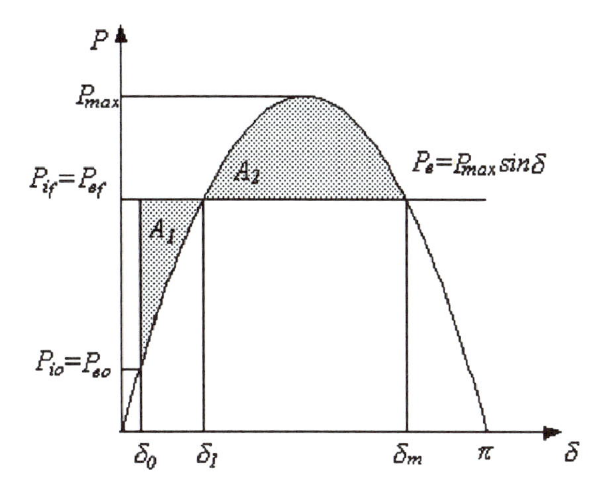

FIGURE 6.4
Example 6.8.

Example 6.8 A synchronous generator, capable of developing 1500 MW of power, operates at a power angle of 8°. By how much can the input shaft power be increased suddenly without loss of stability? Assume that the generator field current and terminal voltages do not change such that P_{max} will remain constant.

Solution
Initially, at $\delta_0 = 8°$, the electromagnetic power being developed is

$$P_{e0} = P_{max} \sin\delta_0 = 500 \sin 8° = 69.6 \text{ MW per phase}$$

Let δ_m (Figure 6.4) be the power angle to which the rotor can swing before losing synchronism. Note that if this angle is exceeded, P_i will again become greater than P_e and the rotor will once again be accelerated and synchronism will be lost as shown in Figure 6.2. Then the equal-area criterion requires that (6.20) be satisfied (with δ_m replacing δ_2). From Figure 6.4, $\delta_m = \pi - \delta_1$, so (6.22) yields

$$(\pi - \delta_1 - \delta_0)\sin\delta_1 + \cos(\pi - \delta_1) - \cos\delta_0 = 0$$

or

$$(\pi - \delta_1 - \delta_0)\sin\delta_1 - \cos\delta_1 - \cos\delta_0 = 0 \tag{6.23}$$

Substituting $\delta_0 = 8° = 0.139$ rad in (6.23) gives

$$(3 - \delta_1)\sin\delta_1 - \cos\delta_1 - 0.99 = 0$$

This yields $\delta_1 = 50°$, for which the corresponding electromagnetic power is

$$P_{ef} = P_{\max} \sin \delta_1 = 500 \sin 50° = 383.02 \text{ MW/phase}$$

The initial power developed by the machine was 69.6 MW. Hence, without loss of stability, the system can accommodate a sudden increase of

$$P_{ef} - P_{e0} = 383.02 - 69.6 = 313.42 \text{ MW/phase}$$

or

$$313.42 \times 3 = 940.3 \text{ MW of input shaft power}$$

6.5 MACHINE FAULTS AND THE CRITICAL CLEARING ANGLE

If a disturbance (or fault) occurs in a system, δ begins to increase under the influence of positive accelerating power, and the system will become unstable if δ becomes very large. There is a critical angle within which the fault must be cleared if the system is to remain stable and the equal-area criterion is to be satisfied. This angle is known as the *critical clearing angle* δ_{cr}. As an example, consider a system that normally operates along curve A in Figure 6.5. If a three-phase short circuit occurs across the line, the terminal voltage goes to zero and the curve of power versus power angle will correspond to the horizontal axis. For stability, the clearing angle, δ_c, must be such that area A_1 = area A_2.

Expressing $A_1 = A_2$ mathematically, we have

$$P_i(\delta_c - \delta_0) = \int_{\delta_c}^{\delta_1} P_{\max} \sin \delta \, d\delta - P_i(\delta_1 - \delta_c) \tag{6.24}$$

Furthermore,

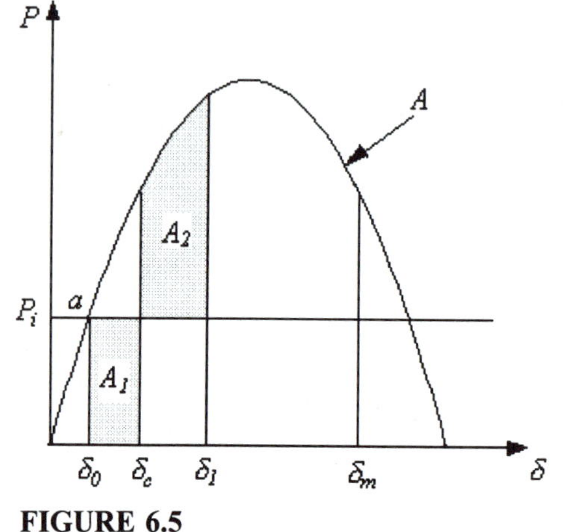

FIGURE 6.5
Clearing angle δ_c.

$$P_i = P_{max} \sin\delta_0 \tag{6.25}$$

Expanding (6.24), combining with (6.25), and upon some simplification we obtain

$$\cos\delta_c = (\delta_1 - \delta_0) \sin\delta_0 + \cos\delta_1 \tag{6.26}$$

To reiterate, with reference to Figure 6.5, the various angles in (6.26) are: δ_c = clearing angle; δ_0 = initial power angle; and δ_1 = power angle to which the rotor advances (or overshoots) beyond δ_c.

In order to determine the clearing time, we re-write (6.8), with $P_e = 0$ since we have a three-phase short-circuit,

$$\frac{d^2\delta}{dt^2} = \frac{180f}{H} P_i$$

Integrating twice and utilizing the fact that $\dot{\delta} = 0$ when $t = 0$ yields

$$\delta = \frac{180fP_i}{2H} t^2 + \delta_0 \tag{6.27}$$

If δ_c is the clearing angle corresponding to a clearing time t_c, we obtain from (6.27)

$$t_c = \sqrt{\frac{2H(\delta_c - \delta_0)}{180fP_i}} \tag{6.28}$$

where δ_c is obtained from (6.26).

If we now wish to determine the critical clearing angle δ_{cr}, we note from Figure 6.5 that the maximum allowable value of δ_1, the overshoot angle, is δ_m. Should δ reach δ_m, the accelerating power will again become positive and synchronism will be lost. Since $\delta_m = \delta_{1(max)} = \pi - \delta_o$, we have upon substitution into (6.26) that

$$\cos\delta_{cr} = (\pi - 2\delta_o) \sin\delta_o + \cos(\pi - \delta_o)$$

or

$$\delta_{cr} = \cos^{-1} \left[(\pi - 2\delta_o)\sin\delta_o - \cos\delta_o\right]. \tag{6.29}$$

The critical clearing time, t_{cr} may then be obtained by substitution of δ_{cr} into (6.28).

Let us now consider the system shown in Figure 6.1(a), which essentially represents a double-circuit line. A three-phase short-circuit fault occurs on one of the lines, as shown. Because of the rotor inertia, the power angle cannot change instantly. Also, some power could still be transmitted during the fault because the terminal voltage of the generator will not be zero.

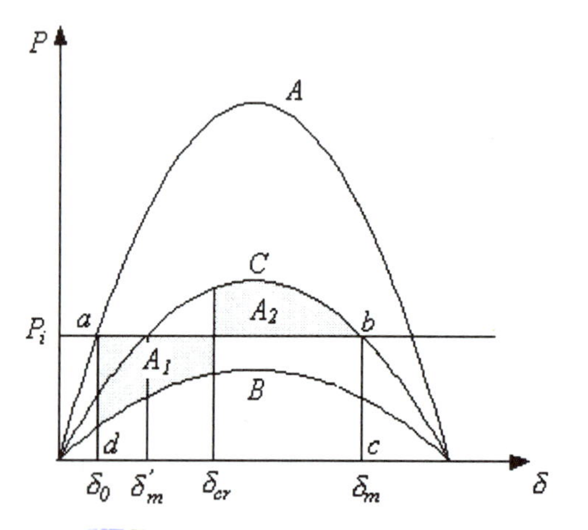

FIGURE 6.6
Determination of δ_c during fault conditions.

In Figure 6.6 we show a power-angle curve A before a fault, B during the fault, and C after the fault such that $A = P_{max}\sin\delta$, $B = k_1 A$, and $C = k_2 A$, with $k_1 < k_2$. For stability, we must have area A_1 = area A_2. Based on Figure 6.6, this condition yields

$$(\delta_m - \delta_0) P_i = \int_{\delta_0}^{\delta_{cr}} B\, d\delta + \int_{\delta_{cr}}^{\delta_m} C\, d\delta \tag{6.30}$$

Substituting for B and C in (6.30), with $P_i = P_{max}\sin\delta_0$, eventually yields

$$\cos\delta_{cr} = \frac{1}{k_2 - k_1}[(\delta_m - \delta_0)\sin\delta_0 - k_1\cos\delta_0 + k_2\cos\delta_m] \tag{6.31}$$

From Figure 6.6, we have

$$P_i = P_m \sin\delta_0 = k_2 P_{max}\sin\delta_m = k_2 P_{max}\sin(\pi - \delta_m) \tag{6.32}$$

Hence, from (6.32),

$$\sin\delta_0 = k_2\sin(\pi - \delta_m) = k_2\sin\delta'_m \tag{6.33}$$

With k_1, k_2, and δ_0 specified, the critical clearing angle may be obtained from (6.31) and (6.33).

We now consider two illustrative examples.

Example 6.9 A 60-Hz synchronous generator capable of supplying 500 MW of power is connected to a large power system and is delivering 100 MW when a three-phase short circuit fault occurs at its terminals. Determine

a) The time in which the fault must be cleared if the maximum power angle is to be 85°. Assume H = 7 MJ/MVA on a 100 MVA base.

b) The critical clearing angle.

Solution

a) From (6.25) and referring to Figure 6.5,

$$\delta_0 = \sin^{-1}\left(\frac{\frac{100}{3}}{\frac{500}{3}}\right) = 11.54° \text{ or } 0.2 \text{ rad}$$

and

$$\delta_1 = 85° \text{ or } 1.48 \text{ rad.}$$

Then from (6.26)

$$\delta_c = \cos^{-1}[(1.48 - 0.2)\sin(0.2) + \cos(1.48)]$$

$$= 69.7° \text{ or } 1.22 \text{ rad}$$

and from (6.28) where P_i = 100 MW or 1.0 per unit

$$t_c = \sqrt{\frac{2(7)(69.7 - 11.54)}{180(60)(1.0)}} = 274 \text{ ms}$$

b) From (6.29)

$$\delta_{cr} = \cos^{-1}[(\pi - 2(.2))\sin(11.54°) - \cos(11.54°)]$$

$$= 115.45°$$

Example 6.10 A synchronous generator is connected to a large power system and is supplying 0.45 pu of its maximum power capacity. A three-phase fault occurs, and the effective terminal voltage of the generator becomes 25% of its value before the fault. The maximum power that can be delivered after the fault is cleared is 70 percent of the original maximum value. Determine the critical clearing angle.

Solution

Let

$$k_1 = \frac{P_{max} \text{ during the fault}}{P_{max} \text{ before the fault}}$$

$$k_2 = \frac{P_{max} \text{ after the fault}}{P_{max} \text{ before the fault}}$$

δ_0 = power angle at the time of the fault
δ_{cr} = critical power angle when fault is cleared
δ_m = maximum angle of swing

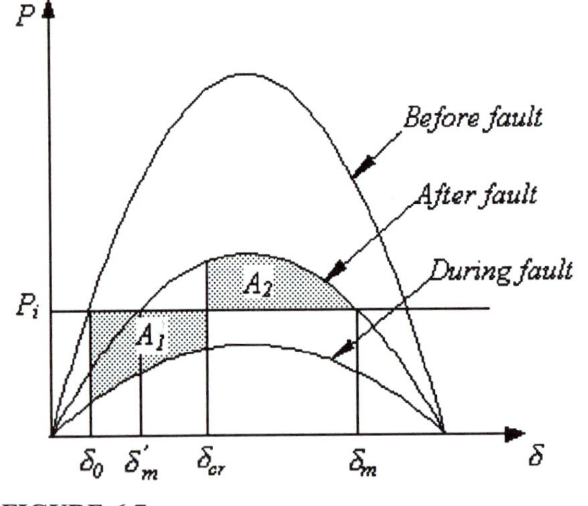

FIGURE 6.7
Example 6.10.

Then the equal-area criterion, $A_1 = A_2$ in Figure 6.7, gives us from (6.31)

$$\cos\delta_{cr} = \frac{1}{k_2 - k_1}\left[\frac{P_i}{P_{max}}(\delta_m - \delta_0) + k_2\cos\delta_m - k_1\cos\delta_0\right]$$

Initially, the generator is supplying 0.45 pu of P_{max}. Thus,

$$P_i = 0.45P_{max} = P_{max}\sin\delta_0$$

from which $\delta_0 = \sin^{-1}0.45 = 26.74°$. Now $P_{max} = |E_g|\ |V_t|/X_d$. When the fault occurs, $|V_t|$ becomes 0.25 $|V_t|$, so that

$$k_1 P_{max}\sin\delta = 0.25\ P_{max}\ \sin\delta$$

and $k_1 = 0.25$.

After the fault, with $k_2 = 0.70$, we have

$$P_i = k_2 P_{max}\sin\delta_m'$$

from which

$$\delta_m' = \sin^{-1}\frac{P_i}{k_2 P_{max}} = \sin^{-1}\frac{0.45P_{max}}{0.70P_{max}} = 40°$$

Then $\delta_m = 180° - \delta_m' = 140°$

$$\delta_m - \delta_0 = 140° - 26.74° = 113.26° \quad \text{or} \quad 1.98 \text{ rad}$$

Hence,

$$\cos\delta_{cr} = \frac{1}{0.70 - 0.25}[0.45(1.98) + 0.70\cos 140° - 0.25\cos 26.74°] = 0.29$$

so that $\delta_{cr} = \cos^{-1}0.29 = 73.2°$.

The preceding example is significant in that in determining the stability, or the critical clearing angle, power flow during fault conditions is taken into account.

6.6 STEP-BY-STEP SOLUTION

The swing equation may be solved iteratively with the step-by-step procedure shown in Figure 6.8. In the solution, it is assumed that the accelerating power P_a and the relative rotor angular velocity ω_r, are constant within each of a succession of intervals (top and middle, Figure 6.8); their values are then used to find the change in δ during each interval.

To begin the iteration, we need $P_a(0+)$, which we evaluate as

$$P_a(0+) = P_i - P_e(0+)$$

Then the swing equation (6.8) may be written as

$$\frac{d^2\delta}{dt^2} = \alpha(0+) = \frac{P_a(0+)(180f)}{H} \tag{6.34}$$

and the change in ω_r, is given (Figure 6.8) by

$$\Delta\omega_r = \alpha(0+)\Delta t \tag{6.35}$$

Then

$$\omega_r = \omega_0 + \Delta\omega_r = \omega_0 + \alpha(0+)\Delta t \tag{6.36}$$

The average value of ω_r during the first interval is then

$$\omega_{r(avg)} = \omega_0 + \frac{\Delta\omega_r}{2} \tag{6.37}$$

Similarly, the change in the power angle for the first interval is

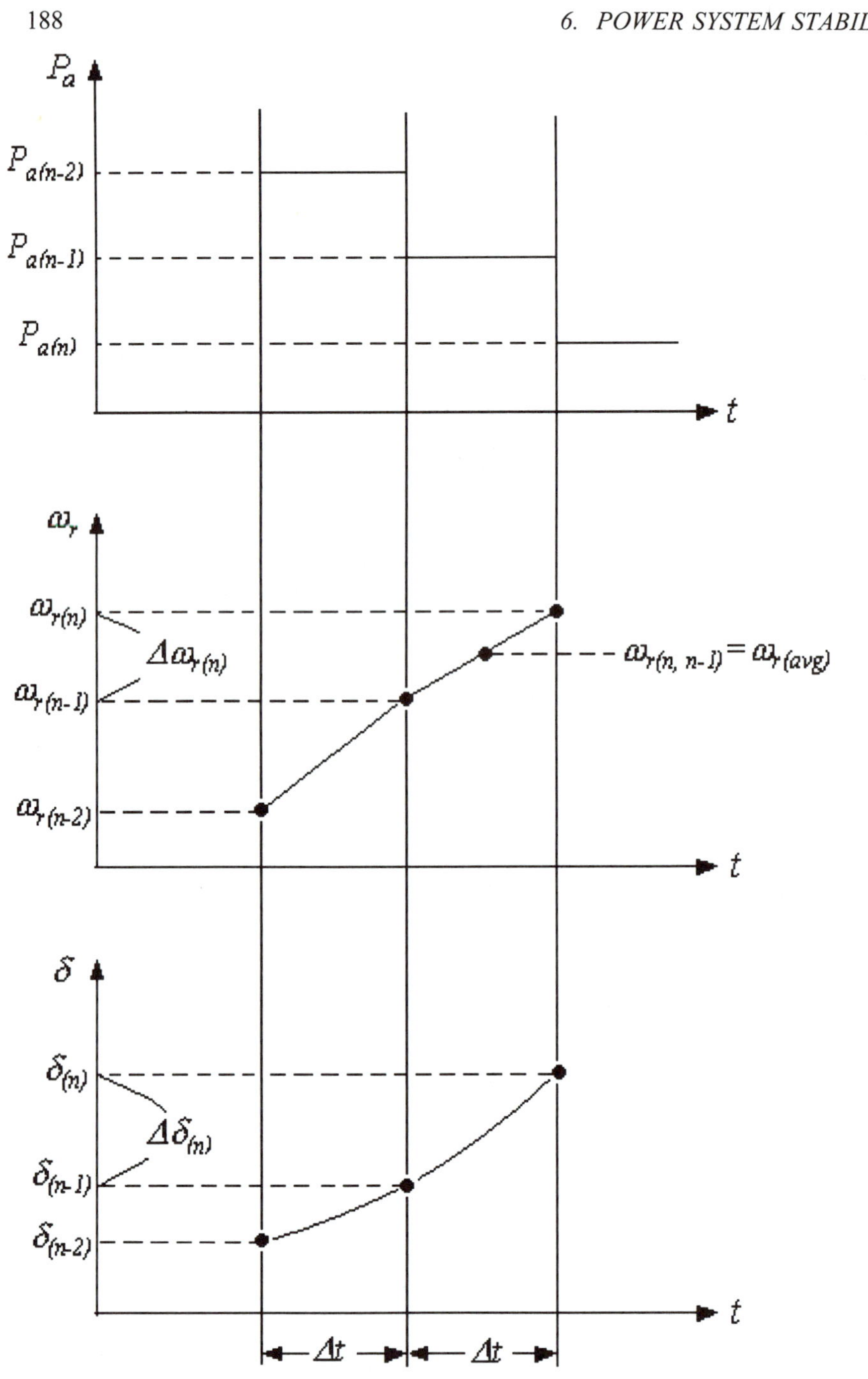

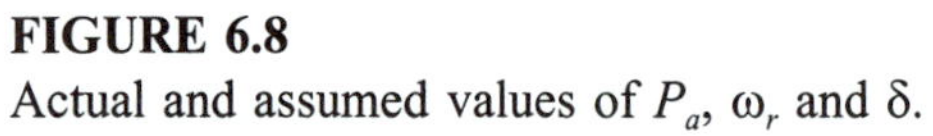

FIGURE 6.8
Actual and assumed values of P_a, ω_r and δ.

$$\Delta\delta_1 = \omega_{r(avg)}\Delta t \tag{6.38}$$

and so

$$\delta_1 = \delta_0 + \Delta\delta_1 \tag{6.39}$$

Evaluation of P_a and $\omega_{r(avg)}$

When using the step-by-step method, the accelerating power P_a is assumed to be constant over the step interval and equal to its value at the beginning of the interval. Thus

$$P_a = P_{a(n-1)+} \tag{6.40}$$

as shown in Figure 6.8. Should a discontinuity occur during a step interval (such as might be caused by the clearing of a fault), the recommended approach is to simply redefine the intervals at that point so that the discontinuity occurs at the end (beginning) of a step interval. Then (6.40) may be used as previously indicated.

When determining the average speed over an interval, the values at the beginning and end of the interval are utilized such that

$$\omega_{r(n,n-1)} = \omega_{r(avg)} = \frac{\omega_{r(n)} + \omega_{r(n-1)}}{2} \tag{6.41}$$

as shown in Figure 6.8. The similarities between (6.41) and (6.37) should be evident.

Algorithm for the Iterations

Returning now to (6.38), we see that δ_1 gives us one point on the swing curve. The algorithm for the iterative process is as follows

$$P_{a(n-1)} = P_i - P_{e(n-1)} \tag{6.42}$$

$$P_{e(n-1)} = \frac{|E|\,|V|}{X}\sin\delta_{(n-1)} \tag{6.43}$$

$$\alpha_{(n-1)} = \frac{P_{a(n-1)}(180f)}{H} \tag{6.44}$$

$$\Delta\omega_{r(n)} = \alpha_{(n-1)}\Delta t \tag{6.45}$$

$$\omega_{r(n,n-1)} = \omega_{r(avg)} = \omega_{r(n-1)} + \frac{\Delta\omega_{r(n)}}{2} \tag{6.46}$$

$$\Delta\delta_{(n)} = \omega_{r(n,n-1)} (\Delta t) \tag{6.47}$$

$$\delta_{(n)} = \delta_{(n-1)} + \Delta\delta_{(n)} \tag{6.48}$$

The use of this algorithm in conjunction with the equal-area criterion provides the critical clearing angle and the corresponding critical clearing time.

Example 6.11 The kinetic energy stored in the rotor of a 60-Hz 50-MVA synchronous machine is 200 MJ. The generator has an internal voltage of 1.2 pu and is connected to an infinite bus operating at a voltage of 1.0 pu through a 0.3 pu reactance. The generator is supplying rated power when a three-phase short circuit occurs on the line. Subsequently, circuit breakers operate and the reactance between the generator and the bus becomes 0.4 pu. Using the step-by-step algorithm, plot the swing curve for the machine for the time before the fault is cleared.

Solution

The per-unit value of the angular momentum, based on the machine rating, is

$$\frac{H}{180f} = \frac{200/50}{180 \times 60} = 3.7 \times 10^{-4}\ \mathrm{s}^2/°$$

From (6.42), we have

$$P_a(0+) = 1.0 - 0.0 = 1.0\ \mathrm{pu}$$

From (6.44),

$$\alpha(0+) = \frac{1.0}{3.7 \times 10^{-4}} = 2702.7°/\mathrm{s}^2$$

From (6.45) with $\Delta t = 0.05$s,

$$\Delta\omega_{r(1)} = 2702.7 \times 0.05 = 131.5°/\mathrm{s}$$

From (6.46),

$$\omega_{r(1,0)} = 0 + \frac{131.5}{2} = 67.55°/\mathrm{s}$$

From (6.47),

$$\Delta\delta_{(1)} = 67.55 \times 0.05 = 3.3775°$$

To complete the first iteration, we determine the initial power angle, δ_0, as follows. Before the fault

$$P_{\max} = \frac{1.2 \times 1.0}{0.3} = 4.0\ \mathrm{pu}$$

Then

$$4\sin\delta_0 = 1.0$$

or,

$$\delta_0 = 14.4775°$$

With this value of the initial power angle, from (6.48) we have

$$\delta_{(1)} = 14.4775 + 3.3775 = 17.855°$$

For the second interval, (6.42) and (6.44) to (6.48) give us

$$P_{a(1)} = 1.0 - 0.0 = 1.0$$

$$\alpha_{(1)} = \frac{1.0}{3.7 \times 10^{-4}} = 2702°/\text{s}$$

$$\Delta\omega_{r(2)} = 2702 \times 0.05 = 135.1°$$

$$\omega_{r(2,1)} = \omega_{r(1)} + \frac{\Delta\omega_{r(2)}}{2} = \omega_{r(0)} + \Delta\omega_{r(1)} + \frac{\Delta\omega_{r(2)}}{2} = 202.65°/\text{s}$$

$$\Delta\delta_{(2)} = \omega_{r(2,1)}\Delta t = 202.65 \times 0.05 = 10.1325°$$

$$\delta_{(2)} = \delta_{(1)} + \Delta\delta_{(2)} = 17.855 + 10.1325 = 27.9875°$$

Since α and $\Delta\omega_r$ do not change during succeeding intervals, we have

$$\omega_{r(3,2)} = \omega_{r(1)} + \Delta\omega_r + \frac{\Delta\omega_r}{2} = 337.75°/\text{s}$$

$$\Delta\delta_{(3)} = \omega_{r(3,2)}\Delta t = 337.75 \times 0.05 = 16.8875°$$

$$\delta_{(3)} = \delta_{(2)} + \Delta\delta_{(3)} = 44.875°$$

and so on. In this way we obtain the following table of values, from which Figure 6.9 is plotted:

t,s	δ, degrees
0.0	14.48
0.05	17.85
0.10	27.99
0.15	44.88
0.20	68.52
0.25	98.92

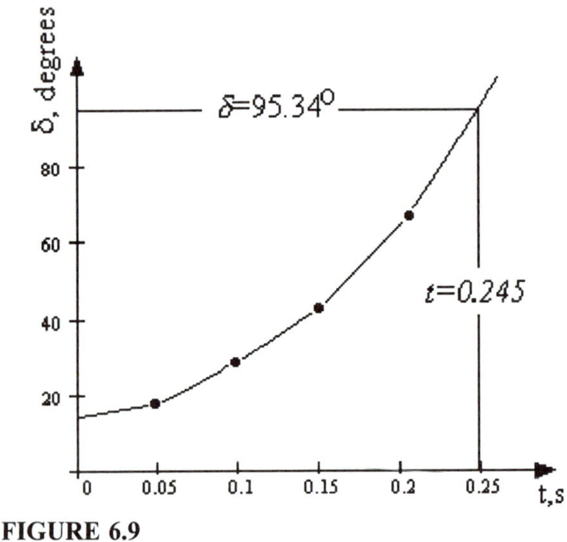

FIGURE 6.9
Example 6.11.

Example 6.12 For the system of Example 6.11, find the critical clearing time in cycles for an appropriately set circuit breaker.

Solution
In order to find the critical clearing time, we must determine the critical clearing angle. In Example 6.11, we have obtained the following results:

Before the fault, P_{max} = 4.0 pu
Initial power angle, δ_0 = 14.4775° = 0.2527 rad.

Now, during the fault, P_{max} = 0 and k_1 = 0 for use in (6.31). After the fault is cleared,

$$P_{max} = \frac{1.2 \times 1.0}{0.4} = 3.0 \text{ pu}$$

and k_2 = 3.0/4.0 = 0.75 for use in (6.31).

The initial power angle δ_0 is given by 4 sin δ_0 − 1.0, from which δ_0 = 0.2527 rad. Define $\delta_m' = \pi - \delta_m$ (see Figure 6.6). The angle δ_m in (6.31) is obtained from

$$\sin\delta_m' = \frac{1}{3.0} \quad \text{and} \quad \delta_m = \pi - \delta_m'$$

from which δ_m = 2.8 rad. Substituting k_1, k_2, δ_0 and δ_m in (6.31) yields

$$\cos\delta_{cr} = \frac{1}{0.75}[(2.8 - 0.2527)\,0.25 - 0 + 0.75\cos 2.8] = -0.093$$

from which δ_{cr} = 95.34°. For this critical clearing angle, Figure 6.9 gives t_{cr} = 0.245s. Hence, the fault must be cleared within 60 × 0.245 = 14.7 cycles.

From the preceding discussions, we conclude that the transient stability of a generator, during and subsequent to fault conditions, depends upon the rotor swing and the critical clearing time. These, in turn, are governed by the machine H constant and the direct-axis transient reactance. The overall system stability may be improved by appropriate control schemes such as turbine valve control, fast fault clearing time, appropriate excitation systems and static VAR compensation.

6.7 MULTI-MACHINE SYSTEMS

Up to this point, we have implied the operation of one machine only. In a system, the rotor of each machine will respond in accordance with (6.8) so that we have

$$\begin{aligned}
\frac{H_1}{180f}\frac{d^2\delta_1}{dt^2} &= P_{i1} - P_{e1} \\
\frac{H_2}{180f}\frac{d^2\delta_2}{dt^2} &= P_{i2} - P_{e2} \\
\vdots \quad \vdots \quad &\vdots \quad \vdots \\
\frac{H_n}{180f}\frac{d^2\delta_n}{dt^2} &= P_{in} - P_{en}
\end{aligned} \tag{6.49}$$

where n = the total number of machines.

From (6.49), we see that there is a separate swing equation for each machine. As a result, it is the relative displacement between the power angles of the machines that is essential in determining the system stability. In multi-machine systems, a single machine is usually chosen as a reference and the rotor swings (changes in power angle) of the remaining machines are determined relative to the reference machine. Stability is maintained if the machine rotors return to a stable operating state relative to each other. These points are illustrated in the example that is presented in the following section.

6.8 COMPUTER ANALYSIS

Step-by-step solution procedures for the analysis of transient stability in a large power system may be conducted using a digital computer. In this way, the transient response to projected disturbances may be evaluated and any potential problems may be corrected.

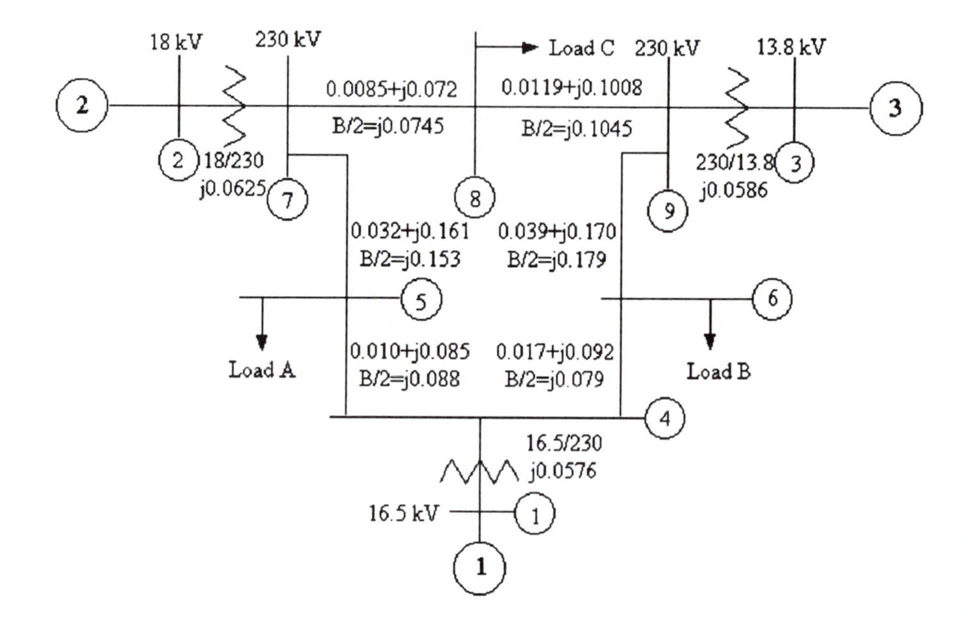

FIGURE 6.10
Power system for Example 6.13.

Example 6.13 An IEEE nine bus sample system taken from Anderson and Fouad[1] is shown in Figure 6.10. All data in this diagram are on a 100 MVA system base and impedances are in per unit. The generator at bus 1 is the swing generator and it is assumed to be operating at a voltage of 1.04 pu. Reactive power from the generators at buses 2 and 3 is used to maintain these bus voltage magnitudes at 1.025 pu. Generator ratings and data on the system base are as given in Table 6.2. Note that transient reactances are generally used to represent machines during the time period of a transient stability analysis.

For this study, *I**SIM software available from SKM Systems Analysis, Inc. was used to simulate the relative generator swings when a three-phase fault occurs at bus 7 and this fault is cleared in 5 cycles (0.0833*s*). Results are shown in Figure 6.11

---

[1]P.M. Anderson and A. A. Fouad, "Power System Control and Stability," IEEE Press Power Engineering Series, 1993.

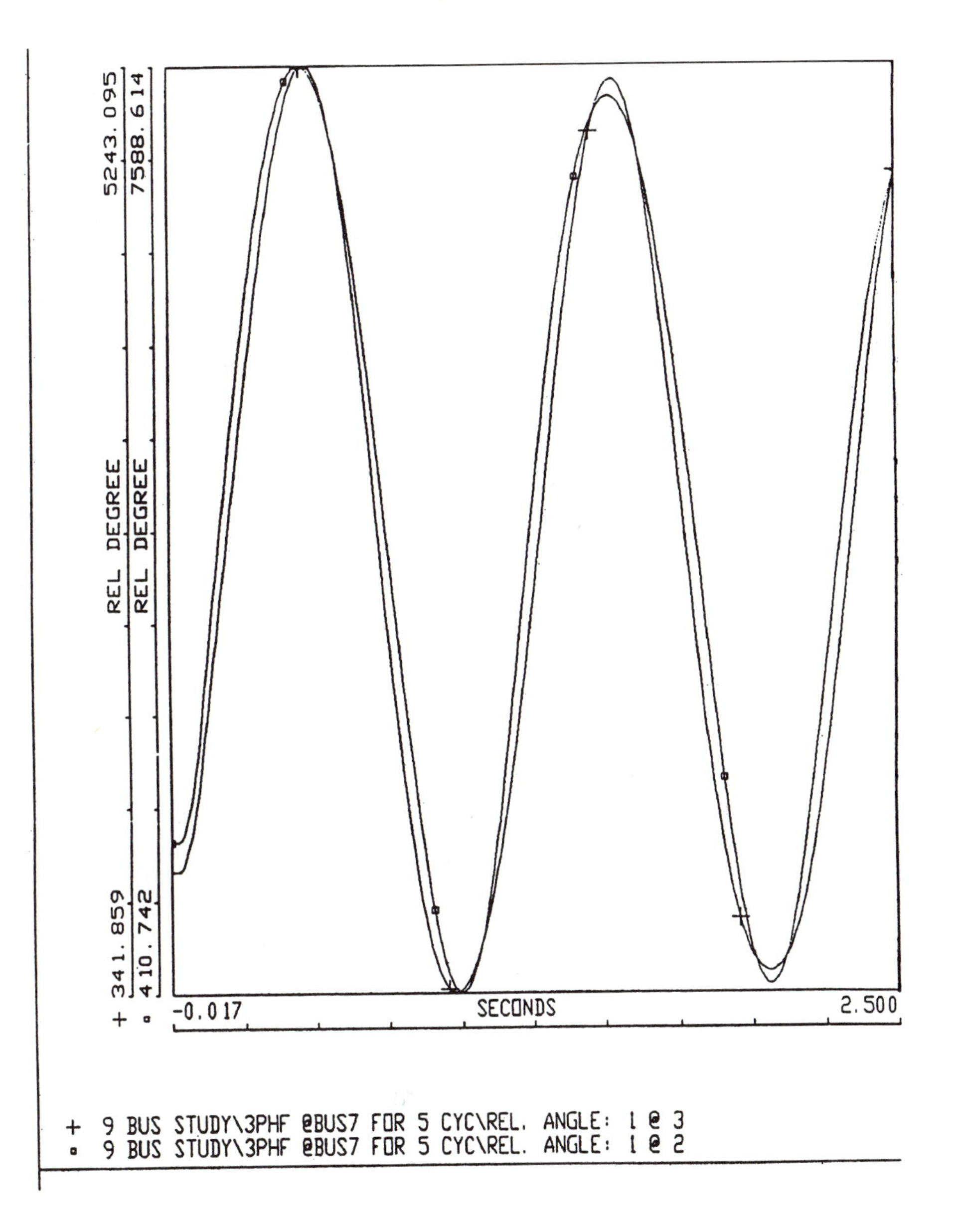

FIGURE 6.11
Generator swings in Example 6.13.

Table 6.2 Machine data for Example 6.13

G_1 (247.5 MVA)	H = 23.64	$X_d' = 0.0608$
G_2 (192.0 MVA)	H = 6.4	$X_d' = 0.1198$
G_3 (128.0 MVA)	H = 3.01	$X_d' = 0.1813$

where the angles of generators 2 and 3 are plotted relative to the angle of the swing generator for a period of 2.5*s*. Note that the system is stable in this instance with maximum relative angles between *G*2 and *G*1 of approximately 84.5° and between *G*3 and *G*1 of approximately 58.2°. Over time, the oscillations shown in Figure 6.11 will decay due to damping and the system will return to a stable steady state.

6.9 SUMMARY

In this chapter, we have discussed the state of equilibrium under which a power system will normally operate as well as the ability of a system to return to a state of equilibrium following a disturbance. Procedures for evaluating the steady-state, dynamic and transient stability of systems have been presented along with examples to illustrate their application using analytical and computer techniques. These procedures allow the practicing engineer to evaluate the response of a power system to various contingencies with the goal of improving system security and achieving a better overall system design.

PROBLEMS

6.1. A 1500-MVA, 1800-rev/min synchronous generator has $WR^2 = 6 \times 10^6$ lb • ft². Find the inertia constant *H* of the machine relative to a 100-MVA base.

6.2. The inertia constant *H* of a 150-MVA, six-pole, 60-Hz synchronous machine is 4.2 MJ/MVA. Determine the value of WR^2 in lb • ft².

6.3. The generator of Problem 6.2 is running at synchronous speed in the steady state. (*a*) What kinetic energy is stored in the rotor? (*b*) If the accelerating power due to a transient change is 28 MW, calculate the rotor acceleration.

6.4. A 300-MVA, 1200-rpm synchronous machine has $WR^2 = 3.6 \times 10^6$ lb • ft². Calculate *H* for the machine (*a*) on its own base and (*b*) on a 100-MVA base.

6.5. A 100-MVA generator has $H = 4.2$ MJ/MVA, and a 250-MVA machine, operating in parallel with the first, has $H = 3.6$ MJ/MVA. Calculate the equivalent inertia constant *H* for the two machines on a 50-MVA base.

6.6. The moment of inertia of a 50-MVA, six-pole, 60-Hz generator is 20×10^3 kg • m². Determine *H* and *M* for the machine.

6.7. A 60-Hz generator, connected directly to an infinite bus operating at a voltage of $1\angle 0°$ pu, has a synchronous reactance of 1.35 pu. The generator no-load voltage is 1.1 pu, and its inertia constant *H* is 4 MJ/MVA. The generator is suddenly loaded to 60 percent of its maximum power limit; determine the required power angle and the initial acceleration of the rotor.

6.8. The kinetic energy stored in the rotor of a 50-MVA, six-pole, 60-Hz synchronous machine is 200 MJ. The input to the machine is increased to 25 MW when the electromagnetic or developed power is 22.5 MW. Calculate the accelerating power and the acceleration.

6.9. If the acceleration of the machine of Problem 6.8 remains constant for ten cycles, what is the power angle at the end of the ten cycles? Assume that the machine is initially operating at synchronous speed and that the initial power angle is 30°.

6.10. A 60 Hz 4-pole lossless synchronous generator having M = 1.2 MJ·s/mech radian is unloaded. The machine is initially operating at synchronous speed with no mechanical power input. The mechanical power supplied to the machine is then increased to 2.4 MW. Assuming a constant acceleration during a period of 10 cycles, calculate the change in power angle during that period in mechanical radians.

6.11. For the machine of Problem 6.10, calculate the rotor speed in rpm at the end of the 10-cycle period.

6.12. A synchronous generator develops 30% of its rated power for a certain load. The mechanical power input to the generator is suddenly increased to 150% of its original value. Neglecting all losses, calculate the maximum power angle on the swing curve.

6.13. A 100 MVA, 60 Hz three-phase synchronous generator is supplying 25% of its maximum power to a large power system when a three-phase short circuit occurs at its terminals. The fault is cleared when the power angle of the machine reaches 150% of its original value. Determine the maximum swing angle of the rotor. Neglect losses.

6.14. For the machine of Problem 6.13, H = 4 MJ/MVA on a 100 MVA base. Determine the critical clearing angle and the critical clearing time if the generator was supplying its rated output power prior to the fault.

6.15. A synchronous generator that can supply a maximum of 100 MW of power is supplying 30 MW when a three-phase short circuit occurs at its terminals. After the fault is cleared, the maximum power that the generator can supply has been reduced to 50 MW. Determine the critical clearing angle.

6.16. Repeat Problem 6.15 if the maximum power that the generator can deliver during the fault is 10 MW.

6.17. The per-unit reactances for a given system are shown in Figure 6.12. A power of 1.0 pu is being delivered to the receiving-end bus of the system at unity power factor and 1.0 pu voltage. A three-phase short circuit occurs at F, the receiving end of one of the lines. Find the critical clearing angle assuming that the receiving end bus voltage remains at 1.0 pu.

6.18. For a given synchronous generator, H/180f is equal to 4.45×10^{-4} s^2/degree. The machine is operating at a steady-state power angle of 24.7° while developing 0.8 pu power. A three-phase short circuit occurs at the terminals of this machine. Use the step-by-step method with Δt = 50 ms to plot the swing curve from t = 0 to t = 100 ms.

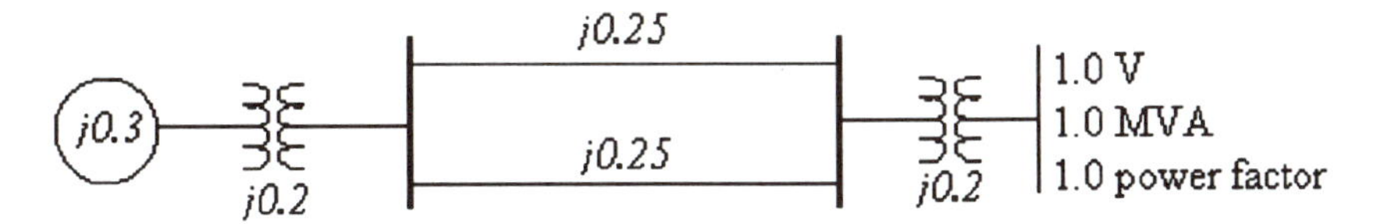

FIGURE 6.12
Problem 6.17.

6.19. The fault on the machine of Problem 6.18 is cleared when t = 100 ms and the generator power angle characteristic returns to its original form. Continue using the step-by-step algorithm to plot the swing curve from 100 ms to 300 ms.

6.20. Using the results of Problems 6.18 and 6.19, plot the speed ω_r vs. time. What can you conclude about the maximum rotor swing and the stability of the system from the angle and speed graphs?

CHAPTER 7

POWER SYSTEM OPERATION AND CONTROL

Numerous aspects of feedback control systems are applied to electric power system operation. For instance, local controls are used at generator units and at controlled buses, whereas central controls are employed at area control centers to monitor area frequency, power flow to interconnected areas, and generating unit outputs. Another very important aspect of a power system operation is economic power dispatch whereby the outputs of controlled units are determined to minimize the total operating cost for a given load on the system.

In this chapter, of the numerous aspects of power system operation and control, we shall consider only the economic operation of power systems and the control of load frequency, generator voltage, and the turbine governor.

7.1 ECONOMIC DISTRIBUTION OF LOAD BETWEEN GENERATORS

Within a power plant, a number of ac generators generally operate in parallel. For the economic operation of the plant, the total load must be appropriately shared by the generating units. Because fuel cost is the major factor in determining economic operation, a curve like that of Figure 7.1 is important to

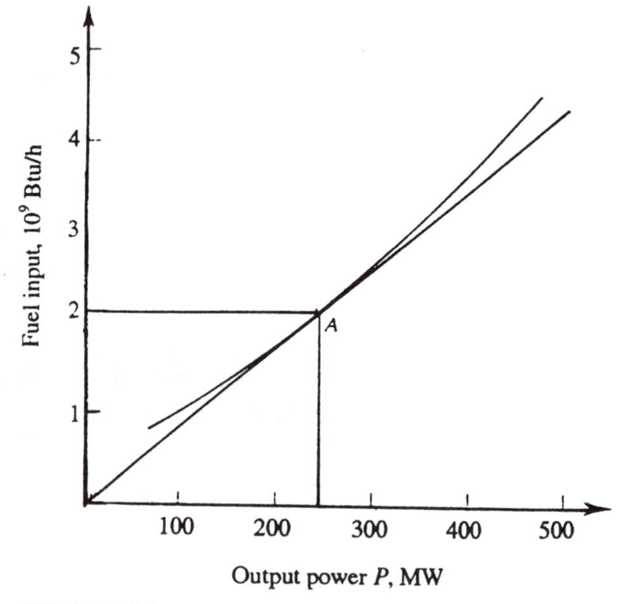

FIGURE 7.1
Fuel input versus output power for a generating unit.

power-plant operation. Note in the figure that the inverse slope of a line drawn from the origin to a point on the curve at any output power is the fuel efficiency (in MWh/Btu) of the generating unit operating at that point. Maximum fuel efficiency occurs at the point at which the line from the origin is tangent to the curve. Point A in Figure 7.1 is such a point for a unit having the input-output characteristic of Figure 7.1; there, an output of 250 MW requires an input of approximately 2.1×10^9 Btu/h. Or, we may say that the *fuel requirement* is 8.4×10^6 Btu/MWh.

To obtain the most economic load distribution between two units, we must determine the incremental cost corresponding to a partial shift of load between the units. We first convert the fuel requirement into a dollar cost per megawatthour. Then the incremental cost is determined from the slopes of the input-output curves (Figure 7.1) for the two units. From Figure 7.1, for each unit,

$$\text{Incremental fuel cost} = \frac{dF}{dP} \quad \text{(in dollars per megawatthour)} \tag{7.1}$$

where F = input cost in dollars per hour, and P = output in megawatts. At a given output, this incremental fuel cost is the additional cost of increasing the output by 1 MW, as illustrated by the following examples.

Example 7.1 Use Figure 7.1 to find the fuel requirements for outputs of (*a*) 100 MW and (*b*) 400 MW. Thus verify that point *A* is probably the maximum fuel-efficiency point.

Solution
(*a*) From Figure 7.1, at 100 MW output, the fuel input is approximately 1×10^9 Btu/h. Hence,

$$\text{Fuel requirement} = \frac{1 \times 10^9}{100} = 10 \times 10^6 \text{ Btu/MWh}$$

(*b*) Similarly, at 400 MW output, the fuel input is approximately 3.6×10^9 Btu/h. Then

$$\text{Fuel requirement} = \frac{3.6 \times 10^9}{400} = 9.0 \times 10^6 \text{ Btu/MWh}$$

Clearly, both values are greater than that for point *A*.

Example 7.2 A certain amount of coal costs \$1.20 and produces 10^6 Btu of energy as fuel for a generating unit. If the input-output characteristic of the unit is that shown in Figure 7.1, determine the incremental fuel cost at point *A*.

Solution

$$\text{Slope at } A = \frac{(2.2 - 2.0)\,10^9}{(260 - 234)} = 7.7 \times 10^6 \text{ Btu/MWh}$$

Thus,

$$\text{Incremental cost} = 7.7 \times 1.20 = \$9.24/\text{MWh}$$

Example 7.3 Convert the curve of Figure 7.1 to a plot of incremental fuel cost versus output power, given a fuel cost of \$1.50 per 10^6 Btu.

Solution
We plot the incremental fuel cost by finding the slope of the input-output curve (Figure 7.1) for several values of output power and plotting slope × cost per Btu against output power. Hence, we obtain the curve shown in Figure 7.2.

Example 7.4 Approximate the curve obtained in Example 7.3 with a straight line, and obtain an equation for the straight line. Use it to determine the incremental cost at 250 MW.

Solution
The approximation is shown in Figure 7.2. The line has an intercept of \$6.25/MWh and a slope of 0.0226. Thus, the required equation is

$$\frac{dF}{dP} = 0.0226P + 6.25 \tag{7.2}$$

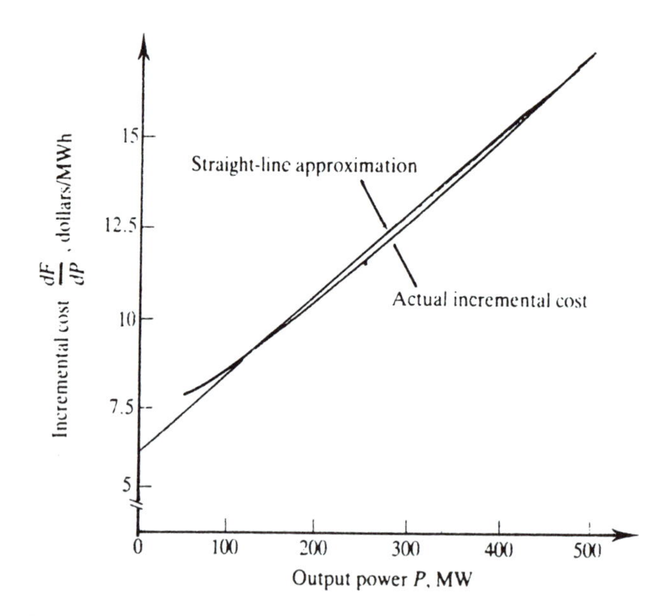

FIGURE 7.2
Example 7.3.

Substituting $P = 250$ in (7.2) yields

$$\text{Incremental cost} = \frac{dF}{dP} = 0.0226 \times 250 + 6.25$$

$$= \$11.9/\text{MWh}$$

In a plant having two operating units, generally the incremental fuel cost of one unit will be higher than that of the other. For the most economic operation, load should be transferred from the unit with the higher incremental cost to the unit with the lower incremental cost, until the incremental costs of the two become equal. In a plant with several units, the criterion for load division is that all units must operate at the same incremental fuel cost. This conclusion may be justified as follows:

If there are n units in the plant, the total input fuel cost F_{total}, in dollars per hour, is given by

$$F_{total} = \sum_{k=1}^{n} F_k \tag{7.3}$$

The total output power P_{total}, in megawatts, may be written as

$$P_{\text{total}} = \sum_{k=1}^{n} P_k \tag{7.4}$$

For a given P_{total}, F_{total} is a minimum when $dF_{\text{total}} = 0$, that is, when

$$dF_{\text{total}} = \sum_{k=1}^{n} \frac{\partial F_{\text{total}}}{\partial P_k} dP_k = 0 \tag{7.5}$$

Since P_{total} is constant, $dP_{\text{total}} = 0$. Then (7.4) yields

$$dP_{\text{total}} = \sum_{k=1}^{n} dP_k = 0 \tag{7.6}$$

Multiplying (7.6) by λ and subtracting the result from (7.5) yields

$$\sum_{k=1}^{n} \left[\left(\frac{\partial F_{\text{total}}}{\partial P_k} - \lambda \right) dP_k \right] = 0 \tag{7.7}$$

The sum in (7.7) will be zero if each term in parentheses is zero. Moreover, for each unit, $\partial F_{\text{total}}/\partial P_k = dF_k/dP_k$, because a change in a unit's power output affects only that unit's fuel cost. Hence

$$\frac{\partial F_{\text{total}}}{\partial P_k} = \frac{dF_k}{dP_k} = \lambda \tag{7.8}$$

and the required condition is

$$\frac{dF_1}{dP_1} = \frac{dF_2}{dP_2} = \ldots = \frac{dF_n}{dP_n} = \lambda \tag{7.9}$$

If a plot of dF_k/dP_k versus P_T for each unit is linear, then λ may be plotted versus P_T to determine the optimum value of λ, where F_k is the input to unit k in dollars per hour, and $\lambda = dF_k/dP_k$ is the incremental fuel cost for unit k in dollars per megawatthour. (λ is also known as the *Lagrange multiplier*).

We now illustrate the preceding discussions by the next example.

Example 7.5 Graphs of the incremental fuel costs (in dollars per megawatthour) for two generating units in a power plant are shown in Figure 7.3. These graphs are linear. The plant output ranges from 240 MW to 1000 MW over a 24-h period. During this period the load on each unit can vary from 120 MW to 600 MW. Determine the incremental fuel cost λ versus plant output for minimum-fuel-cost operation.

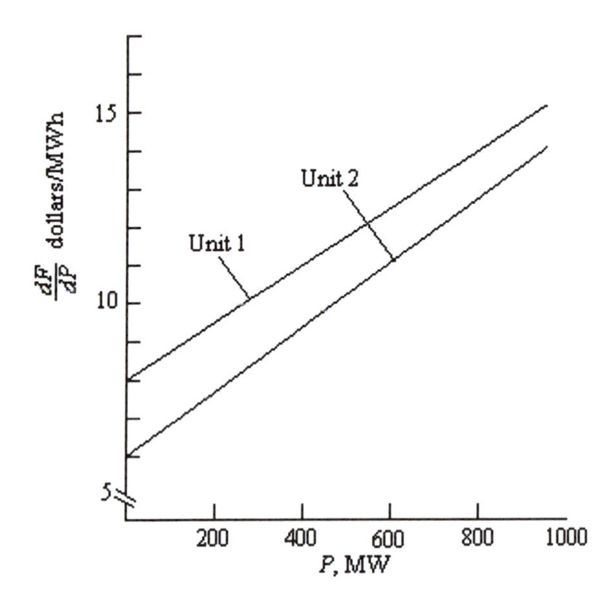

FIGURE 7.3
Example 7.5.

Solution
From Figure 7.3, we obtain

$$\frac{dF_1}{dP_1} = 0.008P_1 + 8 \tag{7.10}$$

and

$$\frac{dF_2}{dP_2} = 0.009P_2 + 6 \tag{7.11}$$

At 120 MW, $dF_1/dP_1 = 8.96$ and $dF_2/dP_2 = 7.08$. Therefore, until dF_2/dP_2 has risen to 8.96, unit 2 should take all the additional load above 120 MW. Using (7.11), we find that dF_2/dP_2 is equal to 8.96 when $P_2 = 328.9$ MW, at which value

$$P_{\text{total}} = P_1 + P_2 = 120 + 328.9 = 448.9 \text{ MW}$$

These values give us the first row of Table 7.1. Similar computations give the remaining rows of the table.

Table 7.1

λ, dollars/MWh	P_1, MW	P_2, MW	P_{total}, MW
8.96	120	328.9	448.9
9.6	200	400	600
10	250	444.4	694.4
10.4	300	488.9	788.9
11.2	400	577.77	977.7

7.2 EFFECT OF TRANSMISSION-LINE LOSS

To include the effect of transmission-line losses on economic operation of a power system, we must express these losses as a function of plant power output. Figure 7.4 shows two plants connected to a three-phase load. The total transmission loss (for all three phases) is given by

$$P_{\text{loss}} = 3(|I_1|^2 R_1 + |I_2|^2 R_2 + |I_3|^2 R_3) \tag{7.12}$$

where the R's are the per-phase resistances of the lines, and where

$$|I_3| = |I_1 + I_2| = |I_1| + |I_2| \tag{7.13}$$

if we assume that I_1 and I_2 are in phase. These currents may be expressed in terms of P_1 and P_2, the respective plant outputs, as

$$|I_1| = \frac{P_1}{\sqrt{3}\,|V_1|\cos\phi_1} \tag{7.14}$$

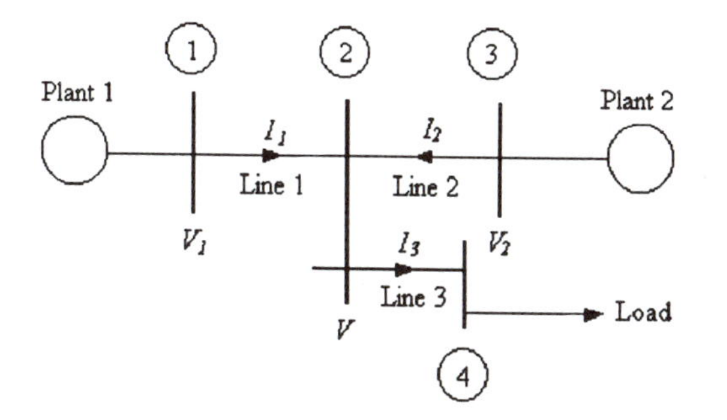

FIGURE 7.4
An interconnected system.

and

$$|I_2| = \frac{P_2}{\sqrt{3}\ |V_2| \cos\phi_2} \tag{7.15}$$

where $\cos\phi_1$ and $\cos\phi_2$ are the power factors at buses 1 and 2, respectively. Equations (7.12) through (7.15) may be combined to yield

$$P_{\text{loss}} = P_1^2 B_{11} + 2P_1 P_2 B_{12} + P_2^2 B_{22} \tag{7.16}$$

where B_{11}, B_{12}, and B_{22} are called *loss coefficients* or *B coefficients* and are given by

$$B_{11} = \frac{R_1 + R_3}{|V_1|^2 \cos^2\phi_1} \tag{7.17}$$

$$B_{12} = \frac{R_3}{|V_1|\,|V_2| \cos\phi_1 \cos\phi_2} \tag{7.18}$$

and

$$B_{22} = \frac{R_2 + R_3}{|V_2|^2 \cos^2\phi_2} \tag{7.19}$$

For a system of n plants, (7.16) may be generalized to

$$P_{\text{loss}} = \sum_{k=1}^{n} \sum_{m=1}^{n} P_k P_m B_{km} \tag{7.20}$$

where $B_{mk} = B_{km}$.

To illustrate the procedure, we consider the following example.

Example 7.6 Find the loss coefficients for the system shown in Figure 7.4 from the following data, in which all numerical quantities are per-unit values: $I_1 = 0.8\angle 0°$, $I_2 = 0.9\angle 0°$, $V_3 = 1.1\angle 0°$, $Z_1 = Z_2 = 0.06 + j0.20$, and $Z_3 = 0.04 + j0.06$. Also calculate the transmission loss in the system of Figure 7.4.

Solution
From Figure 7.4 and the given data,

$$V_1 = 1.1 + (0.8\ \angle 0°)(0.06 + j0.20) = 1.148 + j0.16$$

$$V_2 = 1.1 + (0.9\ \angle 0°)(0.06 + j0.20) = 1.154 + j0.18$$

Hence, $|V_1| \cos\phi_1 = 1.148$ and $|V_2| \cos\phi_2 = 1.154$. Now (7.17) through (7.19) yield

$$B_{11} = \frac{0.06 + 0.04}{(1.148)^2} = 0.0759 \quad \text{pu}$$

$$B_{12} = \frac{0.04}{(1.148)(1.154)} = 0.0302 \quad \text{pu}$$

$$B_{22} = \frac{0.06 + 0.04}{(1.154)^2} = 0.0751 \quad \text{pu}$$

For the losses, we have

$$P_1 = \text{Re}\,[0.8\ \angle 0°)\,(1.148 + j0.16)] = 0.9184 \ \text{pu}$$

$$P_2 = \text{Re}\,[0.9\ \angle 0°)\,(1.154 + j0.18)] = 1.0386 \ \text{pu}$$

Substituting these values and the B coefficients in (7.16) gives

$$P_{\text{loss}} = (0.9184)^2(0.0759) + 2(0.9184)(1.0386)(0.0302) + (1.0386)^2(0.0751) = 0.2026 \text{ pu.}$$

7.3 LOAD DISTRIBUTION BETWEEN PLANTS

In this section we combine the method of Section 7.1 with the results of Section 7.2 to obtain an economic allocation of load among a number of power plants. For a system of n plants, the total cost of fuel per hour is

$$F_{\text{total}} = \sum_{k=1}^{n} F_k \qquad \text{dollars/h} \tag{7.21}$$

and the total power output is

$$P_{\text{total}} = \sum_{k=1}^{n} P_k \qquad \text{MW} \tag{7.22}$$

With transmission losses, we must have

$$P_{\text{total}} = P_R + P_{\text{loss}} \tag{7.23}$$

where P_R and P_{loss} are, respectively, the total power received by the load and lost in transmission. For a given (constant), P_R, $dP_R = 0$. Thus (7.22) and (7.23) yield

$$\sum_{k=1}^{n} dP_k - dP_{\text{loss}} = 0 \tag{7.24}$$

In addition, when the load is allocated among the n plants for minimum fuel cost, $dF_{\text{total}} = 0$. Then

$$dF_{\text{total}} = \sum_{k=1}^{n} \frac{\partial F_{\text{total}}}{\partial P_k} dP_k = 0 \tag{7.25}$$

Also

$$dP_{\text{loss}} = \sum_{k=1}^{n} \frac{\partial P_{\text{loss}}}{\partial P_k} dP_k \tag{7.26}$$

Substituting dP_{loss} from (7.26) into (7.24), multiplying the result by λ, and subtracting that result from the right-hand equality in (7.25) yield

$$\sum_{k=1}^{n} \left[\left(\frac{\partial F_{\text{total}}}{\partial P_k} + \lambda \frac{\partial P_{\text{loss}}}{\partial P_k} - \lambda \right) dP_k \right] = 0 \tag{7.27}$$

This equation holds if

$$\frac{\partial F_{\text{total}}}{\partial P_k} + \lambda \frac{\partial P_{\text{loss}}}{\partial P_k} - \lambda = 0 \qquad \text{for all} \quad k = 1, 2, \ldots, n \tag{7.28}$$

Now, because

$$\frac{\partial F_{\text{total}}}{\partial P_k} = \frac{dF_k}{dP_k} \tag{7.29}$$

condition (7.28) may be written as

$$\frac{dF_k}{dP_k} L_k = \lambda \qquad \text{for} \quad k = 1, 2, \ldots, n \tag{7.30}$$

where L_k, called the *penalty factor* of the kth plant, is given by

$$L_k = \frac{1}{1 - \partial P_{\text{loss}} / \partial P_k} \tag{7.31}$$

Condition (7.30) implies that the system fuel cost is minimized when the incremental fuel cost for each plant, multiplied by its penalty factor, is the same throughout the system, that is, when

$$\frac{dF_1}{dP_1} L_1 = \frac{dF_2}{dP_2} L_2 = \ldots = \frac{dF_n}{dP_n} L_n = \lambda \tag{7.32}$$

To determine the L_k we have, from (7.20)

$$\frac{\partial P_{\text{loss}}}{\partial P_k} = \frac{\partial}{\partial P_k}\left(\sum_{k=1}^{n}\sum_{m=1}^{n} P_k P_m B_{km}\right) = 2\sum_{m=1}^{n} P_m B_{mk} \tag{7.33}$$

The simultaneous equations represented by (7.30) can be solved if a value is assumed for λ.

Example 7.7 In a two-plant system, the entire load is located at plant 2, which is connected to plant 1 by a transmission line. Plant 1 supplies 100 MW of power with a corresponding transmission loss of 5 MW. Calculate the penalty factors for the two plants.

Solution
Since all the load is at plant 2, varying P_2 does not affect the transmission loss P_{loss}. Thus, from (7.16)

$$P_{\text{loss}} = 5 = P_1^2 B_{11} = 10^4 B_{11}$$

so that $B_{11} = 5 \times 10^{-4}\ \text{MW}^{-1}$. Moreover, this expression for P_{loss} yields

$$\frac{\partial P_{\text{loss}}}{\partial P_1} = 2P_1 B_{11} = 2\,(100)\,(5 \times 10^{-4}) = 0.1$$

Then, from (7.31),

$$L_1 = \frac{1}{1 - 0.1} = 1.111$$

Similarly, because $\partial P_{\text{loss}}/\partial P_2 = 0$, we have $L_2 = 1$.

Example 7.8 For the system of Example 7.7, λ = \$15/MWh, and the incremental fuel costs for the two plants are given by

$$\frac{dF_1}{dP_1} = 0.01P_1 + 10 \qquad \text{and} \qquad \frac{dF_2}{dP_2} = 0.02P_2 + 12$$

in dollars per megawatthour, (*a*) How much power should be generated at each plant for minimal total fuel cost? (*b*) For this operating condition, determine the dollar savings that would be realized by coordinating the transmission loss rather than neglecting it.

Solution

(*a*) From (7.32) and from the results of Example 7.7,

$$\frac{dF_1}{dP_1} L_1 = \lambda$$

or

$$(0.01P_1 + 10)\,1.111 = 15$$

from which $P_1 = 350$ MW. Similarly,

$$\frac{dF_2}{dP_2} L_2 = (0.02P_2 + 12)\,1 = 15$$

from which $P_2 = 150$ MW.

With the transmission not coordinated, we would have (based on Section 7.1) $dF_1/dP_1 = dF_2/dP_2$ for economic operation. Hence we would have

$$0.01P_1 + 10 = 0.02P_2 + 12 \tag{7.34}$$

(*b*) Now, we know that the load requires

$$P_1 + P_2 - P_{\text{loss}} = 350 + 150 - P_1^2 B_{11}$$

$$= 500 - 61.25 = 438.75 \text{ MW}$$

Then with the transmission loss *not* coordinated, we would have

$$P_1 + P_2 - 5 \times 10^{-4} P_1^2 = 438.75 \tag{7.35}$$

Solving (7.34) and (7.35) simultaneously yields $P_1 = 417$ MW and $P_2 = 108.5$ MW.

Comparing these results with the results of *(a)* above, we see that the load on plant 1 is increased from 350 MW to 417 MW; hence, its fuel cost increases by

$$\int_{350}^{417} (0.01P_1 + 10)\, dP_1 = \$926.945/\text{h}$$

The load on plant 2 is decreased from 150 MW to 107.5 MW; hence, its fuel cost decreases by

$$-\int_{150}^{108.5} (0.02P_2 + 12)\, dP_2 = \$605.277/\text{h}$$

The savings with loss coordination is thus 926.945 - 605.277 = \$321.67/h.

7.4 POWER SYSTEM CONTROL

A number of automatic controls are used in present-day power systems. These include devices that control the generator voltage, the turbine governor, and the load frequency; there are also computer controls to ensure economic power flow and to control reactive power, among other power-system variables.

Generator voltage control is accomplished by controlling the exciter voltage. The block-diagram representation of a closed-loop automatic voltage regulating system is shown in simplified form in Figure 7.5. (Numerous other forms of generator voltage control also exist.) In Figure 7.5, the open-loop transfer function $G(s)$ is given by

$$G(s) = \frac{k}{(1 + T_a s)(1 + T_e s)(1 + T_f s)} \tag{7.36}$$

where T_a, T_e, and T_f are, respectively, the time constants associated with the amplifier, the exciter, and the generator field, and the open-loop gain k is

$$k = k_a k_e k_f \tag{7.37}$$

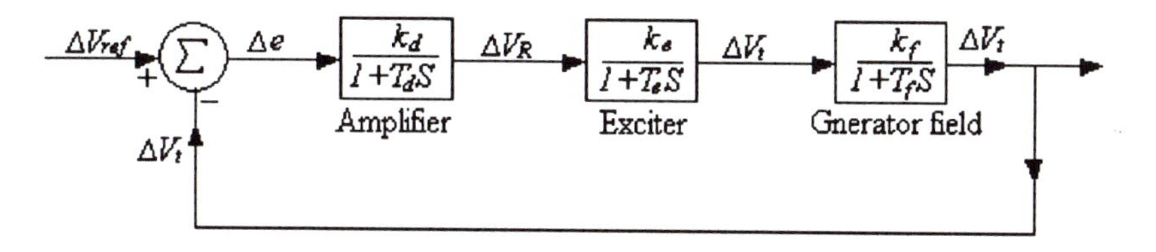

FIGURE 7.5
A voltage regulating system.

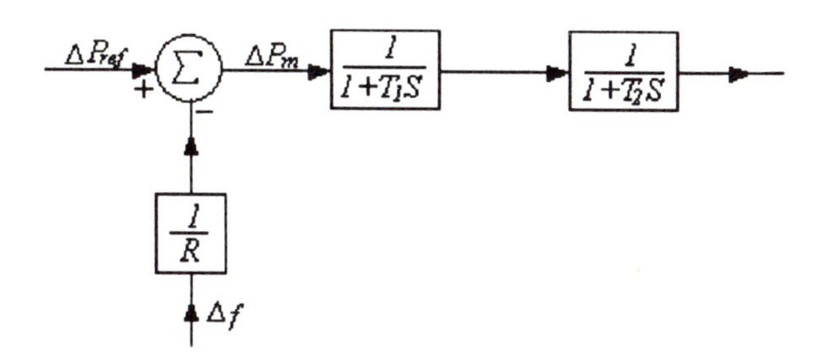

FIGURE 7.6
Turbine-governor control system.

Sudden changes in the load cause the turbine speed and, consequently, the generator frequency to change as well. The change in turbine speed occurs when the generator electromagnetic torque no longer equals the turbine (or other prime mover) mechanical torque. Thus, the change Δf in the generator frequency may be used as a control signal for controlling the turbine mechanical output power. The change in the turbine output power as a function of a change in the generator frequency is given by

$$\Delta P_m = \Delta P_{\text{ref}} - \frac{1}{R}\Delta f \tag{7.38}$$

where ΔP_m and ΔP_{ref} are, respectively, the changes in the turbine output power and the reference power (as determined by the governor setting), and R is known as the *regulation constant.*

Figure 7.6 shows the block diagram for a portion of a turbine-governor control system; we assume the system to be linear, and the governor and turbine-generator to be first-order devices.

In the preceding, we have implied that the accelerations and decelerations of the generator rotor are controlled by the turbine governor. However, the frequency deviation Δf still remains if $\Delta P_{\text{ref}} = 0$. This frequency deviation can be reduced to zero by a process called *load-frequency control* (or LFC). The LFC process then also controls the power flow on the tie line. Thus, via LFC, each interconnected area of a power system maintains the power flow out of that area at its scheduled value, in effect absorbing its own load variations.

To establish the pertinent control strategy for LFC, we define the *area control error* (ACE) as

$$\text{ACE} = \Delta P_{\text{tie}} + B_f \Delta f \tag{7.39}$$

where ΔP_{tie} is the deviation of the tie-line power flow out of the area from the scheduled power flow, Δf is the frequency error, and B_f is known as the *frequency bias constant.*

The change ΔP_{ref} in the reference power setting of the load-frequency-controlled turbine governor is proportional to the integral of ACE. Thus,

$$\Delta P_{\text{ref}} = -K \int \text{ACE}\, dt \tag{7.40}$$

where K is a constant. The minus sign in (7.40) implies that if either the net power flow out of the area or the frequency is low, then ACE is negative and the area should increase its generation.

If an area contains n generating units, we may write (7.38) as

$$\Delta P_{m(\text{total})} = \sum_{k=1}^{n} \Delta P_{mk} = \sum_{k=1}^{n} \Delta P_{\text{ref}k} - \left(\sum_{k=1}^{n} \frac{1}{R_k} \right) \Delta f = \Delta P_{\text{ref(total)}} - \beta \Delta f \quad (7.41)$$

where β is known as the *area frequency-response characteristic* and is given by

$$\beta = \sum_{k=1}^{n} \frac{1}{R_k} \quad (7.42)$$

Also,

$$\Delta P_{\text{ref(total)}} = \sum_{k=1}^{n} \Delta P_{\text{ref}k} \quad (7.43)$$

and Δf remains the same for each unit.

To summarize, (7.39) through (7.41) govern the LFC of the system. To illustrate the procedure we now consider the following examples.

Example 7.9 For the system shown in Figure 7.5, what is the minimum open-loop gain such that the steady-state error Δe_{ss} does not exceed 1 percent?

Solution
From Figure 7.5,

$$\frac{\Delta e}{\Delta V_{\text{ref}}} = \frac{1}{1 + G(s)} \quad (7.44)$$

Substituting (7.36) in (7.44) and setting $s = 0$ (for the steady state) yield

$$\Delta e_{ss} = \frac{(\Delta V_{\text{ref}})_{ss}}{1 + k} \quad \text{or} \quad 1 + k = \frac{(\Delta V_{\text{ref}})_{ss}}{\Delta e_{ss}} \quad (7.45)$$

The condition of the problem implies that the right side of (7.45) is not less than 100. Hence,

$$1 + k \geq 100$$

and $k \geq 99$.

Example 7.10 For a certain turbine-generator set, $R = 0.04$ pu, based on the generator rating of 100 MVA and 60 Hz. The generator frequency decreases by 0.02 Hz, and the system adjusts to steady-state operation. By how much does the turbine output power increase?

Solution
The per-unit frequency change is

$$\text{Per-unit } \Delta f = \frac{\Delta f}{f_{\text{base}}} = \frac{-0.02}{60} = -3.33 \times 10^{-4} \text{ pu}$$

Then (7.38) yields

$$\text{Per-unit } \Delta P_m = -\frac{1}{0.04}(-3.33 \times 10^{-4}) = 8.33 \times 10^{-3} \text{ pu}$$

The actual increase in output power is then

$$\Delta P_m = (8.33 \times 10^{-3})(100) = 0.833 \text{ MW}$$

Example 7.11 An area includes two turbine-generator units, rated at 500 and 750 MVA and 60 Hz, for which $R_1 = 0.04$ pu and $R_2 = 0.05$ pu based on their respective ratings. Each unit carries a 300-MVA steady-state load. The load on the system suddenly increases by 250 MVA. *(a)* Calculate β on a 1000-MVA base. *(b)* Determine Δf on a 60-Hz base and in hertz.

Solution
(*a*) We can change the bases of the R values with the formula

$$R_{\text{new}} = R_{\text{old}} \frac{S_{\text{base(new)}}}{S_{\text{base(old)}}}$$

Thus

$$R_{1\text{(new)}} = (0.04)\frac{1000}{500} = 0.08 \text{ pu}$$

and

$$R_{2\text{(new)}} = (0.05)\frac{1000}{750} = 0.067 \text{ pu}$$

Now, from (7.42)

$$\beta = \frac{1}{R_1} + \frac{1}{R_2} = \frac{1}{0.08} + \frac{1}{0.067} = 27.5 \text{ pu}$$

(*b*) The per-unit increase in the load is 250/1000 = 0.25 pu. From (7.41), with $\Delta P_{\text{ref(total)}} = 0$ for steady-state conditions.

$$\Delta f = \frac{-1}{\beta}\Delta P_m = -\frac{1}{27.5}0.25 = -9.091 \times 10^{-3} \text{ pu}$$

Also,

$$\Delta f = -9.091 \times 10^{-3} \times 60 = -0.545 \text{ Hz}$$

Example 7.12 For areas 1 and 2 in a 60-Hz power system, $\beta_1 = 400$ MW/Hz and $\beta_2 = 250$ MW/Hz. The total power generated in each of these areas is, respectively, 1000 MW, and 750 MW. While each area is generating power at the steady state with $\Delta P_{\text{tie1}} = \Delta P_{\text{tie2}} = 0$, the load in area 1 suddenly increases by 50 MW. Determine the resulting Δf, *(a)* without LFC and *(b)* with LFC. Neglect all losses.

Solution

(*a*) From (7.41), since $\Delta P_{\text{ref(total)}} = 0$ without LFC,

$$50 = -(400 + 250)\Delta f$$

from which $\Delta f = -0.0769$ Hz.

(*b*) With LFC, in the steady state, (7.39) implies that $\text{ACE}_1 = \text{ACE}_2 = 0$; otherwise, the LFC given by (7.39) would be changing the reference power settings of the governors on LFC. Also, the sum of the net tie-line flows, $\Delta P_{\text{tie1}} + \Delta P_{\text{tie2}}$, is zero (neglecting losses). So

$$\text{ACE}_1 + \text{ACE}_2 = 0 = (B_1 + B_2)\Delta f$$

and $\Delta f = 0$, since $B_1 + B_2 \neq 0$.

PROBLEMS

7.1. A graph of fuel input versus power output for a certain plant is given in Figure 7.7. Determine the fuel requirements at *(a)* 120 MW and *(b)* 560 MW output power.

7.2. (*a*) For the plant of Problem 7.1, determine the fuel requirement at the maximum-efficiency operating point. (*b*) What is the power output at that point?

7.3. Assuming a fuel cost of \$1.60 per million Btu for the plant of Problem 7.1, plot the incremental fuel cost versus output power. From the result thus obtained, calculate the incremental fuel cost at the point at which the plant operates at maximum fuel efficiency.

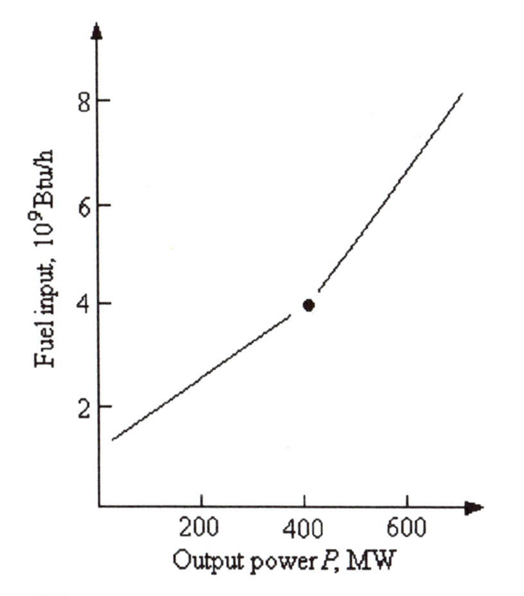

Problem 7.1.

7.4. Approximate the curve obtained in Problem 7.3 with a straight line, and obtain an equation for this line.

7.5. The incremental fuel costs, in dollars per megawatthour, for two units in a plant are given by

$$\frac{dF_1}{dP_1} = 0.007P_1 + 7 \quad \text{and} \quad \frac{dF_2}{dP_2} = 0.009P_2 + 6$$

During a 24-h period the load on each unit varies between 100 MW and 500 MW, whereas the plant output varies from 200 MW to 700 MW. (*a*) At what power level should unit 2 begin to take on all the additional load for the most economic operation of the plant? (*b*) What is the power output of the plant at this point? Neglect losses.

7.6. When the plant of Problem 7.5 is delivering its maximum power output, how should this load be shared between the two units for minimum fuel cost? At what total output should the units share the load equally?

7.7. For maximum demand from the plant of Example 7.5, determine how the load should be shared by the two generating units for minimum fuel cost. Also, determine the incremental fuel cost for each unit.

7.8. For the maximum power output of the plant of Example 7.5, calculate the saving per hour in fuel cost under economic (optimal) operation, as compared to operation with the load equally divided between the two units.

7.9. For the system shown in Figure 7.4, let $I_1 = 1.0\angle 0°$, $I_2 = 0.8\angle 0°$, $V_1 = 1.05\angle 10°$, and $V_2 = 1.07\angle 15°$, all per unit. The line impedances, again per unit, are $Z_1 = 0.05 + j0.20$, $Z_2 = 0.06 + j0.30$, and $Z_3 = 0.06 + j0.40$. Determine the system loss coefficients.

7.10. Calculate the transmission loss for the system of Problem 7.11 with the loss formula (7.16).

7.11. Express the result given by (7.20) in matrix form.

7.12. Calculate the penalty factors for the two plants operating as in Problem 7.9.

7.13. For the system of Problem 7.9, the incremental fuel costs for the two plants are given by

$$\frac{dF_1}{dP_1} = 0.01P_1 + 10 \quad \text{and} \quad \frac{dF_2}{dP_2} = 0.02P_2 + 10$$

The system operates at λ = \$16/MWh for minimum fuel cost. Determine the power generated at each plant.

7.14. Obtain the form of the dynamic response of the system of Figure 7.5 to a step change in the reference input voltage.

7.15. Assume that there are no changes occurring in the reference power setting of a turbine-governor system (that is, the system is operating in the steady state), and the frequency-power relationship of the turbine governor is that represented graphically in Figure 7.8. Determine the regulation constant *R*.

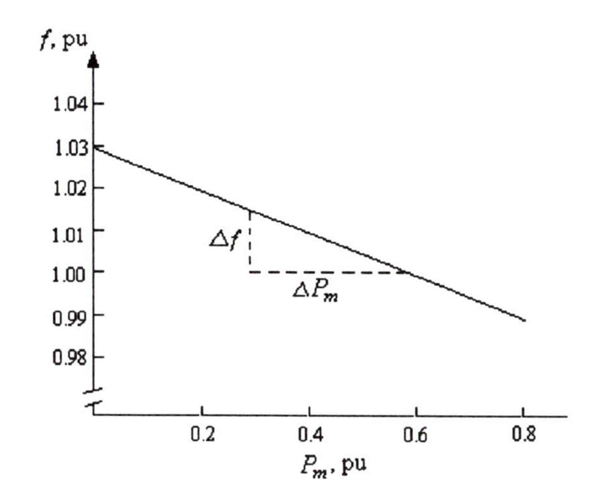

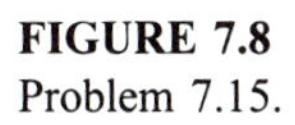

FIGURE 7.8
Problem 7.15.

APPENDIX I

CHARACTERISTICS OF ALUMINUM CONDUCTORS

TABLE I.1
Electrical characteristics of bare aluminum conductors steel-reinforced (ACSR)

Code word	Aluminum area, cmil	Standing Al/St	Layers of aluminum	Outside diameter,in	Resistance			GMR D_s,ft	Resistance per conductor 1-ft spacing, 60 Hz	
					Dc, 20°C, Ω/1,000 ft	Ac, 60 Hz			Inductive X_a Ω/mi	Capacitive X_a' MΩ•mi
						20°C, Ω/mi	50°C, Ω/mi			
Waxwing	266,800	18/1	2	0.609	0.0646	0.3488	0.3831	0.0198	0.476	0.1090
Partridge	266,800	26/7	2	0.642	0.0640	0.3452	0.3792	0.0217	0.465	0.1074
Ostrich	300,000	26/7	2	0.680	0.0569	0.3070	0.3372	0.0229	0.458	0.1057
Merlin	336,400	18/1	2	0.684	0.0512	0.2767	0.3037	0.0222	0.462	0.1055
Linnet	336,400	26/7	2	0.721	0.0507	0.2737	0.3006	0.0243	0.451	0.1040
Oriole	336,400	30/7	2	0.741	0.0504	0.2719	0.2987	0.0253	0.445	0.1032
Chickadee	397,500	18/1	2	0.743	0.0433	0.2342	0.2572	0.0241	0.452	0.1031
Pelican	477,000	18/1	2	0.814	0.361	0.1957	0.2148	0.0264	0.441	0.1004
Flicker	477,000	24/7	2	0.846	0.0359	0.1943	0.2134	0.00284	0.432	0.992
Hawk	477,000	27/7	2	0.858	0.357	0.1931	0.2120	0.0289	0.430	0.0988
Hen	477,000	30/7	2	0.883	0.355	0.1919	2.0107	0.0304	0.424	0.0980
Osprey	556,500	18/1	2	0.879	0.0309	0.1679	0.1843	0.0284	0.432	0.0981
Parakeet	556,500	24/7	2	0.914	0.0308	0.1669	0.1832	0.0306	0.423	0.0969
Dove	556,500	26/7	2	0.927	0.0307	0.1663	0.1826	0.0314	0.420	0.0965
Drake	795,000	26/7	2	1.108	0.0215	0.1172	0.1284	0.0373	0.399	0.0912
Tern	795,000	45/7	3	1.063	0.0217	0.1188	0.1302	0.0352	0.406	0.0925
Cardinal	954,000	54/7	3	1.196	0.0180	0.0988	0.1082	0.0402	0.390	0.0890
Bluejay	1,113,000	45/7	3	1.259	0.0155	0.0861	0.0941	0.0415	0.386	0.0874
Finch	1,113,000	54/19	3	1.293	0.0155	0.0856	0.9937	0.0436	0.380	0.0866
Pheasant	1,272,000	54/19	3	1.382	0.0035	0.0751	0.0821	0.0466	0.372	0.0847
Bluebird	2,156,000	84/19	4	1.762	0.0080	0.0476	0.0515	0.0586	0.344	0.0776

APPENDIX II

POWER TRANSFORMERS

The most common function of a transformer in a power system is to change voltage (and implicitly current) levels on the system. In this appendix we will review briefly the construction, operation, and analysis of single-phase and three-phase power transformers.

II.1 CONSTRUCTION

A transformer consists of one or more electrical windings linked or coupled together magnetically by a magnetic circuit or core. The magnetic circuit of a power transformer is constructed of a magnetic material. Transformers may have one, two, three or more windings on the same core. Since all transformers perform similar functions, the theory of two-winding transformers may easily be extended to multiwinding transformers and one-winding transformers. Transformers with a single tapped winding are normally called autotransformers. The two basic windings of a transformer are often called the primary and the secondary. The meaning usually attached to this nomenclature is that the input or source energy is applied to the primary windings and the output energy is taken from the secondary winding. However, since a transformer is a bilateral

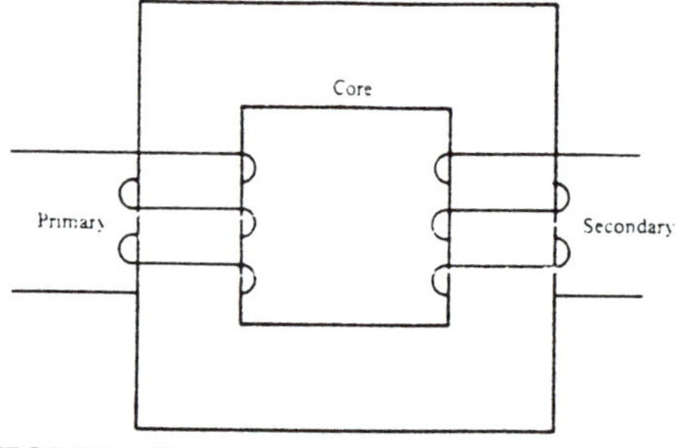

FIGURE II.1
Elementary model of a transformer.

device and is often operated bilaterally, this meaning is not very significant and these words are used more as a way to distinguish the two windings. It is morecommon to designate the windings by numbered subscripts or as high-voltage and low-voltage windings. A simple two-winding transformer model is shown in Figure II.1.

The electromagnetic structure of a transformer is contained within a housing or case for safety and protection. In several types of transformers, the space surrounding the electromagnetic structure is filled with an electrically insulating material to prevent damage to the windings or core and to prevent their movement or facilitate heat transfer between the electromagnetic structure and the case. In many power transformers, a nonflammable insulating oil called transformer oil is used. Transformer oil serves an added function of improving the insulation characteristics of the transformer, since it has a higher dielectric strength than air. In most oil-filled transformers, the oil is permitted to circulate through cooling fins or tubes on the outside of the case to improve further the heat transfer characteristics. The fins or tubes are often cooled by forced air. In large transformers operating at high voltage and current levels, there are other important structural components. Such components include porcelain bushings, through which the winding leads are brought for external connection, oil pressure and temperature gauges, and internal structural supports to prevent movement of the leads or windings caused by electromagnetic forces resulting from high current levels.

The magnetic core of a transformer must be constructed in a manner to minimize the magnetic losses. Power transformer cores are generally constructed from soft magnetic materials in the form of punched laminations or wound tapes. Lamination or tape thickness is a function of the transformer frequency. The most common lamination materials are silicon-iron, nickel-iron, and cobalt-iron alloys.

Transformer windings are constructed of solid or stranded copper or aluminum conductors. The windings of large power transformers generally use conductors with heavier insulation than magnet wire insulation. The windings are assembled with much greater mechanical support, and winding layers are insulated from each other. Larger, high-power windings are often preformed, and the transformer is assembled by stacking the laminations within the preformed coils.

II.2 OPERATION

In order to understand the operation of a transformer, and to obtain certain basic relationships for an ideal transformer, consider the simple model shown as Figure II.2.

A fundamental parameter of the transformer is the turns ratio, defined as

$$a = \frac{N_1}{N_2} \tag{II.1}$$

The actual number of turns in each winding is generally known only to the transformer manufacturer or the person winding a laboratory transformer. The ratio can be measured in the laboratory by measuring the induced voltages in the two windings. The turns ratio is frequently given as part of the nameplate data by the manufacturer in both electronics and power systems transformers, normally as rated voltages for the primary and secondary windings.

The transformer of Figure II.2 is ideal in the sense that its core is lossless, it is infinitely permeable, there are no leakage fluxes, and the windings have no losses. Absence of leakage flux implies that the entire flux links with both windings completely. In Figure II.2 the basic components are the *core*, the *primary winding* having N_1 turns, and the *secondary winding* having N_2 turns. If ϕ is the mutual (or core) flux linking N_1 and N_2, then according to Faraday's law of electromagnetic induction, emf's e_1 and e_2 are induced in N_1 and N_2 due to a time rate of change of ϕ such that

$$e_1 = N_1 \frac{d\phi}{dt} \tag{II.2}$$

and

$$e_2 = N_2 \frac{d\phi}{dt} \tag{II.3}$$

The direction is such as to oppose the flux change, according to Lenz's law. The transformer being ideal, $E_1 = V_1$ where V_1 is the phasor value of the input terminal voltage and E_1 is the phasor value of the induced voltage e_1 (Figure II.2). The dots shown on each winding denote positive voltage terminals in accordance with the winding polarities. From (II.2) and (II.3), we have

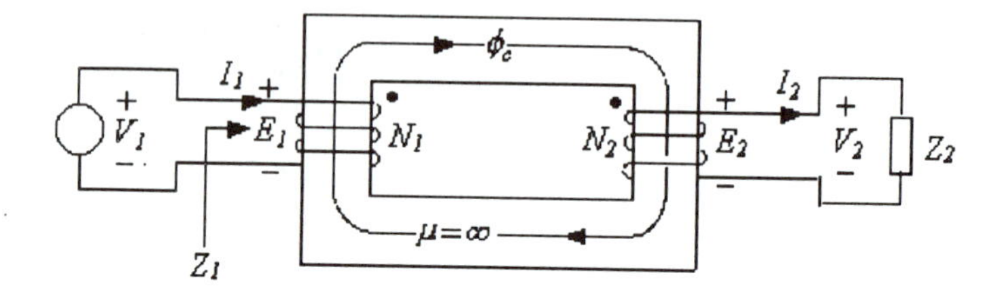

FIGURE II.2
An ideal transformer model showing polarities and dot convention.

$$\frac{e_1}{e_2} = \frac{N_1}{N_2}$$

which may also be written in terms of rms values as

$$\frac{|E_1|}{|E_2|} = \frac{N_1}{N_2} = a$$

or in terms of phasor values as

$$\frac{E_1}{E_2} = \frac{N_1}{N_2} = a \tag{II.4}$$

where a is the *turns ratio.*

For a sinusoidal applied voltage v_1, the flux varies sinusoidally, such that

$$\phi = \phi_m \sin\omega t. \tag{II.5}$$

Then, from (II.2) and (II.5), the corresponding induced voltage, e_1, linking an N-turn winding is given by

$$v_1 = e_1 = \omega N_1 \phi_m \cos\omega t \tag{II.6}$$

From (II.6), the rms value of the induced voltage is

$$|V_1| = |E_1| = \frac{\omega N_1 \phi_m}{\sqrt{2}} = 4.44 f N_1 \phi_m \tag{II.7}$$

In (II.7), $f = \omega/2\pi$ is the frequency in hertz.

Equation (II.7) is known as the *emf equation* of a transformer.

II.3 VOLTAGE, CURRENT, AND IMPEDANCE TRANSFORMATIONS

The voltage transformation property, mentioned in the preceding section, of an ideal transformer is expressed as

$$\frac{V_1}{V_2} = \frac{E_1}{E_2} = a \tag{II.8}$$

where the subscripts 1 and 2 correspond to the primary and secondary sides, respectively.

For an ideal transformer, the net mmf around its magnetic circuit must be zero. Consequently,

$$\mathscr{F} = \mathscr{F}_1 + \mathscr{F}_2 = N_1 |I_1| - N_2 |I_2| = 0 \tag{II.9}$$

where $|I_1|$ and $|I_2|$ are the rms values of the primary and the secondary currents, respectively. From (II.4) and (II.9) we get

$$\frac{|I_2|}{|I_1|} = \frac{N_1}{N_2} = a$$

or in terms of phasor quantities

$$\frac{I_2}{I_1} = \frac{N_1}{N_2} = a \tag{II.10}$$

From (II.8) and (II.10), the impedance transformation relationships can be obtained. If Z_1 is the impedance seen at the primary terminals, then

$$\begin{aligned} Z_1 &= \frac{V_1}{I_1} = \frac{E_1}{I_1} = \frac{aE_2}{I_2/a} = a^2 \frac{E_2}{I_2} \\ &= \frac{a^2 V_2}{I_2} = a^2 Z_2 \end{aligned} \tag{II.11}$$

II.4 THE NONIDEAL TRANSFORMER

A nonideal (or an actual) transformer differs from an ideal transformer in that the former has hysteresis and eddy-current (or core) losses, and has resistive

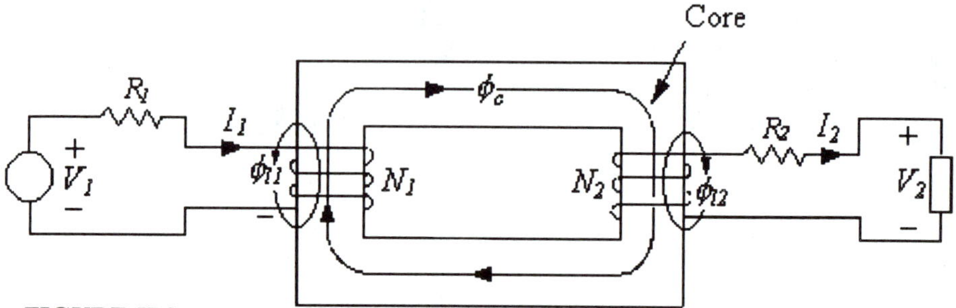

FIGURE II.3
A nonideal transformer showing the leakage fluxes and the winding resistances.

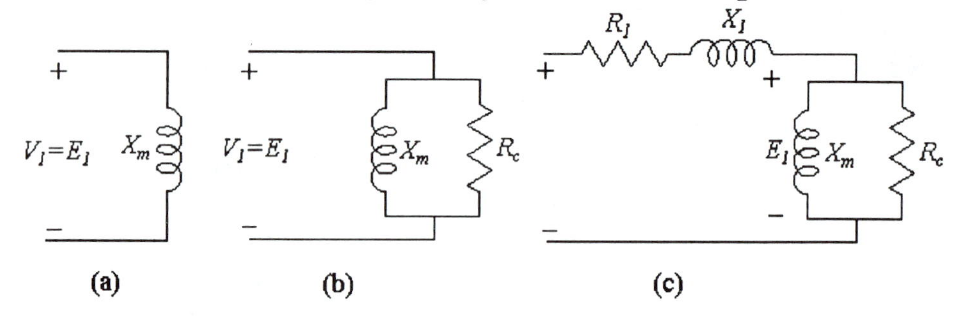

FIGURE II.4
Equivalent circuits of an iron core excited by an ac mmf.

(I^2R) losses in its primary and secondary windings. Furthermore, the core of a nonideal transformer is not perfectly permeable, and the transformer core requires a finite mmf for its magnetization. Also, not all fluxes link with the primary and secondary windings simultaneously because of leakages. Referring to Figure II.3, we observe that R_1 and R_2 are the respective resistances of the primary and secondary windings. The flux ϕ_c, which replaces the flux ϕ of Figure II.2, is called the *core flux* or *mutual flux*, as it links both the primary and secondary windings. The primary and secondary leakage fluxes are shown as ϕ_{l1} and ϕ_{l2}, respectively.

Considering the primary winding and the nonideal core, first, we assume that the core has a finite and constant permeability, but no core losses. We also temporarily assume that the windings have no resistance. In such an idealized case, the magnetic circuit and the primary winding can be represented just by an inductance L_m, as given in Figure II.4*(a),* which corresponds to the core flux, ϕ_c. If the coil is excited by a sinusoidal ac voltage $v_1 = V_m \sin \omega t$, it is conventional to express L_m as an inductive reactance $X_m = \omega L_m$. This reactance is known as the *magnetizing reactance*. Thus, in Figure II.4*(a)* we have X_m across which we show the terminal voltage V_1, equal to the induced voltage E_1. Note that V_1 and E_1 are phasor values.

Next, we include the hysteresis and eddy-current losses by the resistor R_c in parallel with X_m, such that the voltage V_1 appears across R_c also, the core losses being directly dependent on V_1. We thus obtain Figure II.4*(b)*, where again $V_1 = E_1$. Finally, we include the series resistance R_1, the resistance of the primary winding, and X_1, its *leakage reactance*, which arises from leakage fluxes. Hence, we obtain the electrical equivalent shown in Figure II.4*(c)*, which represents an iron core excited by an ac voltage. We will use this basic equivalent circuit to obtain the complete equivalent circuit of a transformer.

Continuing with the development of an equivalent circuit of a nonideal transformer, the leakage fluxes ϕ_{l1} and ϕ_{l2} give rise to the leakage reactances, X_1 and X_2, respectively. The mutual flux ϕ_c is represented by the magnetizing reactance X_m and the core losses are represented by a resistance, R_c, in parallel with X_m. Winding resistances may be represented by the parameters R_1 and R_2. Consequently, the equivalent circuit of an ideal transformer shown in Figure II.5(*a*) modifies to that shown in Figure II.5*(b)*, which is the equivalent circuit of a nonideal transformer. In Figure II.5*(b)*, notice that the circuit components denoting the imperfections of the transformer are coupled by an ideal transformer

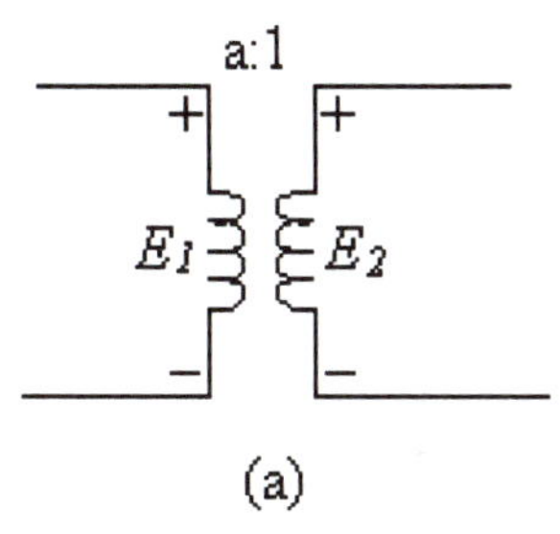

(a)

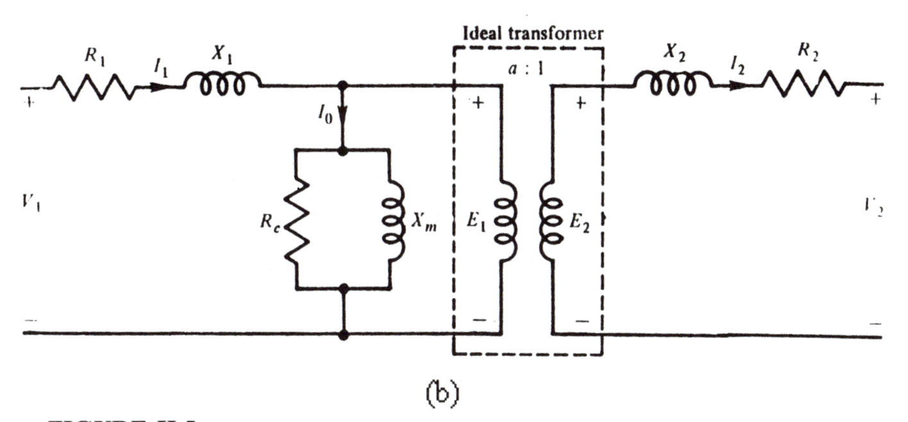

(b)

FIGURE II.5

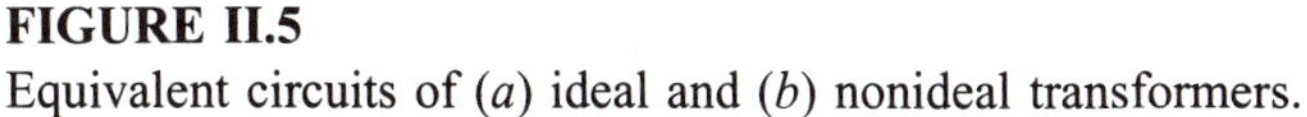

Equivalent circuits of (*a*) ideal and (*b*) nonideal transformers.

of proper turns ratio. This ideal transformer may be removed and the entire equivalent circuit may be referred either to the primary or to the secondary of the transformer by using its transformation properties, as given by (II.8), (II.10), and (II.11).

Referring to the primary side, according to (II.11) an impedance Z_2, on the secondary side, will appear as a^2Z_2 on the primary side. In accordance with (II.10), a current I_2 on the secondary side corresponds to a current I_2/a on the primary side. Finally, to eliminate the ideal transformer from Figure II.5*(b)*, and refer the secondary quantities to the primary, we use (II.8) to replace E_2 by $aE_2 = E_1$. The circuit then becomes as shown in Figure II.6. Of course, the primary quantities (V_1, I_1, R_1, X_1, X_m and R_c) remain unchanged.

To obtain the equivalent circuit referred to the secondary, we again use (II.8), (II.10), and (II.11). Thus, we get the equivalent circuit shown in Figure II.7.

A phasor diagram for the circuit of Figure II.6, for lagging power factor, is shown in Figure II.8. In Figs. II.6 to II.8, the various symbols are as follows:

a = turns ratio
E_1 = primary voltage induced by ϕ_c
E_2 = secondary voltage induced by ϕ_c
V_1 = primary terminal voltage
V_2 = secondary terminal voltage
I_1 = primary current
I_2 = secondary current
I_0 = no-load (primary) current or exciting current
R_1 = resistance of the primary winding
R_2 = resistance of the secondary winding
X_1 = primary leakage reactance
X_2 = secondary leakage reactance
I_m, X_m = magnetizing current and reactance
I_c, R_c = current and resistance accounting for the core losses.

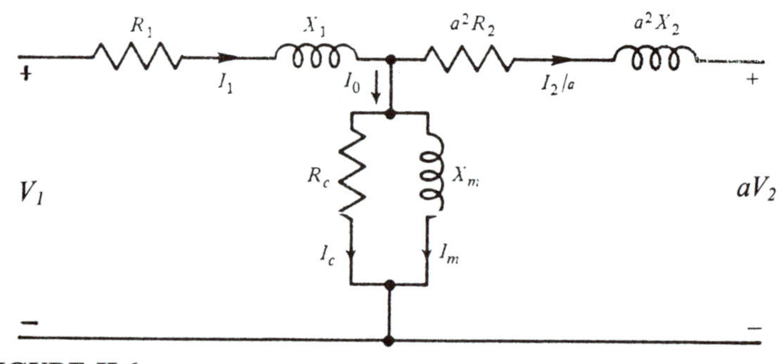

FIGURE II.6
Equivalent circuit referred to primary.

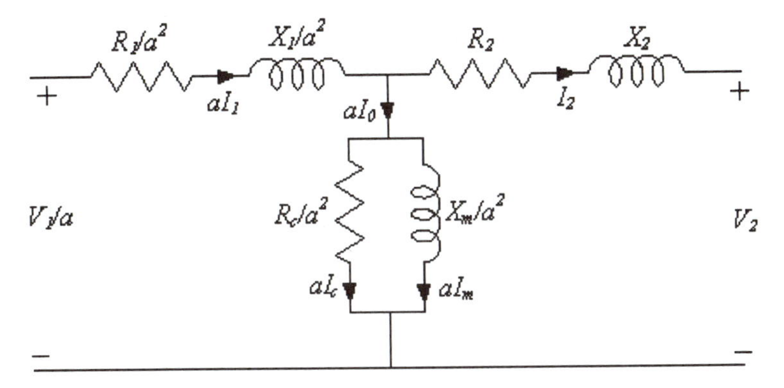

FIGURE II.7
Equivalent circuit referred to secondary.

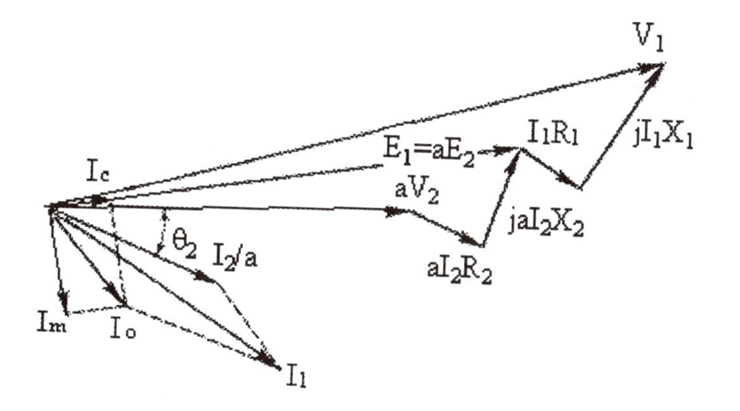

FIGURE II.8
Phasor diagram corresponding to Figure II.6.

II.5 PER-UNIT IMPEDANCES

In the equivalent circuits presented above, the impedances are generally expressed in ohms. However, in the text we have demonstrated the usefulness of expressing various quantities related to a power system in per unit (pu). In fact, much of the analysis has been carried out in per unit. Specifically, for a transformer, the per-unit system simplifies the equivalent circuit. Referring to Figure II.5, the ideal transformer shown to couple the primary and secondary windings may be eliminated, since the per-unit parameters do not change when they are referred from one side of the transformer to the other. This is a result of the standard practice of assuming base voltages in accordance with the turns ratio of the transformer. This fact is demonstrated by the following example.

Example II.1 A 150-kVA 2400/240-V transformer has the following parameters: $R_1 = 0.2\ \Omega$, $R_2 = 0.002\ \Omega$, $X_1 = 0.45\ \Omega$, $X_2 = 0.0045\ \Omega$, $R_c = 10{,}000\ \Omega$, and $X_m = 1550\ \Omega$, where the symbols are shown in Figure II.5(*b*). Express the secondary leakage impedance referred to the primary and the secondary, in per unit.

Solution

We choose the base values as the transformer ratings. Thus,

$$S_{\text{base}} = 150\ \text{kVA};\quad V_{\text{base1}} = 2400\ \text{V};\quad V_{\text{base2}} = 240\ \text{V}$$

$$Z_{\text{base2}} = \frac{V_{\text{base2}}^2}{S_{\text{base}}} = \frac{240^2}{150{,}000} = 0.384\ \Omega$$

Now, $Z_2 = R_2 + jX_2 = 0.002 + j0.0045 = 0.00492\angle 66°\Omega$. Referred to the secondary winding, this secondary leakage impedance in per unit becomes:

$$Z_2 = \frac{0.00492\angle 66°}{0.384} = 0.0128\angle 66°\ \text{pu}$$

Next, referring to the primary winding, the ohmic value of the secondary leakage impedance is

$$a^2 Z_2 = 10^2 \times 0.00492\angle 66° = 0.492\angle 66°\ \Omega$$

To express this value in per unit, we have

$$Z_{\text{base1}} = \frac{V_{\text{base1}}^2}{S_{\text{base}}} = \frac{2400^2}{150{,}000} = 38.4\ \Omega$$

In per unit, therefore, we obtain

$$a^2 Z_2 = \frac{0.492\angle 66°}{38.4} = 0.0128\angle 66°\ \text{pu}$$

which is the same as the per-unit secondary impedance, as expected. Thus, $a^2 Z_2 = Z_2$ when these quantities are expressed in per unit.

In a power system, the magnetizing branch of the equivalent circuit of a transformer is generally neglected. So the transformer is represented by its total series impedance ($R + jX$), the per-unit value of which does not change when referred to one winding or the other. As a result, we have the transformer model that has been previously given in Figure 3.2.

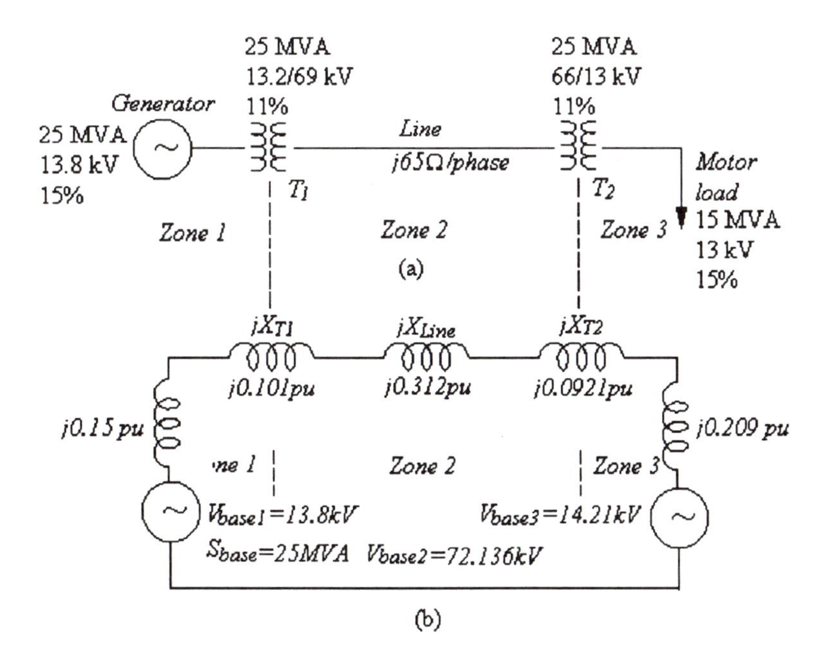

FIGURE II.9
(*a*) Portion of a power system and (*b*) its reactance diagram.

As we know, transformers are extensively used as components of power systems to change voltage levels. In Figure II.9 we show a portion of a three phase power system having three levels of voltages resulting from the transformers T_1 and T_2. Notice that in this diagram all resistances are neglected as is a common practice. In order to draw the reactance diagram, we divide the system into zones 1 through 3, each corresponding to a voltage level. Notice that these voltage levels are due to the transformers. The reactance diagram is shown in Figure II.9(b). The numerical values are calculated as follows.

Let S_{base} = 25 MVA be the base value of the entire network. Considering the zones in sequence, we have

For zone 1:

$$V_{\text{base1}} = 13.8 \text{ kV}$$

$$\text{generator reactance} = 0.15 \text{ pu}$$

For zone 2:

$$V_{\text{base2}} = 13.8 \times \frac{69}{13.2} = 72.136 \text{ kV}$$

From (3.6)

$$\text{transformer reactance} = \left(\frac{13.2}{13.8}\right)^2 \times 0.11 = 0.101 \text{ pu}$$

or

$$\text{transformer reactance} = \left(\frac{69}{72.136}\right)^2 \times 0.11 = 0.101 \text{ pu}$$

From (3.9):

$$\text{line reactance} = 65 \times \frac{25}{(72.136)^2} = 0.312 \text{ pu}$$

For zone 3:

$$V_{\text{base3}} = 72.136 \times \frac{13}{66} = 14.21 \text{ kV}$$

$$\text{transformer reactance} = \left(\frac{13}{14.21}\right)^2 \times 0.11 = 0.0921 \text{ pu}$$

$$\text{motor reactance} = \left(\frac{25}{15}\right)\left(\frac{13}{14.21}\right)^2 \times 0.15 = 0.209 \text{ pu}$$

Consequently, we obtain the reactance diagram shown in Figure II.9(b).

II.6 PERFORMANCE CHARACTERISTICS

As mentioned earlier, the major use of the equivalent circuit of a transformer is in determining its characteristics. The characteristics of most interest to power engineers are voltage regulation and efficiency. *Voltage regulation* is a measure of the change in the terminal voltage of the transformer with load. From Figs. II.6 and II.8, it is clear that the rms terminal voltage aV_2 is load-dependent. Specifically, we define voltage regulation as

$$\text{percent regulation} = \frac{|V|_{\text{no-load}} - |V|_{\text{load}}}{|V|_{\text{load}}} \times 100 \qquad \text{(II.12)}$$

With reference to Figure II.8, we may rewrite (II.12) as

$$\text{percent regulation} = \frac{a\,|V_2| - a\,|V_2'|}{a\,|V_2'|} \times 100$$

where $|V_2'|$ is the actual rms secondary voltage when a load is connected to the secondary terminals and $|V_2|$ is the actual rms secondary voltage when no load is connected.

There are two kinds of efficiencies of transformers of interest to us, known as *power efficiency* and *energy efficiency*. These are defined as follows:

$$\text{power efficiency} = \frac{\text{output power}}{\text{input power}} \tag{II.13}$$

$$\text{energy efficiency} = \frac{\text{output energy for a given period}}{\text{input energy for the same period}} \tag{II.14}$$

Generally, energy efficiency is taken over a 24-hour (h) period and is called *all-day efficiency*. In such a case (II.14) becomes

$$\text{all-day efficiency} = \frac{\text{output energy for 24 h}}{\text{input energy for 24 h}} \tag{II.15}$$

II.7 AUTOTRANSFORMERS

In contrast to the two-winding transformers considered so far, the autotransformer is a single-winding transformer having a tap brought out at an intermediate point. Thus, as shown in Figure II.10, *ac* is the single winding (wound on a laminated core) and *b* is the intermediate point where the tap is brought out. The autotransformer may be used as either a step-up or a step-down operation, like a two-winding transformer. Considering a step-down arrangement, let the primary applied (terminal) voltage be V_1, resulting in a magnetizing current and a core flux, ϕ_c. Let the secondary be open-circuited. Then the primary and secondary voltages obey the same rules as in a two-winding ideal transformer, and we have

$$\frac{V_1}{V_2} = \frac{E_1}{E_2} = \frac{N_1}{N_2} = a \tag{II.16}$$

with $a > 1$ for step-down. In (II.16) N_1 is the total turns from a to c and N_2 is the number of turns from b to c.

Furthermore, from conservation of energy

$$V_1 I_1 = V_2 I_2$$

and

$$\frac{V_1}{V_2} = \frac{I_2}{I_1} = a$$

For an ideal transformer, we must have the mmf balance equation as

$$N_2 I_3 = (N_1 - N_2) I_1$$

or

$$I_3 = \frac{N_1 - N_2}{N_2} I_1 = (a - 1) I_1 = I_2 - I_1$$

Example II.2. The autotransformer shown in Figure 11.10 has N_1 = 100 turns and N_2 = 90 turns. Assuming that the transformer is ideal, determine the input real power and the real power delivered to an impedance of 3 + j4 Ω connected to the secondary terminals if the primary voltage is 200 V rms.

Solution

Assume

$$V_1 = 200\angle 0° \text{ V}$$

then

$$V_2 = \frac{N_2}{N_1} V_1 = \frac{90}{100}(200\angle 0°) = 180\angle 0° \text{ V}$$

and

$$I_2 = \frac{V_2}{3 + j4} = \frac{180\angle 0°}{5\angle 53.1°} = 36\angle -53.1° \text{ A}$$

$$P_2 = (180)(36) \cos (53.1°) = 3.9 \text{ kW}$$

also

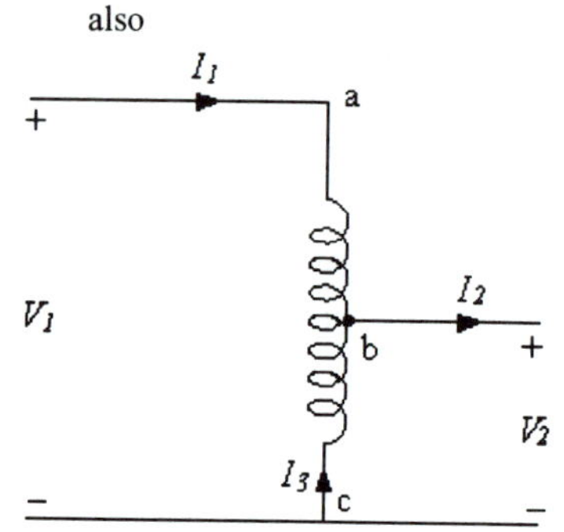

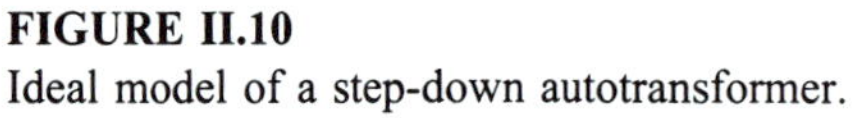

FIGURE II.10
Ideal model of a step-down autotransformer.

$$I_1 = \frac{N_2}{N_1} I_2 = \frac{90}{100} (36\angle -53.1°) = 32.4 \angle -53.1° \text{ A}$$

and

$$P_1 = (200)(32.4) \cos (53.1°) = 3.9 \text{ kW} = P_2$$

Finally,

$$I_3 = I_2 - I_1 = 36 \angle -53.1° - 32.4 \angle -53.1° = 3.6 \angle -53.1° \text{ A}$$

and

$$N_2 I_3 = 90(3.6\angle -53.1°) = (N_1 - N_2)I_1 = (100 - 90)(32.4\angle -53.1°)$$

$$= 324\angle -53.1° \text{ At}$$

as required.

II.8 THREE-PHASE TRANSFORMER CONNECTIONS

As we have seen in the text, generators and transmission lines of a power system are linked by three-phase transformers. The primary and secondary windings of single-phase transformers may be interconnected to obtain three-phase transformer banks.

Some of the factors governing the choice of connections are as follows:

1. Availability of a neutral connection for grounding, protection, or load connections.

2. Insulation to ground and voltage stresses.

3. Availability of a path for the flow of third-harmonic (exciting) currents and zero-sequence (fault) currents.

4. Need for partial capacity with one unit out of service.

5. Parallel operation with other transformers.

6. Operation under fault conditions.

7. Economic considerations.

Keeping these factors in mind, we now consider some of the three-phase transformer connections.

There are four primary possibilities for the connection of three-phase transformers using wye or delta configurations. These are the wye-wye, wye-delta, delta-wye, and delta-delta connections as shown in Figure II.11(*a*)-(*d*). In each case, these are simply three single-phase transformers (although they may be wound on a common core) connected with the polarities as indicated by the dots. Correspondingly lettered phases are assumed to be coupled. Each single-

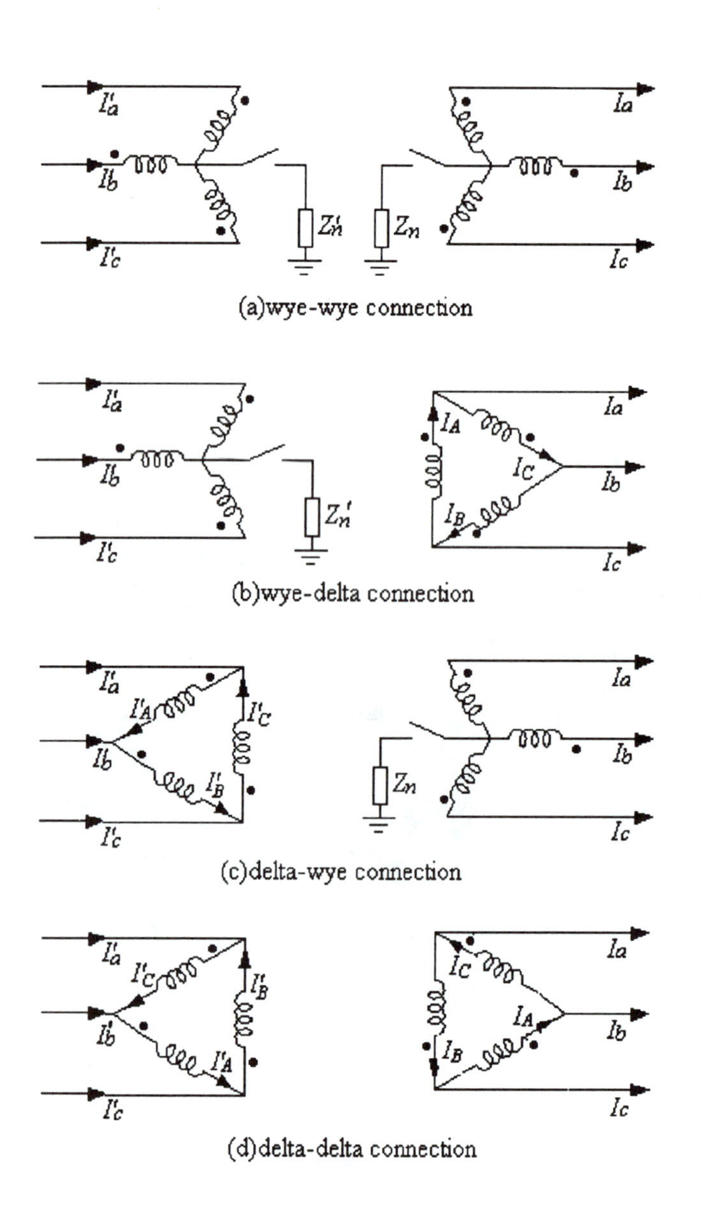

FIGURE II.11
Three-phase transformer connections.

phase transformer has its own kVA and voltage ratings (these are normally the same for each transformer) or the three-phase bank may be rated as a whole. In this latter situation, it is conventional to specify the total kVA rating (3 × the kVA rating of each phase) and the line-to-line voltage ratings (regardless of the connection). Thus a bank consisting of three 100 kVA 4160:13,800 V transformers connected in wye-delta would be equivalent to a three-phase transformer rated at 300 kVA 4160$\sqrt{3}$:13,800 V.

The junctions points of all wye-connected transformers may be ungrounded, solidly grounded, or connected to ground through an impedance Z_n. For balanced operation (positive- or negative-sequence voltages and currents), these connections may be ignored. However, they do provide a path for zero-sequence currents which must be considered in the analysis of unbalanced faults involving ground (see Chapter 5).

When working in the per-unit system, the turns ratios of the transformers do not have to be considered as impedances are the same referred to either side of the transformer. The magnitudes of voltages and currents are also independent of the turns ratio and are the same even if they are line or phase quantities. Thus, for example, the currents I_a, I_A and I_a' in Figure II.11(b) have equal magnitudes in per unit. The actual differences in these currents when specified in amperes is taken into account by the differences in base current values at each location. There are, however, phase shifts in the voltages and currents each time we change from line currents to phase currents of a delta (or vice versa) that are not provided for in the per-unit system. These phase shifts are of importance when referring unbalanced currents or voltages across a wye-delta or delta-wye transformer. Since the positive- and negative-sequence components of these quantities (Chapter 5) are shifted differently, this phase shift should be taken into account, even when working in per unit.

Following is a summary of the properties of zero-, positive- and negative-sequence currents flowing through the three-phase transformer connections *for the notations given.* Different notation schemes will yield different but equivalent results. Similar relationships may also be developed for positive- and negative-sequence voltages. In each case, I_{a0}, I_{a1} and I_{a2} are the symmetrical components of I_a, I_b, and I_c, I_{a0}', I_{a1}' and I_{a2}' are the symmetrical components of I_a', I_b' and I_c', etc. All quantities are assumed to be in per unit.

Wye-wye

If either Z_n or $Z_n' \to \infty$ (ungrounded wye), $I_{a0} = I_{a0}' = 0$ and this connection blocks zero-sequence current (the sum of three equal currents will equal zero only if the individual currents are zero). Otherwise, a zero-sequence current may flow through $3Z_n'$, $3Z_n$ and the zero-sequence transformer impedance. Also, $I_{a1}' = I_{a1}$ and $I_{a2}' = I_{a2}$ in per unit.

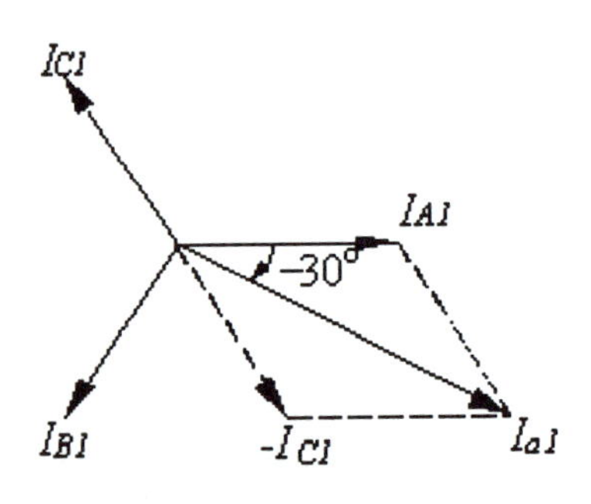

(a) Positive sequence

IB2 -IC2 Ia2 30° IA2 IC2

(b) Negative sequence

FIGURE II.12
Sequence currents.

Wye-delta

If $Z_n' \to \infty$, $I_{a0} = I_{a0}' = 0$ and this connection blocks zero-sequence current. Otherwise, zero-sequence current may flow through the transformer impedance and $3Z_n'$ to reference and $I_{a0}' = I_{A0}$ may have a non-zero value. I_{a0} must still equal zero. For positive- and negative-sequence currents, consider the phasor diagrams shown in Figure II.12(a) and (b). Since $I_{a1}' = I_{A1}$ and $I_{a2}' = I_{A2}$ in per unit, the delta currents only are shown on these diagrams. Now since $I_{a1} = I_{A1} - I_{C1}$, it may be seen that $I_{a1} = 1\ \angle{-30°}\ I_{A1} = 1\ \angle{-30°}\ I_{a1}'$ (remembering that the difference in magnitude is accounted for by the base values in the per-unit scheme). Similarly, $I_{a2} = 1\angle 30°\ I_{A2} = 1\ \angle 30°\ I_{a2}'$. We may therefore conclude that:

$$\begin{aligned} I_{a1}' &= 1\angle 30°\ I_{a1} \\ I_{a2}' &= 1\angle{-30°}\ I_{a2} \end{aligned} \tag{II.17}$$

Delta-wye

If $Z_n \to \infty$, $I_{a0}' = I_{a0} = 0$ and zero-sequence currents are blocked. Otherwise, zero-sequence current may flow through the transformer impedance and $3Z_n$ such that $I_{a0} = I_{A0}'$ may not be zero. In either case, $I_{a0}' = 0$. By a development similar to that for the wye-delta transformer, it may be shown that

$$\begin{aligned} I_{a1}' &= 1\ \angle{-30°}\ I_{a1} \\ I_{a2}' &= 1\ \angle 30°\ I_{a2} \end{aligned} \tag{II.18}$$

Delta-delta

In this case, zero-sequence current is blocked in the lines and $I_{a0}' = I_{a0} = 0$. It is possible that $I_{A1}' = I_{A1} \neq 0$. For positive- and negative-sequence currents, the delta-delta connection may always be labeled such that $I_{a1}' = I_{a1}$ and $I_{a2}' = I_{a2}$ and the total phase shift is zero.

Example II.3 A 60 MVA, 13,800 V wye-connected generator is connected to a three-phase transformer bank consisting of three 20 MVA, 13,800 to 240,000 V transformers connected in delta-wye. The transformer is unloaded when a single-line-to-ground short circuit having a fault current of 1800 A occurs on phase *a* of the high-voltage side of the transformer. The generator and wye side of the transformer are solidly grounded.

a) Determine the rating of this bank as a three-phase transformer.

b) Determine the current in phase *b* of the generator.

Solution

a) The rating of an equivalent three-phase transformer is 20 × 3 = 60 MVA with line voltages of 13,800:240,000 $\sqrt{3} \approx$ 13.8: 416 kV.

b) For a SLG fault,

$$I_{a0} = I_{a1} = I_{a2} = \frac{I_f}{3} = \frac{1800}{3} = 600 \text{ A}$$

Selecting a 60 MVA 13,800:416,000 base for this problem:

$$I_{\text{base}} = \frac{60{,}000{,}000}{\sqrt{3}\ (416{,}000)} = 83.3 \text{ A}$$

and

$$I_{a0} = I_{a1} = I_{a2} = \frac{600}{83.3} = 7.2 \text{ pu}$$

If a reference angle of −90° is now assumed for the currents

$$I_{a0} = I_{a1} = I_{a2} = -j7.2 \text{ pu}$$

Referring these currents across the delta-wye transformer (See Figure II.11(*c*) and Eqs. II.18)

$$I_{a0}' = 0$$

$$I_{a1}' = (1\angle -30°)\, I_{a1} = (1\angle -30°)(-j7.2) = 7.2\angle -120°$$

$$I_{a2}' = (1\angle 30°)\, I_{a2} = (1\angle 30°)(-j7.2) = 7.2\angle -60°$$

$$I_b' = I_{a0}' + a^2 I_{a1}' + a I_{a2}' = 0 + (1\angle 240°)(7.2\angle -120°) + (1\angle 120°)(7.2\angle -60°)$$

$$I_b' = 12.5\angle 90°$$

On the low voltage side of the three phase transformer

$$I_{\text{Base}} = \frac{60{,}000{,}000}{\sqrt{3}\,(13{,}800)} = 2510 \text{ A}$$

and

$$|I_b'| = (12.5)(2510) \approx 31.3 \text{ kA}$$

This result may be checked as follows (Figure II.11(c))

$$I_a = -j1800 \text{ A}\,, \quad I_b = I_c = 0$$

and

$$a = \frac{13{,}800}{240{,}000} = 0.0575$$

so that

$$I_A' = \frac{-j1800}{0.0575} = -j31.3 \text{ kA}$$

$$I_B' = I_C' = 0$$

and

$$I_a' = I_A' - I_C' = -j31.3\text{kA}$$

$$I_b' = I_B' - I_A' = j31.3 \text{ kA}$$

$$I_c' = I_C' - I_B' = 0$$

as before.

II.9 TAP-CHANGING TRANSFORMERS

As was mentioned in Chapter 3, tap-changing or regulating transformers may be used to control the voltage levels and power flows in a network. One side of the transformer has incremental tap positions which might be used to vary the voltage to perhaps ±10% of the nominal setting. Since base values in the per-unit system would normally be chosen using the 100% tap setting, any deviation from this value creates a transformer that does not conform to the chosen base voltages. As a result, this deviation must be accounted for by using a transformer turns ratio as shown in Figure II.13. In the Figure, the off-nominal

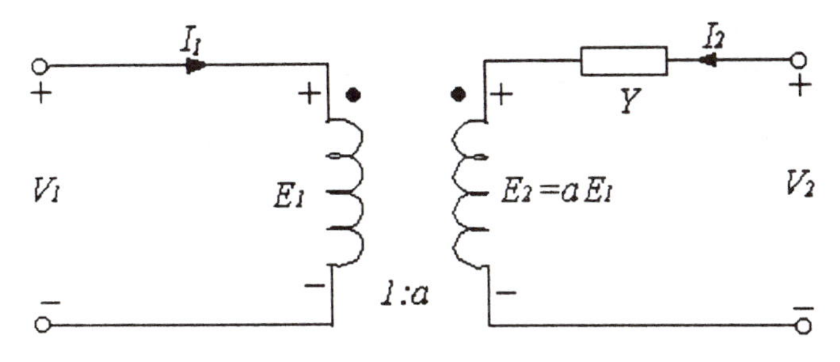

FIGURE II.13
Tap-changing transformer.

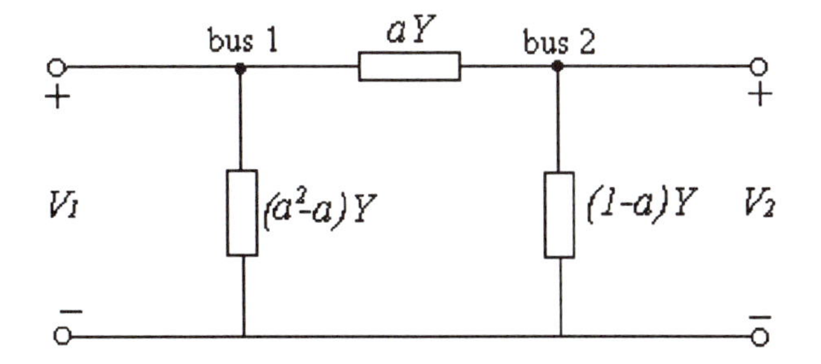

FIGURE II.14
Equivalent circuit for a voltage magnitude tap-changing transformer.

turns ratio is a ($a = 1$ for a 100% tap) and the reciprocal of the transformer series impedance is designated by the transformer admittance Y. The secondary (side 2) is the tap-changing side. In general, the turns ratio may be real (voltage magnitude changing transformer) or complex (changes in phase and magnitude). The development which follows pertains to transformers which are used to change voltage magnitude only such that a is real. In this case:

$$I_2 = (V_2 - aE_1)Y = YV_2 - aYV_1$$

and

$$I_1 = -aI_2 = a^2YV_1 - aYV_2$$

Now consider the $\mathbf{Y}_{\mathbf{bus}}$ matrix that relates I_1 and I_2 to V_1 and V_2

$$\begin{bmatrix} I_1 \\ I_2 \end{bmatrix} = \begin{bmatrix} a^2Y & -aY \\ -aY & Y \end{bmatrix} \begin{bmatrix} V_1 \\ V_2 \end{bmatrix}$$

Using the standard means for formation of $\mathbf{Y}_{bus}$, we arrive at the π-equivalent circuit for the voltage magnitude tap-changing transformer as shown in Figure II.14.

We now have an equivalent circuit that may be used to relate the input and output voltages of a voltage magnitude tap-changing transformer in a power flow or other analysis. The shunt admittances in Figure II.14 may be positive or negative depending upon whether $a < 1$ or $a > 1$. Note that when $a = 1$, this model reduces to the standard equivalent circuit.

APPENDIX III

SYNCHRONOUS MACHINES

We have seen in the text that synchronous machines are an essential component of a power system. They are among the three most common types of electric machines, the other two being dc commutator machines and polyphase induction machines. The bulk of electric power for everyday use is produced by three-phase synchronous generators, which are the largest single-unit electric machines in production. For instance, synchronous generators with power ratings of several hundred megavolt-amperes (MVA) are fairly common. These are called synchronous machines because they operate at constant speeds and constant frequencies under steady-state conditions. Like most rotating machines, synchronous machines are capable of operating both as a motor and as a generator. They are used as motors in constant-speed drives and, where a variable speed drive is required, a synchronous motor may be used with an appropriate frequency changer. As generators, several synchronous machines often operate in parallel in a power station. While operating in parallel, the generators share the load with each other. At a given time, one of the generators may be allowed to "float" on the line as a synchronous motor on no-load.

The operation of a synchronous generator is based on Faraday's law of electromagnetic induction, and in an ac synchronous generator the generation of emf's is by the relative motion of conductors and magnetic flux. The two basic parts of a synchronous machine are the magnetic field structure, carrying a dc-excited winding, and the armature. The armature has a three-phase winding in which the ac emf is generated. Almost all modern synchronous machines have stationary armatures and rotating field structures. The dc winding on the rotating field structure is often connected to an external source through slip rings and brushes. Some field structures do not have brushes but, instead, have brushless excitation by rotating diodes.

III.1 OPERATION OF A SYNCHRONOUS GENERATOR

As mentioned above, the operation of the synchronous generator is based on Faraday's law of electromagnetic induction. The law states that an emf is induced in a circuit placed in a magnetic field if either (1) the magnetic flux linking the circuit is time varying, or (2) there is a relative motion between the circuit and the magnetic field such that the conductors comprising the circuit cut across the magnetic flux lines. The first form of the law, stated as (1), is the basis of operation of transformers. The second form, stated as (2), is the basic principle of operation of electric generators.

We may apply Faraday's law in the form of voltage-induced = time rate of change of flux linkage. Consider a three-phase round-rotor 2-pole machine shown in Figure III.1*(a)*. We assume that the flux-density distribution due to the dc field excitation is sinusoidally distributed in the airgap. This implies that the flux per pole [Figure III.1(*a*)] is given by

$$\phi = \int_0^{\pi} B_m \sin\theta l \frac{D}{2} \, d\theta = B_m l D \tag{III.1}$$

where

B_m = maximum value of the airgap flux density
ϕ = flux per pole produced by the field winding
D = diameter of the coil aa'
l = axial length of the machine core.

As the rotor is moved by an angle ωt (Figure III.1), the flux linking the coil is $\lambda = N\phi \cos \omega t$. If the rotor rotates at a constant angular velocity ω, the emf induced in aa' is

$$\begin{aligned} e_a &= -\frac{d\lambda}{dt} = \omega N\phi \sin\omega t = 2\pi f N\phi \sin\omega t \\ &= E_m \sin\omega t \end{aligned} \tag{III.2}$$

where $E_m = 2\pi f N\phi$ and $f = \omega/2\pi$ is the frequency of the induced voltage.

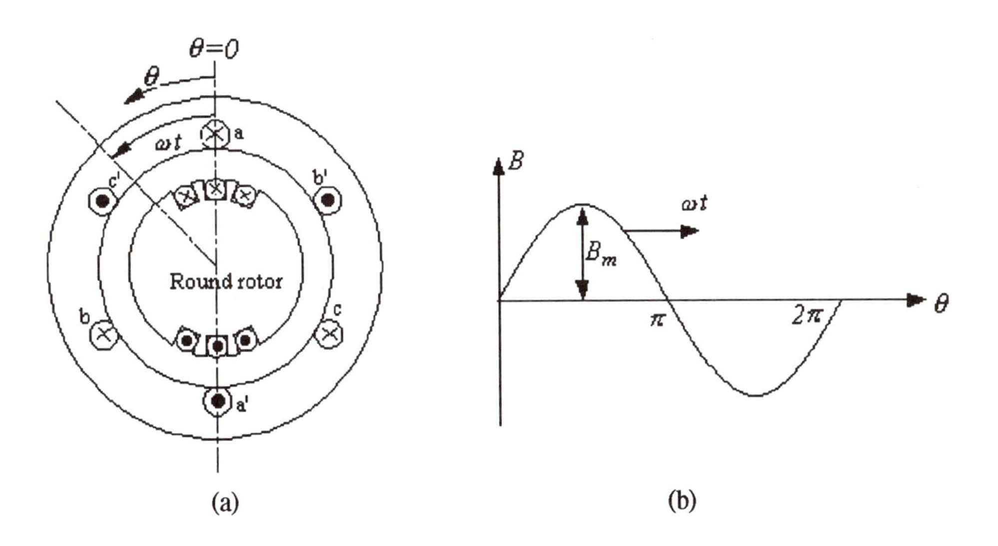

FIGURE III.1
(*a*) A round or cylindrical-rotor 2-pole synchronous machine. (*b*) Flux-density distribution due to the field excitation.

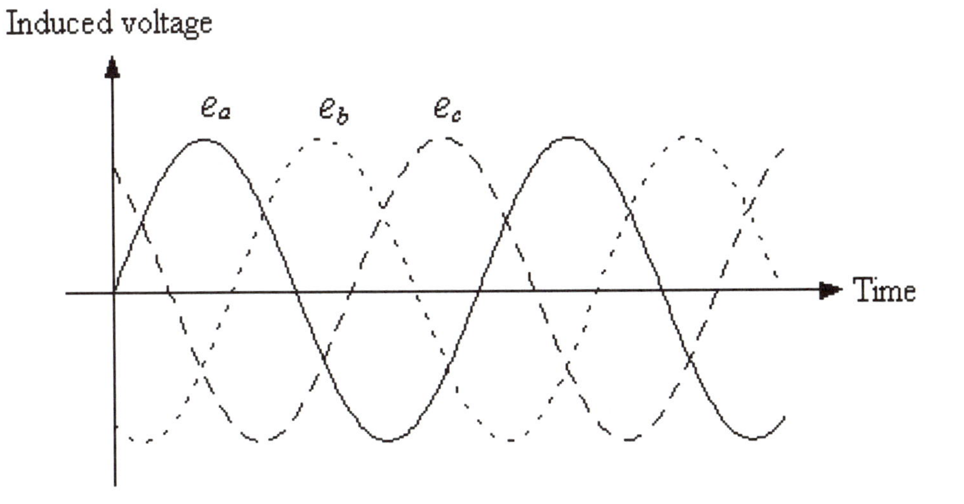

FIGURE III.2
Three-phase voltages produced by a three-phase synchronous generator.

Because phases b and c are displaced from phase a by ±120° [Figure III.1(a)], the corresponding voltages may be written as

$$e_b = E_m \sin(\omega t - 120°)$$

$$e_c = E_m \sin(\omega t + 120°)$$

These voltages are sketched in Figure III.2.

A similar analysis may be performed on a machine with any even number of poles with the major difference that the electrical radian frequency $\omega = 2\pi f$ is no longer equal to the rotor mechanical speed in radians/s which we may now call $\omega_m = (2/P)\omega$ where P = the number of poles in the machine. In addition, we must also realize that the mechanical angle $\theta_m = (2/P)\theta$ where θ is the electrical angle as in Figure III.1.

III.2 CONSTRUCTION FEATURES

From the preceding discussion we conclude that a synchronous machine must have (in principle) a load (or armature) winding and a source for the magnetic flux. This source of flux may be either a permanent magnet or an electromagnet (provided by the dc field excitation). Some of the factors that dictate the form of construction of a synchronous machine follow.

1. *Form of Excitation.* We recall from the preceding remarks that the field structure is usually the rotating member of a synchronous machine and is supplied with a dc-excited winding to produce the magnetic flux. This dc excitation may be provided by a self-excited dc generator mounted on the same shaft as the rotor of synchronous machine. Such a generator is known as the *exciter.* The direct current thus generated is fed to the synchronous machine field winding. Brushes and slip rings are required in such machines. In slow-speed machines with large ratings, such as hydroelectric generators, the exciter may not be self-excited. Instead, a pilot exciter, which may be self-excited or may have a permanent magnet, activates the exciter. The maintenance problems of direct-coupled dc generators impose a limit on this form of excitation at about a 100-MW rating.

An alternative form of excitation is provided by silicon diodes and thyristors. The two types of solid-state excitation systems are:

(a) Static systems that have stationary diode or thyristor rectifiers, in which the current is fed to the rotor through slip rings.

(b) Brushless systems that have shaft-mounted rectifiers that rotate with the rotor, thus avoiding the need for brushes and slip rings.

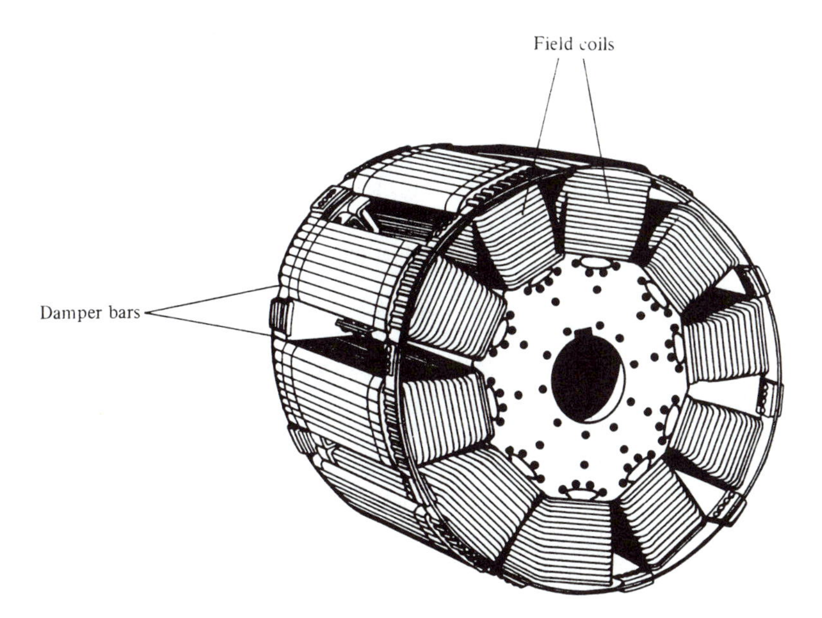

FIGURE III.3
A salient rotor showing the field windings and damper bars (shaft not shown).

2. *Field Structure and Speed of the Machine.* We have already mentioned that the synchronous machine is a constant-speed machine. This mechanical speed, known as synchronous speed, n_s in rpm, is given by

$$n_s = \frac{120f}{P} \tag{III.3a}$$

If (III.3*a*) is expressed in mechanical radians/s

$$\omega_m = \left(2\pi \ \frac{\text{mech rad}}{\text{rev}}\right)\left(\frac{1}{60} \ \frac{\text{min}}{\text{s}}\right)\left(\frac{120f}{P}\right) = \frac{4\pi f}{P} \tag{III.3b}$$

Thus a 60-Hz, two-pole synchronous machine must run at 3600 rpm, whereas the synchronous speed of a 12-pole, 60-Hz machine is only 600 rpm. The rotor field structure consequently depends on the speed rating of the machine. Therefore, turbogenerators, which are high-speed machines, have *round* or *cylindrical rotors.* Hydroelectric and diesel-electric generators are low-speed machines and have *salient pole rotors.* (Figure III.3). Such rotors are less expensive to fabricate than round rotors. They are not suitable for large, high-speed machines, however, because of the excessive centrifugal forces and mechanical stresses that develop at speeds around 3600 rpm.

Another feature in the construction of a synchronous machine stems from the mounting of the rotor. For example, a round-rotor, turbine-driven machine or a salient-rotor, diesel, engine-driven machine has a horizontally mounted rotor. A waterwheel-driven machine invariably has a vertically mounted, salient pole rotor.

3. *Stator.* The stator of a synchronous machine has distributed slot-embedded three-phase armature windings. There is essentially no difference between the stator of a round-rotor machine and that of a salient-rotor machine. The stators of waterwheel generators, however, usually have a large-diameter armature compared to other types of generators. The stator core consists of punchings of high-quality laminations having slot-embedded lap windings.

4. *Cooling.* Because synchronous machines are often built in extremely large sizes, they are designed to carry very large currents. A typical armature current density may be of the order of 10 A/mm^2 in a well-designed machine. Also, the magnetic loading of the core is such that it reaches saturation in many regions. The severe electric and magnetic loadings in a synchronous machine produce heat that must be appropriately dissipated. Thus, the manner in which the active parts of a machine are cooled determines its overall physical structure. In addition to air, some of the coolants used in synchronous machines include water, hydrogen, and helium.

5. *Damper Bars.* So far, we have mentioned only two electrical windings of a synchronous machine: the three-phase armature winding and the field winding. We also pointed out that, under steady-state, the machine runs at a constant speed, that is, at the synchronous speed. However, like other electric machines, a synchronous machine undergoes transients during starting and under abnormal conditions. During transients, the rotor may undergo mechanical oscillations and its speed deviates from the synchronous speed, which is an undesirable phenomenon. To overcome this, an additional set of windings, resembling the cage of an induction motor, is mounted on the rotor. This winding is called the damper winding and is shown in Figure III.3. When the rotor speed is different from the synchronous speed, currents are induced in the damper windings. The damper winding acts like the cage rotor of an induction motor, producing a torque to restore the synchronous speed. Also, the damper bars provide a means of starting the machine when it is run as a motor. Otherwise, it would not be self-starting.

III.3 STEADY-STATE ANALYSIS OF SYNCHRONOUS MACHINES

We now consider the operation of a synchronous machine under steady state. The analysis of a round-rotor machine is somewhat different from the procedure for a salient pole machine. In all cases, the analysis makes use of the machine parameters, which must be identified before proceeding with the analytical details.

Since, for now, we are considering only the steady-state behavior of the machine, circuit constants of the field and damper windings need not be considered. The presence of the field winding will be denoted by the flux produced by the field excitation. Turning to the armature winding, we will represent it on a per-phase basis. Obviously, the armature winding has a resistance. But the ohmic value of this resistance must include the effects of the operating temperature and the alternating currents flowing in the armature conductors (causing skin effect, for instance). As a consequence, the value of the armature resistance becomes larger compared to its dc resistance. The larger value of the resistance is known as the *effective resistance* of the armature and is denoted by R_a. An approximate value of R_a is 1.6 times the dc resistance.

Next, we consider the reactances pertaining to the armature winding. First, the leakage reactance is caused by the leakage fluxes linking the armature conductors only because of the currents in the conductors. These fluxes do not link with the field winding and are therefore not mutual fluxes. As in an induction motor, for convenience in calculation, the leakage reactance is divided into (1) end-connection leakage reactance, (2) slot-leakage reactance, (3) tooth-top and zigzag leakage reactance, and (4) belt-leakage reactance. All of these components are not significant in every synchronous machine. In most large machines the last two reactances are a small portion of the total leakage reactance.

Flux paths contributing to the end-connection leakage and slot-leakage reactances are shown in Figure III.4*(a)* and *(b)*, respectively. We denote the total leakage reactance of the armature winding per phase by X_l. Now, to proceed with the analysis, let us first consider a round-rotor synchronous generator. We will also introduce the concept of *synchronous reactance*, the most important parameter in determining the steady-state characteristics of a synchronous machine.

PERFORMANCE OF A ROUND-ROTOR SYNCHRONOUS GENERATOR. At the outset, we wish to point out that we will study the machine on a per-phase basis, implying a balanced operation. Thus, let us consider a round-rotor machine operating as a generator on no-load. Let the open-circuit phase voltage be E_g for a certain field current I_{fld}. Here, E_g is the

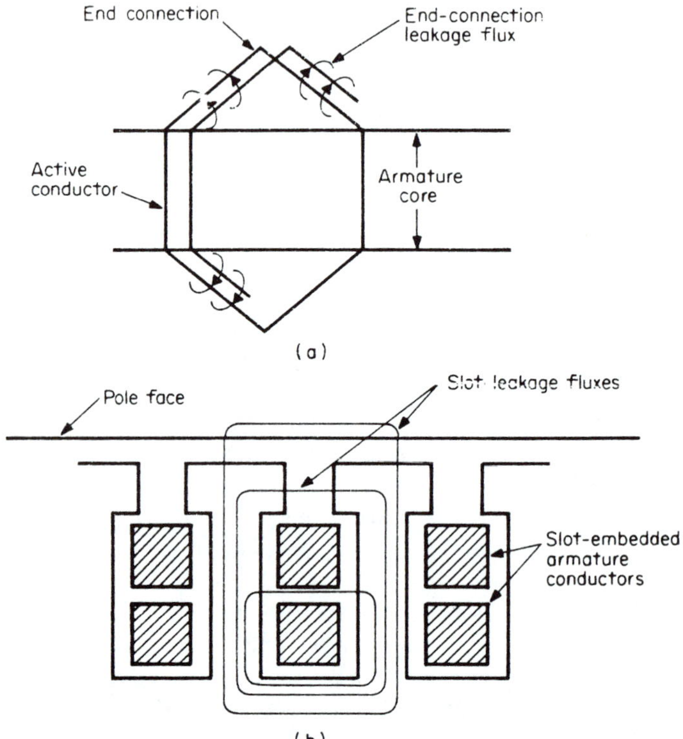

FIGURE III.4
(a) End-connection leakage flux path. *(b)* Slot-leakage flux paths.

phasor value at the internal voltage of the generator which would be obtained from (III.2). We assume that I_{fld} is such that the machine is operating in an unsaturated condition. Next, we short-circuit the armature at the terminals, keeping the field current unchanged (at I_{fld}), and measure the armature phase current I_a after any transients have subsided. In this case, the entire internal voltage E_g is dropped across the internal impedance of the machine. In mathematical terms,

$$E_g = I_a Z_s$$

and Z_s is known as the *synchronous impedance*. One portion of Z_s is R_a and the other a reactance, X_s, known as *synchronous reactance*; that is,

$$Z_s = R_a + jX_s = R_a + j(X_a + X_l) \tag{III.4}$$

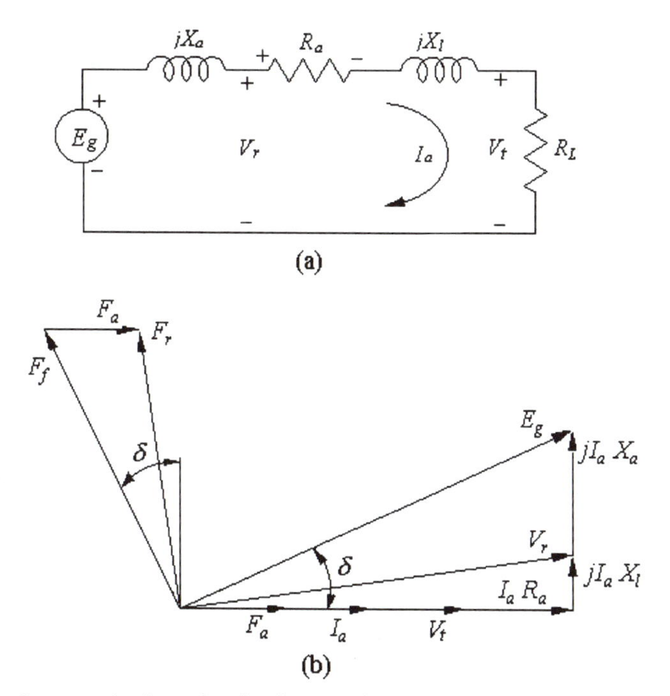

(*a*) Per phase equivalent circuit of a round rotor generator at unity power factor. (*b*) Phasor diagram for a round-rotor generator at unity power factor (to define X_S).

FIGURE III.5

In (III.4) X_s is greater than the armature leakage reactance X_l, mentioned earlier. Where does the additional reactance X_a come from? We will answer this question in the following discussion.

Let the generator supply a phase current I_a to a load at unity power factor and at a terminal voltage V_t V/phase. This is shown in the per-phase equivalent circuit of Figure III.5(*a*) and the phasor diagram of Figure III.5(*b*). These figures show that the voltage V_r is the phasor sum of V_t and the drop due to the armature resistance and armature leakage reactance. Remembering that mmf $\mathcal{F}$ = NI where N is the number of turns in a coil, we now have two mmf's—$\mathcal{F}_a$ attri-butable to the armature current and $\mathcal{F}_f$ attributable to the field current present in the machine. To find the mmf $\mathcal{F}_r$ that produces the voltage V_r, refer to the magnetization characteristic of the generator, in Figure III.6, which shows $\mathcal{F}_r$ corresponding to V_r.

The mmf and flux produced by a current is in phase with the current. As dictated by $e = Nd\phi/dt$, however, the voltage induced by a certain flux is behind the mmf by 90°. Therefore, we lay $\mathcal{F}_r$ ahead of V_r by 90° and $\mathcal{F}_a$ in phase with I_a, as depicted in Figure III.5(*b*). The mmf $\mathcal{F}_a$ is known as the *armature reaction mmf.* Sufficient mmf must be supplied by the field to overcome $\mathscr{F}_a$ such that we have a net $\mathscr{F}_r$, to produce V_r. The mmf $\mathscr{F}_r$ must therefore be the vector sum of $\mathscr{F}_f$ and $\mathscr{F}_a$ as shown in Figure III.5(*b*). Corresponding to this field mmf, $\mathscr{F}_f$, the internal voltage of the generator is E_g, as shown in Figure III.5(*b*).

From Figures III.5(*a*) and (*b*), it is clear that

$$E_g = V_t + I_a(R_a + jX_s) \qquad \text{(III.5)}$$

where X_s, the synchronous reactance, is defined in (III.4). The "extra" reactance, in addition to X_l (Figure III.5), is introduced by the armature reaction. Therefore, the synchronous reactance is the sum of the armature leakage reactance and armature reaction reactance.

In an actual synchronous machine, except in very small ones, we almost always have $X_s >> R_a$, in which case $Z_s \cong jX_s$. With this restriction, the similarity of (III.5) to (5.4) and (5.5) should be evident. The only differences are that we have used I_a to emphasize that the load current I_L is actually the armature current and we are representing steady-state conditions instead of subtransient or transient operation. (See Section III.4). The equivalence of X_s to the direct axis reactance X_d will become apparent when we discuss salient pole machines later in this appendix.

Among the steady-state characteristics of a synchronous generator, its voltage regulation and power-angle characteristics are the most important ones. As for a transformer, we define the voltage regulation of a synchronous generator at a given load as

$$\text{percent voltage regulation} = \frac{|E_{g0}| - |V_t|}{|V_t|} \times 100 \qquad \text{(III.6)}$$

where $|V_t|$ is the terminal voltage on load and $|E_{g0}|$ is the no-load terminal voltage. Clearly, for a given V_t, we can find E_g from (III.5) and, hence, the voltage regulation.

Neglecting R_a, phasor diagrams shown in Figure III.7 are drawn for lagging and leading power factors of a synchronous generator.

POWER ANGLE CHARACTERISTICS. The angle by which E_g lends V_t in Figure III.7 is defined as the electrical power angle, δ. To justify this definition, we reconsider Figure III.7(*a*), from which we obtain

$$|I_a|\,|X_s|\cos\phi = |E_g|\,\sin\delta \qquad \text{(III.7)}$$

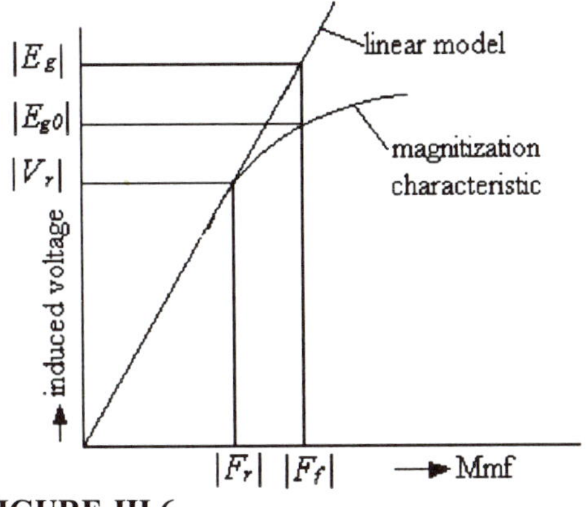

FIGURE III.6
Magnetization characteristic of a synchronous generator.

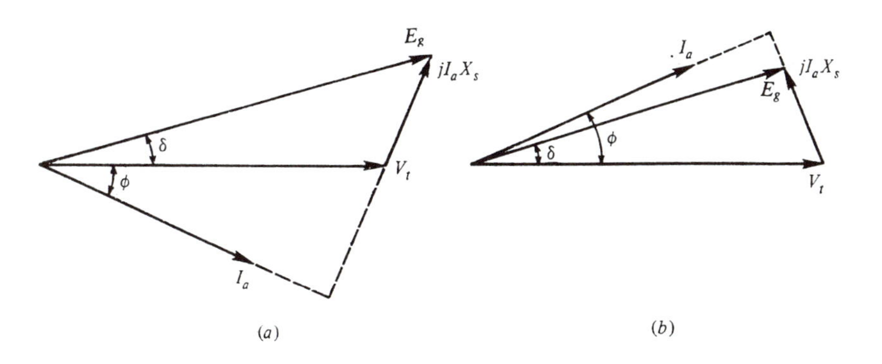

FIGURE III.7
Phasor diagrams. (*a*) Lagging power factor. (*b*) Leading power factor.

Continuing to neglect R_a, the internally developed power, P_e is equal to the output (or developed) power $P_d = |V_t||I_a| \cos \phi$. Hence, in conjunction with (III.7), we get

$$P_e = P_d = \frac{|E_g||V_t|}{X_s} \sin\delta = P_{\max} \sin\delta \text{ W/phase} \tag{III.8}$$

which shows that the internal power of the machine is proportional to sin δ. Equation (III.8) is often said to represent the power-angle characteristic of a synchronous machine. The assumption of neglecting the armature resistance is a valid assumption in most power system studies. Thus the equivalent circuit of a round-rotor synchronous machine under steady-state becomes as shown in Figure III.8(*a*) and the corresponding power angle characteristic is shown in Figure III.8(*b*). In Figure III.8(*a*), we have shown the terminal voltage V_t, the

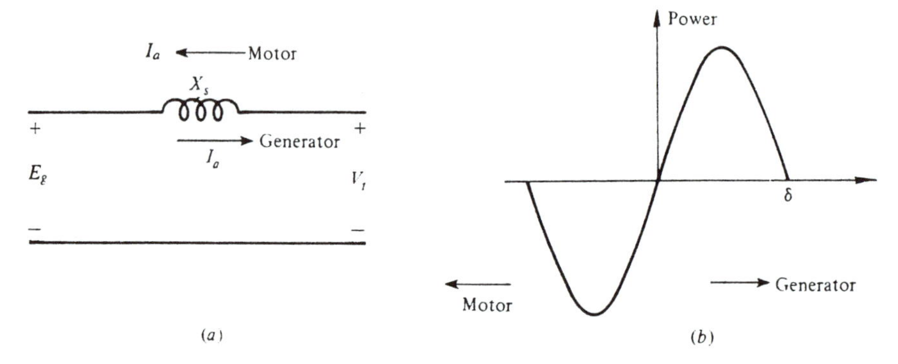

FIGURE III.8
(*a*) An approximate equivalent circuit. (*b*) Power-angle characteristics of a synchronous machine.

internal excitation voltage E_g, and the armature current I_a going "into" the machine or "out of" it, depending on the mode of operation—"into" for motor and "out of" for generator. In Figure III.8(*b*) we show the power-angle characteristics as given by (III.8). Here positive power and positive δ imply generator operation, while a negative δ corresponds to a motor operation. Because δ is the angle by which E_g leads V_t, E_g is ahead of V_t in a generator, whereas in a motor, V_t is ahead of E_g.

SALIENT POLE SYNCHRONOUS MACHINES. In the preceding discussion we analyzed the round-rotor machine and made extensive use of the machine parameter, which we defined as synchronous reactance. Because of saliency, the reactance measured at the terminals of a salient-rotor machine will vary as a function of the rotor position. This is not so in a round-rotor machine.

To overcome this difficulty, we use the *two-reaction theory* proposed by André Blondel. The theory proposes to resolve the given armature mmf's into two mutually perpendicular components, with one located along the axis of the rotor salient pole, known as the *direct* (or *d*) axis and with the other in quadrature and known as the *quadrature* (or *q*) axis. The *d*-axis component of the mmf, $\mathscr{F}_d$, is either magnetizing or demagnetizing; the *q*-axis component, $\mathscr{F}_q$, results in a cross-magnetizing effect. Thus, if the amplitude of the armature mmf is $\mathscr{F}_a$, then

$$\mathscr{F}_d = \mathscr{F}_a \sin\psi$$

and

$$\mathscr{F}_q = \mathscr{F}_a \cos\psi$$

where ψ is the phase angle between the armature current I_a and the internal (or excitation) voltage E_g. In terms of space distribution, the mmf's and ψ are shown in Figure III.9. The effects of $\mathscr{F}_d$ and $\mathscr{F}_q$ are that they produce magnetic fluxes which will induce voltages. They may therefore be associated with reactances X_d and X_q, respectively. To illustrate, we consider a salient pole generator having a terminal voltage V_t, supplying a load of lagging power factor (cos ϕ), and drawing a phase current I_a. For the operating condition (given field current), we also know the no-load voltage, E_g, from the magnetization characteristic. These parameters (V_t, E_g, I_a, and ϕ) are shown in Figure II.10(*a*). To construct this diagram, we choose E_g as the reference phasor and neglect the armature resistance. We can resolve I_a into its *d*- and *q*-axis components, I_d and I_q, respectively, so also the I_aX_s drop shown in Figure III.7(*a*) is replaced by the I_dX_d and I_qX_q drops as shown in Figure III.10(*a*). Thus, the phasor diagram is complete. We can associate physical meanings to the reactances X_d and X_q. These are, respectively, the *direct-axis* and *quadrature-axis* reactance of a salient pole machine. These reactances can be measured experimentally. The preceding discussions are valid on a per-phase basis for a balanced machine. The phasor diagram (Figure III.10(*a*)) depicts the complete steady-state performance characteristics of the machine. For example, to obtain the power-angle characteristics of a salient pole machine, operating either as a generator or as a motor, we refer to Figure III.10(*a*). From this figure we have, neglecting R_a) and the internal losses,

$$\text{power output} = |V_t|\,|I_a|\cos\phi = \text{developed power} = P_d \tag{III.9}$$

and

$$|I_q|X_q = |V_t|\sin\delta \tag{III.10}$$

$$|I_d|X_d = |E_g| - |V_t|\cos\delta \tag{III.11}$$

Also,

$$|I_d| = |I_a|\sin(\delta + \phi) \tag{III.12}$$

and

$$|I_q| = |I_a|\cos(\delta + \phi) \tag{III.13}$$

Combining (III.10) - (III.13) and solving for $|I_a|$ cos ϕ yields

$$|I_a|\cos\phi = \frac{|E_g|}{X_d}\sin\delta + \frac{|V_t|}{2X_q}\sin 2\delta - \frac{|V_t|}{2X_d}\sin 2\delta \tag{III.14}$$

Finally, combining (III.14) and (III.9) gives

$$P_d = \frac{|V_t|\,|E_g|}{X_d}\sin\delta + \frac{|V_t|^2}{2}\left(\frac{1}{X_q} - \frac{1}{X_d}\right)\sin 2\delta \tag{III.15}$$

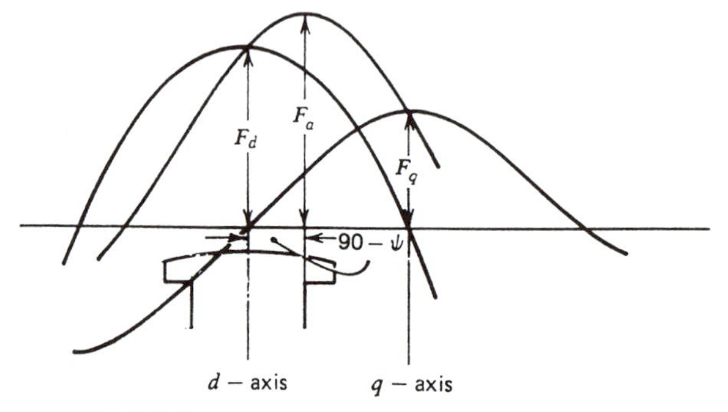

FIGURE III.9
Armature mmf and its d and q components.

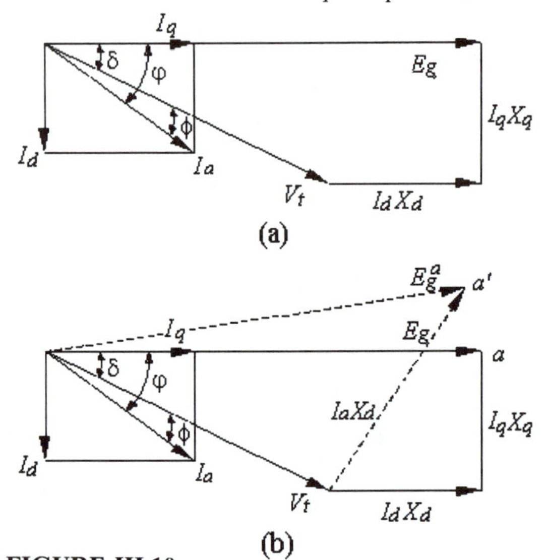

FIGURE III.10
Phasor diagram of a salient pole machine.

This variation of the developed power, P_d, as a function of the power angle δ is shown in Figure III.11. Notice that the resulting power is composed of power due to saliency—the second term in (III.15)—and of power due to field excitation—the first term in (III.15). Clearly, when $X_d = X_q$, the machine has no saliency and only the first term in (III.15) is nonzero, which represents the power-angle characteristic of a round-rotor machine. On the other hand, if there is no field excitation, implying $|E_g| = 0$, the first term in (III.15) reduces to zero. We then have the power-angle characteristics of a reluctance machine as given by the second term. As in a round-rotor machine (discussed earlier), the power-angle characteristics given by (III.15) reflect generator as well as motor operation. The term δ is positive for the former and negative for the latter.

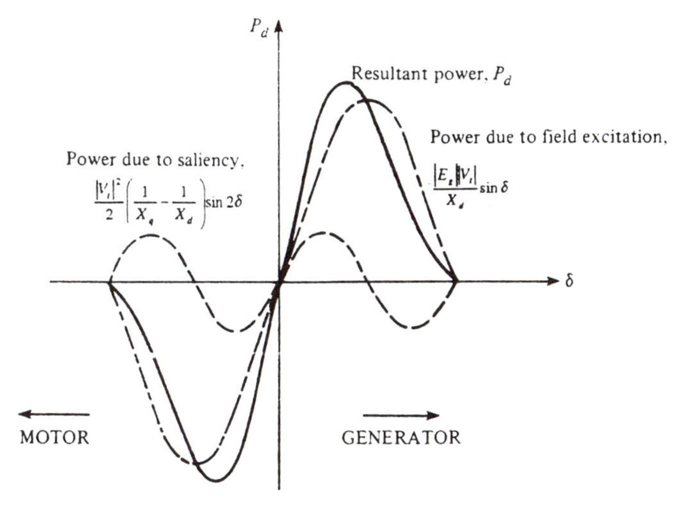

FIGURE III.11
Power-angle characteristics of salient pole machine.

For the case of the round rotor synchronous machine, it was possible (with armature resistance neglected) to represent the internal impedance of the machine by its synchronous reactance as shown in Figure III.8(*a*). This is not strictly possible for the salient pole situation as this reactance has been broken down into direct and quadrature axis components. However, it seems clear that the direct axis component, X_d, is along the axis of easy magnetization and should be the major factor. To investigate this possibility, consider the changes that would occur in the phasor diagram of Figure III.10(*a*) if we were to represent the salient-pole machine by X_d alone. In this case, $E_g = V_t + jI_aX_d$ and we obtain the E_g^a instead of E_g as shown in Figure III.10(*b*). Intuitively, it may be seen that the magnitude of E_g^a is close to the magnitude of E_g because aa' is very nearly perpendicular to E_g. In practice and for normal operating conditions, the effects of salient poles are normally of minor importance when evaluating power, current and voltage interrelationships, and the salient pole synchronous machine may be represented by X_d just as the round-rotor machine is represented by X_s. Thus, Figure III.8(*a*) is unchanged with the exception that X_s is replaced by X_d.

III.4 TRANSIENTS IN SYNCHRONOUS MACHINES

In the preceding sections we focused our attention on the steady-state behavior of synchronous machines. In this section we will briefly review some cases involving transients in synchronous machines. Of particular interest is a sudden short circuit at the armature terminals of a synchronous generator. There are numerous other cases involving transients in synchronous machines, but these will not be considered at this point.

We know from earlier considerations that the performance of a machine, for a given operating condition, can be determined if the machine parameters are known. For instance, we have already expressed the steady-state power-angle characteristics of a salient pole synchronous machine in terms of the *d*-axis and *q*-axis reactances. Similarly, the constants by which transient behavior of a synchronous machine is known are the transient and subtransient reactances and pertinent time constants. We now define these quantities while relating them to the study of an armature short circuit similar to those considered in Chapter 5.

Sudden Short Circuit at the Armature Terminals

At the outset, we assume no saturation and neglect the resistances of all the windings—the armature, field, and damper windings. Thus, only the inductances remain, implying that the flux linking a closed circuit (or winding) cannot change instantaneously, as dictated by the constant flux linkage theorem. Stated differently, the sum of the flux linkages is initially constant for each winding. With these assumptions in mind, we assume that a three-phase short circuit suddenly occurs on the armature of an unloaded generator having a dc field current I_{fld}. The generator is rotating at synchronous speed prior to the fault so there are no prefault voltages induced in the damper winding and the damper currents are zero (see Section III.2). Upon occurrence of the short circuit, large 3-phase armature currents will begin to flow creating an mmf $\mathscr{F}_a$ as shown in Figure III.5(*b*). This suddenly produced mmf will act to change the mutual flux linkages between the armature and both the field and the damper windings in violation of the constant flux linkage theorem. Currents must therefore be induced in the field and damper circuits which will oppose the action to immediately change these flux linkages. An approximate representation of the field and damper currents that might flow during this transition period is shown in Figure III.12.

Note that in Figure III.12 the induced currents in both rotor circuits are shown to decay exponentially to zero as the time from inception of the fault increases. This is a natural occurrence as there is no force to maintain the induced currents and they will subside in accordance with the time constant of an R-L circuit. Normal design practices for synchronous machines dictate that the time constant for the damper windings will be smaller than that of the field circuit. These current decays, when plotted on a logarithmic scale, relate to the decrements shown in Figure 5.4 of the text. Note also that while we have neglected resistances in the determination of the initially induced currents, the field and damper winding resistances play an important role in the decay time associated with the respective currents.

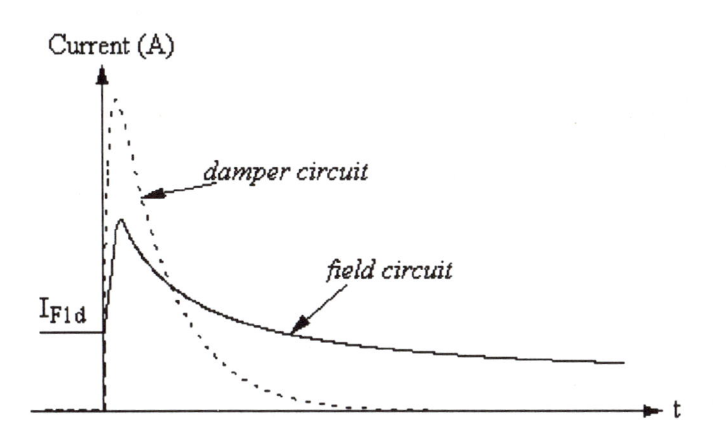

FIGURE III.12
Approximate field and damper currents. Three-phase armature short-circuit at $t = 0$.

The result of these temporarily induced rotor winding currents is that larger armature short circuit currents will flow than would be predicted by our steady-state machine model of Figure III.8(*a*). To include this observation in our machine model, and remembering that the major phenomenon in a salient-pole machine occurs in the direct axis, we may simply replace the synchronous reactance (X_s or X_d) in Figure III.8(*a*) with a reactance X_d'' during the time when both damper and field winding induced currents are flowing and with X_d' after the damper current has decayed and the induced field current is still significant. Of course, we will be back to using X_d (or X_s) when the induced currents in both rotor windings have returned to zero. These reactances are the X_d'', X_d' and X_d defined in (5.1) - (5.3) and it seems evident that $X_d'' < X_d' < X_d$. Should the generator be loaded prior to the occurrence of the short circuit, (5.4) or (5.5) may be used to compute the internal voltage in the machine model. Typical values for the machine reactances are given in Table III.1.

TABLE III.1
Per-unit synchronous machine reactances

Reactance	Salient Pole Machine	Round-Rotor Machine
X_d	1.0 to 1.25	1.0 to 1.2
X_d'	0.35 to 0.40	0.15 to 0.25
X_d''	0.20 to 0.30	0.10 to 0.15

As a final note, it should be remembered that the constant flux linkage theorem pertains to the individual armature windings as well as to the mutual effects between the armature and rotor windings. Thus an exponentially decaying component of dc current will normally appear on these windings also as discussed in Chapter 5 and shown for one armature phase in Figure 5.27. These dc currents will create ripples in the induced rotor circuit currents shown in Figure III.12. This is not a significant effect for model development and we may continue to handle these dc currents separately as discussed in Chapter 5.

APPENDIX IV

POWER TRANSMISSION LINES

In Chapter 3 we considered the basic power system components: transformers, rotating electric machines, and transmission lines. Transmission lines physically integrate the output of generating plants and the requirements of customers by providing pathways for the flow of electric power between various circuits in the system. These circuits include those between generating units, between utilities, and between generating units and substations (or load centers). For our purposes, we consider a transmission line to have a sending end and a receiving end, and to have a series resistance and inductance and shunt capacitance and conductance as primary parameters. We will classify transmission lines as short, medium-length, and long lines. In a short line, the shunt effects (conductance and capacitance) are negligible. Often, this approximation is valid for lines up to 80 km long. In a medium-length line, the shunt capacitances are lumped at a few predetermined locations along the line. A medium-length line may be anywhere between 80 and 240 km in length. Lines longer than 240 km are considered as long lines which are represented by (uniformly) distributed parameters. The operating voltage of a transmission line often depends on the length of the line. Some of the common operating voltages for transmission lines are 115, 138, 230, 345, 500, and 765 kV. Whereas most transmission lines operate on ac, with the advent of high-power solid-state conversion equipment, high-voltage dc trans-

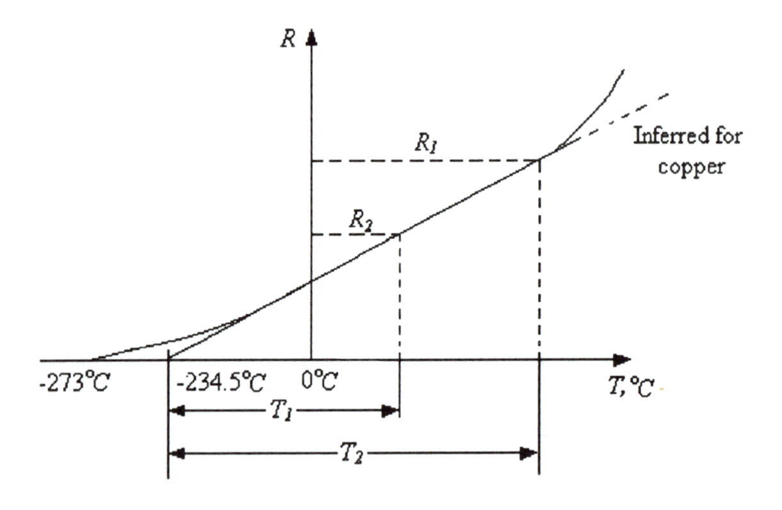

FIGURE IV.1
Variation of resistance of copper conductor with temperature.

mission lines are sometimes used for long distances exceeding 600 km (375 miles).

IV.1 TRANSMISSION-LINE PARAMETERS

Resistance

The first transmission-line parameter to be considered is the resistance of the conductors. Resistance is the cause of I^2R loss in the line, and also results in an IR-type voltage drop. The dc resistance R of a conductor of length l (in m) and area of cross section A (in m^2) is given by

$$R = \rho \frac{l}{A} \quad \Omega \tag{IV.1}$$

where ρ is the *resistivity* of the conductor in Ω-m. Whereas in (IV.1) we have used SI units, sometimes English units are also used. In these units the area A is expressed in circular mils (cmil). One inch equals 1000 mils, and a circular mil is the area of a circle of 1 mil in diameter. The resistivity in English units is given in Ω-cmil/ft.

The dc resistance of a conductor is affected only by the operating temperature of the conductor, linearly increasing with temperature. However, when operating on ac, the current-density distribution across the conductor cross section becomes nonuniform, and is a function of the ac frequency. This phenomenon is known as the *skin effect*, and as a consequence, the ac resistance of a conductor is higher than its dc resistance. Approximately at 60 Hz, the ac resistance of a transmission line conductor may be 5 to 10 percent higher than its dc resistance.

The variation of conductor resistance with temperature is expressed by the temperature coefficient of resistance α. Explicitly, the resistance R_T at a temperature T°C is related to the resistance R_0 at 0°C by

$$R_T = R_0 (1 + \alpha_0 T) \tag{IV.2}$$

where α_0 is the temperature coefficient at 0°C. This relation is depicted for copper in Figure IV.1, which also shows the inferred absolute zero for copper.

Alternatively, the temperature dependence of a resistance is given by

$$R_2 = R_1 [1 + \alpha(T_2 - T_1)] \tag{IV.3}$$

where R_1 and R_2 are the resistances at temperatures T_1 and T_2, respectively, and α is the temperature coefficient of resistance. The resistivities and temperature coefficients of certain materials are given in Table IV.1.

In practice, transmission-line conductors contain a stranded steel core surrounded by aluminum conductors (for electrical conduction). Such a conductor is designated ACSR—aluminum conductor steel reinforced. Whereas it is difficult to calculate the exact ac resistance of such a conductor, measured values of resistance of various types of stranded conductors are tabulated in the literature. Because stranded conductors are spiraled, the length of the spiraled conductors becomes 1 to 2 percent greater than the apparent length of the transmission line. The cross section of a stranded conductor is shown in Figure IV.2.

TABLE IV.1
Values of ρ and α

Material	Resistivity, ρ, at 20°C (μΩ-cm)	Temperature Coefficient, α, at 20°C
Aluminum	2.83	0.0039
Brass	6.4-8.4	0.0020
Copper		
Hard-drawn	1.77	0.00382
Annealed	1.72	0.00393
Iron	10.0	0.0050
Silver	1.59	0.0038
Steel	12-88	0.001-0.005

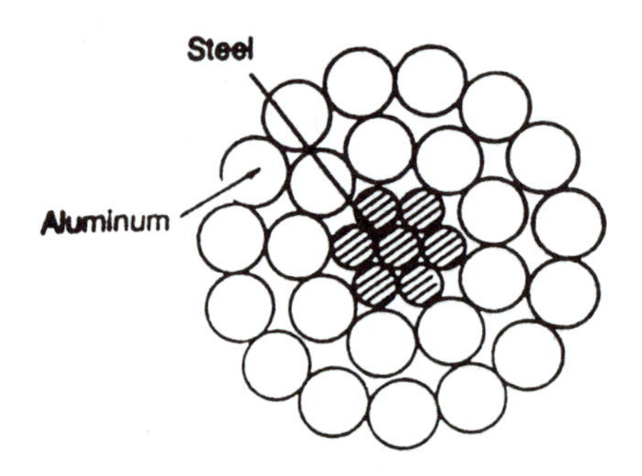

FIGURE IV.2
Cross section of a conductor having 7 steel strands and 24 aluminum strands.

Example IV.1 A piece of copper wire has a resistance of 20 Ω at 0°C. What is its resistance at −20°C?

Solution
From Figure IV.1, we have $\alpha_0 = 1/234.5$. From the given data, and from (IV.2) we obtain

$$R_2 = R_0(1 + \alpha_0 T_2) = 20\left(1 + \frac{-20}{234.5}\right) = 18.29\ \Omega$$

Example IV.2 A sample of copper wire has a resistance of 50 Ω at 10°C. What must be the maximum operating temperature of the wire if its resistance is to increase by at most 10 percent? Take the temperature coefficient at 10°C to be $\alpha = 0.00409°C^{-1}$

Solution
Here we have $R_1 = 50\ \Omega$ and $R_2 = 50 + 0.1 \times 50 = 55\ \Omega$. Also, $T_1 = 10°C$, and we require T_2. From (IV.3) we obtain

$$55 = 50[1 + 0.00409(T_2 - 10)] \quad \text{or} \quad T_2 = 34.45°C$$

Example IV.3 A transmission-line cable consists of 19 strands of identical copper conductors, each 1.5 mm in diameter. The length of the cable is 2 km but, because of the twist of the strands, the actual length of each conductor is increased by 2 percent. What is the resistance of the cable? Take the resistivity of copper to be $1.72 \times 10^{-8}\ \Omega$–m.

Solution
Allowing for twist, we find that l = (1.02)(2000) = 2040 m. The cross-sectional area of all 19 strands is $19(\pi/4)(1.5 \times 10^{-3})^2 = 33.576 \times 10^{-6}\text{m}^2$. Then, from (IV.1)

$$R = \frac{\rho l}{A} = \frac{1.72 \times 10^{-8} \times 2040}{33.576 \times 10^{-6}} = 1.045\ \Omega$$

In summary, (IV.1) gives the dc resistance of a conductor. Correction to account for a temperature variation is made by (IV.3). Spiraling increases the length of the conductor by about 2 percent and skin-effect is obtained from tabulated results. See, for instance, Appendix I, where a conductor is identified by a code word. The use of this table is illustrated by the following example.

Example IV.4 For *Partridge*, from Appendix I, we have: R_{dc} = 0.0640 Ω/1000 ft at 20°C and R_{ac} = 0.3792 Ω/mile at 50°C. The conductor cross section is 266,800 cmil. Verify the value of the dc resistance and evaluate R_{ac}/R_{dc}. The conductors are made of aluminum which has a resistivity of 2.83×10^{-8} Ωm or 17 Ωcmil/ft.

Solution
From (IV.1) we have

$$R_{dc} = \frac{17.0 \times 1000}{266{,}800} = 0.0637\ \Omega/1000\ \text{ft}$$

Allowing for a 1 percent increase in length due to spiraling, we finally obtain

$$R_{dc} = 1.01 \times 0.0637 = 0.0643\ \Omega/1000\ \text{ft}$$

which agrees with the tabulated value.

Now, from Table IV.1, for aluminum, α = 0.0039. Thus, from (IV.3) we obtain:

$$R_{50} = 0.0643\ [1 + 0.0039(50 - 20)] = 0.0718\ \Omega/1000\ \text{ft}$$

Since 5280 ft = 1 mile, we have

$$\frac{R_{ac}}{R_{dc}} = \frac{0.3792}{0.0718 \times 5.28} = 1.00025$$

which does not show much increase in the dc resistance.

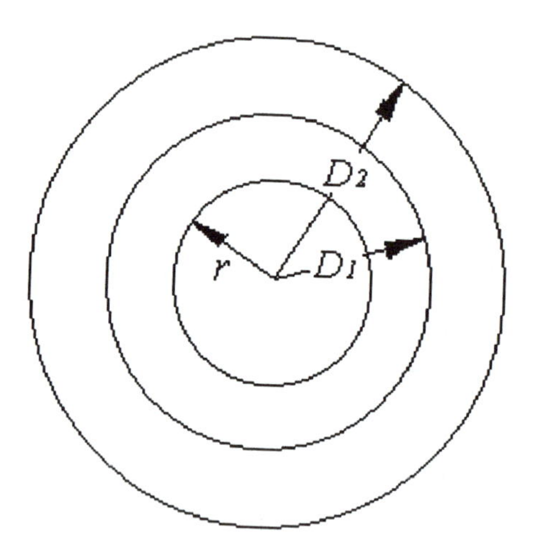

FIGURE IV.3
Magnetic fields internal and external to a conductor.

Inductance

The next parameter of interest is the transmission line inductance. The inductance may be determined either by finding the flux linkage or from magnetic field energy storage concept. We will use the former. In the following, we first determine the inductance of a solid conductor and that of a loop comprised of two conductors. Finally, we will evaluate the inductances of an array of conductors.

We begin our study with a single round conductor, the cross section of which is shown in Figure IV.3. We proceed to determine the inductance of this conductor by applying the defining equation

$$L \equiv \frac{\lambda}{i} \tag{IV.4}$$

according to which inductance = flux linkage per ampere.

We determine the magnetic fields within the conductor and external to the conductor by applying Ampere's law stated as follows:

$$\oint \boldsymbol{H} \cdot d\boldsymbol{l} = \int \boldsymbol{J} \cdot d\boldsymbol{s} = I \tag{IV.5}$$

Stated in words, the line integral of the magnetic field intensity ***H*** around a closed path equals the total current I enclosed by the path. Accordingly, referring to the conductor under consideration, if r is the radius of the conductor and it carries a current I, the field at some radius x is given by:

$$H_\phi \, (2\pi x) = I_x \qquad 0 < x < r \tag{IV.6}$$

where I_x is the total current through a cross section πx^2. In terms of I, we may write I_x as

$$I_x = \frac{I}{\pi r^2} \left(\pi x^2\right) \tag{IV.7}$$

where we have assumed a uniform current density in the conductor. Combining (IV.6) and (IV.7) yields:

$$H_\phi = \frac{Ix}{2\pi r^2} \qquad 0 < x < r \tag{IV.8}$$

Since **B** and **H** are related by $\mathbf{B} = \mu\mathbf{H} = \mu_0\mathbf{H}$ (the conductor being non-magnetic), (IV.8) becomes

$$B_\phi = \frac{\mu_0 Ix}{2\pi r^2} \qquad 0 < x < r \tag{IV.9}$$

The circumferential flux within the annular cylinder (Figure IV.3) of thickness dx per-unit length of the cylinder is

$$d\phi = B_\phi \, dx = \frac{\mu_0 I}{2\pi r^2} \, x\, dx \tag{IV.10}$$

The flux linkage within the conductor is the product of the flux and the fraction of the current linked. Thus,

$$\lambda_{in} = \int_0^r \frac{\pi x^2}{\pi r^2} \, d\phi = \frac{\mu_0 I}{2\pi r^4} \int_0^r x^3 \, dx = \frac{\mu_0 I}{8\pi} \tag{IV.11}$$

Combining (IV.4) and (IV.11) finally yields

$$L = \frac{\lambda_{in}}{I} = \frac{\mu_0}{8\pi} = \frac{1}{2} \times 10^{-7} \quad \text{H/m} \tag{IV.12}$$

In order to find the inductance due to the flux linkage external to the conductor, we first determine the circumferential flux within a cylinder bounded by the outer radius D_2 and inner radius D_1. External to the conductor, Ampere's law yields:

$$H_\phi = \frac{I}{2\pi x} \qquad r < x < \infty \tag{IV.13}$$

And

$$B_\phi = \mu_0 H_\phi = \frac{\mu_0 I}{2\pi x} \qquad r < x < \infty \tag{IV.14}$$

Thus

$$d\phi = B_\phi \, dx = \frac{\mu_0 I}{2\pi} \frac{dx}{x} \tag{IV.15}$$

The flux linkage between the radii D_2 and D_1 then becomes

$$\lambda_{ex} = \frac{\mu_0 I}{2\pi} \int_{D_1}^{D_2} \frac{dx}{x} = \frac{\mu_0 I}{2\pi} \ln \frac{D_2}{D_1} \tag{IV.16}$$

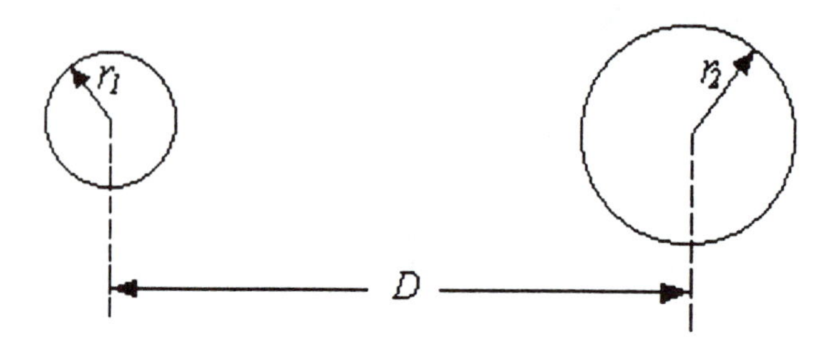

FIGURE IV.4
A transmission line having two conductors.

Setting $D_1 = r$ and $D_2 = D_1$ greater than r (but finite) in (IV.16), we finally obtain

$$\lambda_{ex} = \frac{\mu_0 I}{2\pi} \ln \frac{D}{r}$$

and

$$L_{ex} = \frac{\mu_0}{2\pi} \ln \frac{D}{r} = 2 \times 10^{-7} \ln \frac{D}{r} \quad \text{H/m} \tag{IV.17}$$

We now extend the preceding results, (IV.12) and (IV.17) to obtain the inductance of a loop of two conductors shown in Figure IV.4. From the dimensions shown in Figure IV.4, we substitute $D = D$ and $r = r_1$ in (IV.17) to obtain

$$L_{ex,1} = 2 \times 10^{-7} \ln \frac{D}{r_1} \quad \text{H/m} \tag{IV.18}$$

Combining this with (IV.12) yields

$$L_1 = \left(\frac{1}{2} + 2 \ln \frac{D}{r_1} \right) \times 10^{-7} \quad \text{H/m} \tag{IV.19}$$

Because $\ln (e^{1/4}) = \frac{1}{4}$, we may rewrite (IV.19) as

$$L_1 = 2 \times 10^{-7} \left(\ln e^{1/4} + \frac{D}{r_1} \right) = 2 \times 10^{-7} \ln \left(\frac{D}{r_1 e^{-1/4}} \right)$$

Or

$$L_1 = 2 \times 10^{-7} \ln \frac{D}{r_1'} \quad \text{H/m} \tag{IV.20}$$

where $r_1' = r_1 e^{-1/4}$, which may be taken as the radius of a (hollow) conductor with no internal flux. Similarly, for the second conductor we have

$$L_2 = 2 \times 10^{-7} \ln \frac{D}{r_2'} \quad \text{H/m} \tag{IV.21}$$

The loop inductance, per-unit length, then becomes

$$L = L_1 + L_2 = 4 \times 10^{-7}\, ln\, \frac{D}{\sqrt{r_1' r_2'}} \quad \text{H/m} \tag{IV.22}$$

If the conductors have the same radius, $r = r_1 = r_2$, then (IV.22) reduces to

$$L = 4 \times 10^{-7}\, ln\, \frac{D}{r'} \quad \text{H/m} \tag{IV.23}$$

In the preceding, we have presented the basic approach to determining the inductance of a line consisting of two solid conductors of cylindrical cross section. Whereas in practice a transmission line is seldom constructed of a solid conductor, the previous discussions may be used to determine the inductance of a line, each phase of which has more than one conductor; that is, such a line may be made of stranded conductors.

We consider an arbitrary arrangement of n conductors as shown in Figure IV.5. First, we determine the flux linkage with one conductor only, assuming that the sum of the conductor currents is zero. Expressed mathematically,

$$I_1 + I_2 + \ldots + I_n = \sum_{k=1}^{n} I_k = 0 \tag{IV.24}$$

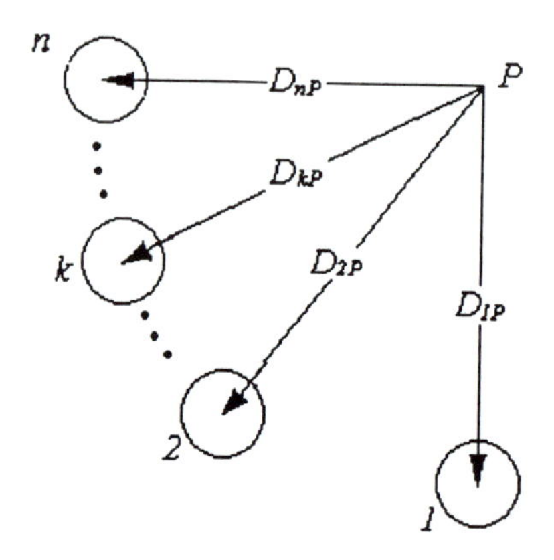

FIGURE IV.5
A line consisting of n conductors per phase.

Now, the flux linking conductor 1 out to P (but excluding the flux beyond P) due to the current I_1, from (IV.16) and (IV.20), is given by:

$$\lambda_{1P1} = 2 \times 10^{-7}\ I_1\ ln\ \frac{D_{1P}}{r_1^{/}} \qquad \text{(IV.25)}$$

where D_{1P} is the distance between P and conductor 1, and $r_1' = r_1 e^{-1/4}$, r_1 being the radius of the conductor.

Similarly, the flux linking conductor 1, due to the current I_2, out to P is the flux produced between conductor 1 and point P such that

$$\lambda_{1P2} = 2 \times 10^{-7}\ I_2\ ln\ \frac{D_{2P}}{D_{1P}} \qquad \text{(IV.26)}$$

Finally, the flux linkage λ_{1P} with conductor 1, due to all n conductors, out to P becomes

$$\lambda_{1P} = 2 \times 10^{-7} \left(I_1\ ln\ \frac{D_{1P}}{r_1^{/}} + I_2\ ln\ \frac{D_{2P}}{D_{12}} + I_3\ ln\ \frac{D_{3P}}{D_{13}} + \ldots + I_n\ ln\ \frac{D_{nP}}{D_{1n}} \right) \qquad \text{(IV.27)}$$

In compact notation, (IV.27) may also be written as

$$\lambda_{1P} = 2 \times 10^{-7} \sum_{k=1}^{n} I_k\ ln\ \frac{D_{kP}}{D_{1k}} \qquad \text{(IV.28)}$$

where $r_1' = D_{11}$. In general, we define

$$r_k^{/} = D_{kk} \qquad \text{(IV.29)}$$

We rewrite (IV.28) as

$$\lambda_{1P} = 2 \times 10^{-7} \left(\sum_{k=1}^{n} I_k\ ln\ \frac{1}{D_{1k}} + \sum_{k=1}^{n} I_k\ ln\ D_{kP} \right) \qquad \text{(IV.30)}$$

If we now solve for $I_n = -(I_1 + I_2 + \ldots + I_{n-1})$ from (IV.24) and substitute in the second term of (IV.30), we obtain

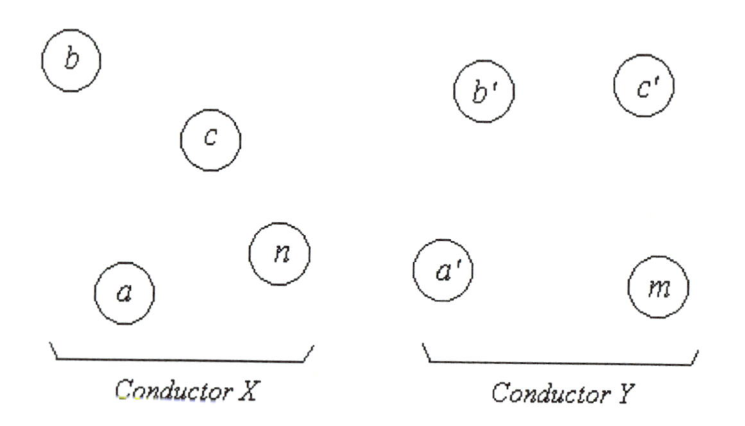

FIGURE IV.6
1-phase two-conductor line, each consisting of multiple filaments.

$$\lambda_{1P} = 2 \times 10^{-7} \left(\sum_{k=1}^{n} I_k \, ln \frac{1}{D_{1k}} + \sum_{k=1}^{n-1} I_k \, ln \frac{D_{kP}}{D_{nP}} \right) \tag{IV.31}$$

To include all the fluxes, we let P tend to infinity. Thus, $ln(D_{kP}/D_{nP}) \rightarrow 0$ as $P \rightarrow \infty$ and the second term in (IV.31) goes to zero. The total flux linking conductor 1, out to infinity, finally becomes:

$$\lambda_{1P} = 2 \times 10^{-7} \sum_{k=1}^{n} I_k \, ln \frac{1}{D_{1k}} \tag{IV.32}$$

We may apply this result to obtain the inductance of a line, each conductor of which consists of filaments, sub-conductors or bundled conductors. We assume that all filaments constituting a conductor are identical. Consider a line consisting of conductors X with n filaments and Y with m filaments as shown in Figure IV.6. The filament currents in conductors X and Y, respectively, are:

$$I_{filX} = \frac{I}{n} \; ; \quad I_{filY} = - \frac{I}{m} \tag{IV.33}$$

where I = total current, n = number of filaments in conductor X, and m = number of filaments in conductor Y. The negative sign in I_{filY} is for the return current. Thus, from (IV.32) the total flux linking filament a is

$$\lambda_a = 2 \times 10^{-7} I \left(\frac{1}{n} \sum_{k=a}^{n} ln \frac{1}{D_{ak}} - \frac{1}{m} \sum_{l=a'}^{m} ln \frac{1}{D_{al}} \right) \tag{IV.34}$$

In product form, (IV.34) may be written as

$$\lambda_a = 2 \times 10^{-7} I \, ln \frac{\sqrt[m]{(D_{aa'} D_{ab'} D_{ac'} \ldots D_{am})}}{\sqrt[n]{(D_{aa} D_{ab} D_{ac} \ldots D_{an})}} \tag{IV.35}$$

Alternatively, in compact notation (using Π for product), (IV.35) may be expressed as

$$\lambda_a = 2 \times 10^{-7} I \, ln \frac{\sqrt[m]{\prod_{l=a'}^{m} D_{al}}}{\sqrt[n]{\prod_{k=1}^{n} D_{ak}}} \tag{IV.36}$$

The inductance of filament a, from (IV.33) and (IV.36), is given by

$$L_a = \frac{\lambda_a}{I/n} = 2n \times 10^{-7} \, ln \frac{\sqrt[m]{\prod_{l=a'}^{m} D_{al}}}{\sqrt[n]{\prod_{k=a}^{n} D_{ak}}} \quad \text{H/m} \tag{IV.37}$$

Similarly, the inductance of filament b is

$$L_b = 2n \times 10^{-7} \, ln \frac{\sqrt[m]{\left(\prod_{l=a'}^{m} D_{bl}\right)}}{\sqrt[n]{\left(\prod_{k=a}^{n} D_{bk}\right)}} \quad \text{H/m} \tag{IV.38}$$

and so on for filaments c through n.

We may now write the average of the filaments constituting conductor X as

$$L_{av} = \frac{1}{n} \sum_{k=a}^{n} L_k \quad \text{H/m} \tag{IV.39}$$

To form conductor X, n filaments each having an inductance given by (IV.39) are connected in parallel. Consequently, the inductance of conductor X becomes:

$$L_x = \frac{1}{n} L_{av} = \frac{1}{n^2} \sum_{k=a}^{n} L_k \quad \text{H/m} \tag{IV.40}$$

Substituting (IV.37), (IV.38), etc., in (IV.40) and combining terms yields

$$L_x = 2 \times 10^{-7} \frac{\sqrt[mn]{(D_{aa'} D_{ab'} \ldots D_{am})(D_{ba'} D_{bb'} \ldots D_{bm}) \ldots (D_{na'} D_{nb'} \ldots D_{nm})}}{\sqrt[n^2]{(D_{aa} D_{ab} \ldots D_{an})(D_{ba} D_{bb} \ldots D_{bn}) \ldots (D_{na} D_{nb} \ldots D_{nn})}} \quad \text{H/m} \tag{IV.41}$$

Again, we may rewrite (IV.41) in compact form as:

$$L_x = 2 \times 10^{-7}\, ln \frac{\left(\prod_{k=a}^{n} \prod_{l=a'}^{m} D_{kl} \right)^{1/mn}}{\left(\prod_{k=a}^{n} \prod_{l=a}^{m} D_{kl} \right)^{1/n^2}} = 2 \times 10^{-7}\, ln \frac{D_m}{D_s} \quad \text{H/m} \tag{IV.42}$$

where we define:

D_m = mutual geometric mean distance between X and Y,

$$= GMD = \sqrt[mn]{\prod_{k=a}^{n} \prod_{l=a'}^{m} D_{kl}} \tag{IV.43}$$

and

D_s = self geometric mean distance of X

$$= \text{geometric mean radius} = GMR = \sqrt[n^2]{\prod_{k=a}^{n} \prod_{l=a}^{m} D_{kl}} \tag{IV.44}$$

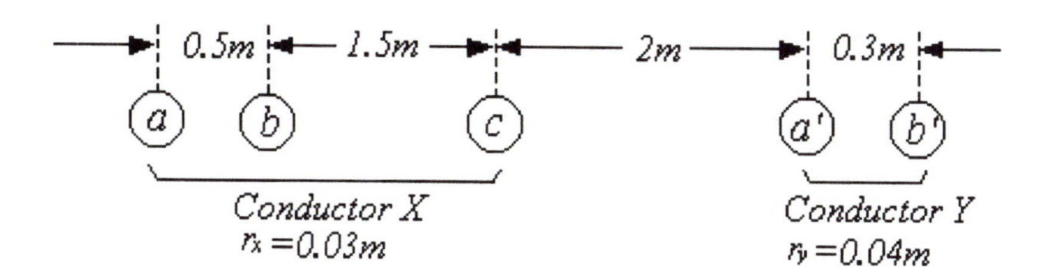

FIGURE IV.7
Example IV.5.

Example IV.5 Conductors X and Y are made of filaments as shown in Figure IV.7. The radii of filaments constituting X and Y are 3.0 cm and 4.0 cm respectively. Calculate the loop inductance of the line if it is 100 m long.

Solution
From (IV.29) we have, for conductor X,

$$D_{aa} = D_{bb} = D_{cc} = r_x' = e^{-0.25}\,(0.03) = 0.02336 \text{ m}$$

From (IV.43) and the given data, we obtain

$$D_{mx} = \sqrt[6]{\prod_{k=a}^{c}\prod_{l=a'}^{b'} D_{kl}}$$

$$= \sqrt[6]{\prod_{k=a}^{c}\left(D_{ka'}D_{kb'}\right)}$$

$$= \sqrt[6]{(D_{aa'}D_{ab'})(D_{ba'}D_{bb'})(D_{ca'}D_{cb'})}$$

$$= \sqrt[6]{4 \times 4.3 \times 3.5 \times 3.8 \times 2 \times 2.3} = 3.189 \text{ m}$$

Similarly, (IV.44) gives

$$D_{sx} = \sqrt[9]{\prod_{k=a}^{c} \prod_{k=a}^{c} D_{kl}}$$

$$= \sqrt[9]{\prod_{k=a}^{c} (D_{ka} D_{kb} D_{kc})}$$

$$= \sqrt[9]{(D_{aa} D_{ab} D_{ac})(D_{ba} D_{bb} D_{bc})(D_{ca} D_{cb} D_{cc})}$$

$$= \sqrt[9]{(0.02336)^3 (0.5)^2 \times 2^2 \times 1.5^2} = 0.3128 \text{ m}$$

Consequently

$$L_X = 2 \times 10^{-7} \, ln \, \frac{D_{mx}}{D_{sx}} = 2 \times 10^{-7} \, ln \, \frac{3.189}{0.3128}$$

$$= 4.644 \times 10^{-7} \quad \text{H/m}$$

Similarly,

$$L_Y = 2 \times 10^{-7} \, ln \, \frac{3.189}{0.09667} = 6.992 \times 10^{-7} \quad \text{H/m}$$

And, the total loop inductance for 100 m becomes:

$$L = 100 \, (L_X + L_Y) = 100 \, (4.644 + 6.992) \, 10^{-7}$$

$$= 0.1164 \quad \text{mH}$$

Whereas the preceding example illustrates the procedure of finding the inductance of a multi-filament conductor, tables are available which list GMRs and inductive reactances of standard conductors.

Example IV.6 Find the per-phase per-unit length inductance of a transmission line, the conductors of which are located at the corners of an equilateral triangle. The conductors are separated from each other by a distance D and each is of radius r

Solution
From (IV.43),

$$GMD = D_m = \sqrt[3]{D_{ab} D_{bc} D_{ca}} = \sqrt[3]{D^3} = D$$

Since each phase consists of one conductor, its $GMR = D_s = r' = re^{-1/4}$

Thus, applying (IV.42) yields

$$L_a = L_b = L_c = 2 \times 10^{-7}\, ln \frac{D}{r'} = 2 \times 10^{-7}\, ln \frac{D}{D_s} \quad \text{H/m}$$

In the above example, we determine the per phase, or line-to-neutral, inductance of a line with equilaterally spaced conductors. In practice, the three conductors of a three-phase line are seldom equilaterally spaced. Such an unsymmetrical spacing results in unequal inductances in the three phases, leading to unequal voltage drops and an imbalance in the line. To offset this difficulty, the positions of the conductors are interchanged at regular intervals along the line. This practice is known as *transposition* and is illustrated in Figure IV.8,

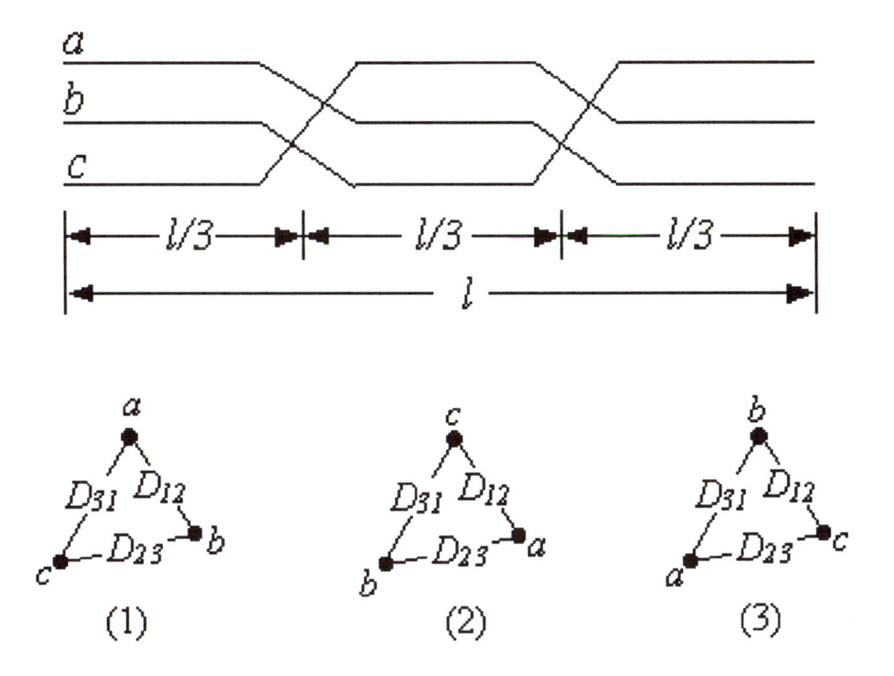

FIGURE IV.8
Transposition of unequally spaced three-phase transmission line conductors.

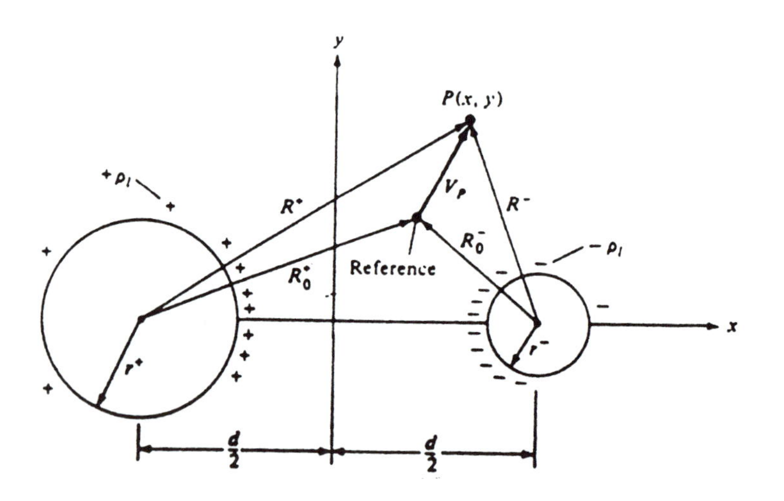

FIGURE IV.9
A two-conductor transmission line.

which also shows the unequal spacings between the conductors. The average per phase inductance for such a case is still given by the end result of Example IV.6 except that the spacing D in the equation is replaced by the equivalent spacing D_e obtained from

$$D_e = \sqrt[3]{D_{12}D_{23}D_{31}} \tag{IV.45}$$

where the distances D_{12}, and so on, are shown in Figure IV.8.

Capacitance

The third and last parameter of importance, especially in long and medium-length lines, is the capacitance between various conductors of the line. Similar to (IV.4) defining inductance as flux linkage per ampere, we define capacitance (in farad) as charge per-unit volt; that is,

$$C = \frac{Q}{V} \quad \text{F} \tag{IV.46}$$

Specifically, with reference to the two-conductor line shown in Figure IV.9, Q is the charge (in coulomb) on one of the conductors and V is the voltage (or potential difference) between the two conductors. In the following, we will use (IV.46) as the basis for the determination of capacitances of various line configurations.

If we relate the charge to the voltage, then we can obtain the capacitance as given by (IV.46). To do so, we use Gauss' law to determine the electric field due to the charge and then find the voltage from the electric field. According to Gauss' law, "the total outward electric flux from a closed surface equals the charge enclosed by the surface," where the electric flux density **D** is related to the electric field **E** by

$$\mathbf{D} = \varepsilon\, \mathbf{E} \tag{IV.47}$$

In (IV.47) ε is defined as the permittivity of the material (in F/m) in which **D** and **E** exist. Referring to Figure IV.9, we imagine a cylinder of unit length and of radius R^+, concentric with the conductor a. The total outward flux from the surface of this cylinder is given by

$$\psi_e = D_r 2\pi R^+ \tag{IV.48}$$

where D_r is the flux density at the surface of the cylinder. Now, if Q is the total charge on a unit length of the conductor, then Gauss' law combined with (IV.47) and (IV.48) yields

$$Q = \psi_e = D_r 2\pi R^+ = 2\pi\varepsilon_0 E_r R^+ \tag{IV.49}$$

where ε_0 = permittivity of free space = $10^{-9}/36\pi$ F/m. Finally, solving for the electric field in terms of the charge, we obtain

$$E_r = \frac{Q}{2\pi\varepsilon_0 R^+} \quad \text{V/m} \tag{IV.50}$$

Now, we determine the voltage V_{OP} (Figure IV.9) from (IV.50) such that

$$\begin{aligned} V_{OP}^+ &= -\int_{R_0^+}^{R^+} E_r\, dr = -\frac{Q}{2\pi\varepsilon_0}\int_{R_0^+}^{R^+} \frac{1}{R^+}\, dR^+ \\ &= \frac{Q}{2\pi\varepsilon_0}\, ln\, \frac{R_0^+}{R^+} \end{aligned} \tag{IV.51}$$

This voltage is due to the charge Q on conductor a. Repeating the process for the charge $-Q$ on conductor b, we obtain

$$V_{OP}^{-} = -\frac{Q}{2\pi\varepsilon_0} \ln \frac{R_0^{-}}{R^{-}} \tag{IV.52}$$

Combining (IV.51) and (IV.52) yields

$$V_{OP} = V_{OP}^{+} + V_{OP}^{-} = \frac{Q}{2\pi\varepsilon_0} \ln \frac{R_0^{+}R^{-}}{R^{+}R_0^{-}} \tag{IV.53}$$

We now let P come to the surface of a and O to the surface of b, such that $R^{+} = r_a$, $R_0^{-} = r_b$, $R_0^{+} \simeq D \simeq R^{-}$. Substituting these in (IV.53) gives:

$$V_{OP} = \frac{Q}{\pi\varepsilon_0} \ln \left(\frac{D}{\sqrt{r_a r_b}} \right) \tag{IV.54}$$

Hence, from (IV.46) and (IV.54), the capacitance per-unit length of the line is given by

$$C = \frac{\pi\varepsilon_0}{\ln\left(D/\sqrt{r_a r_b}\right)} \quad \text{F/m} \tag{IV.55}$$

As a special case, if the two conductors are of the same radius, $r = r_a = r_b$, then (IV.55) becomes

$$C = \frac{\pi\varepsilon_0}{\ln(D/r)} \quad \text{F/m} \tag{IV.56}$$

The preceding results may be generalized for a multiconductor system shown in Figure IV.10. Thus, from (IV.51), the voltage between conductors k and i due to a charge Q_m on conductor m is

$$V_{kim} = \frac{Q_m}{2\pi\varepsilon_0} \ln \frac{D_{im}}{D_{km}} \tag{IV.57}$$

Or, due to all charges, we have

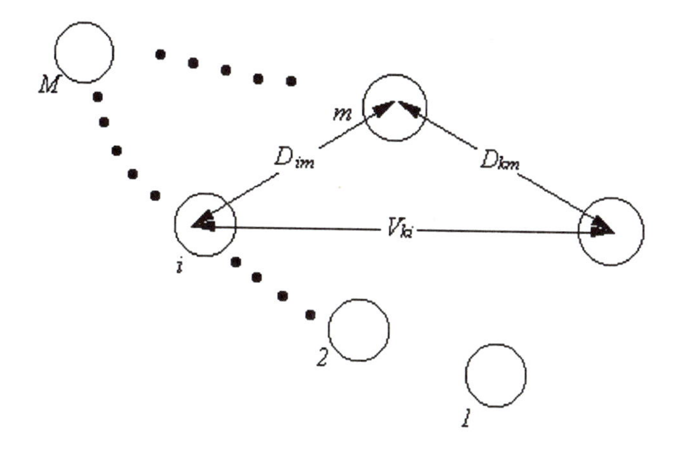

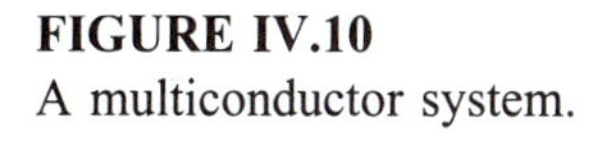

FIGURE IV.10
A multiconductor system.

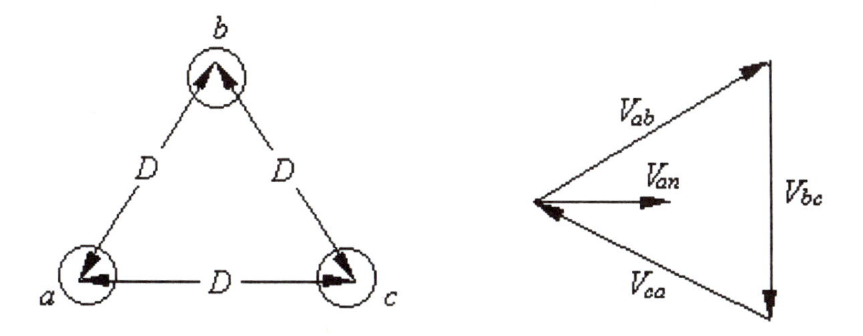

FIGURE IV.11
A 3-phase symmetrical line.

$$V_{ki} = \frac{1}{2\pi\varepsilon_0} \sum_{m=1}^{M} Q_m \ln \frac{D_{im}}{D_{km}} \qquad \text{(IV.58)}$$

We may apply this result to obtain the per phase capacitance of a three-phase line. For instance, for a three-phase symmetrical line shown in Figure IV.11 which also shows the various voltage phasors, we use (IV.58) to write

$$V_{ab} = \frac{1}{2\pi\varepsilon_0} \sum_{m=a,b,c} Q_m \ ln \frac{D_{bm}}{D_{am}}$$

$$= \frac{1}{2\pi\varepsilon_0} \left(Q_a \ ln \frac{D_{ba}}{D_{aa}} + Q_b \ ln \frac{D_{bb}}{D_{ab}} + Q_c \ ln \frac{D_{bc}}{D_{ac}} \right) \tag{IV.59}$$

Now, $D_{aa} = D_{bb}$ = radius of the conductor, r, and $D_{ab} = D_{ac} = D$ = separation between the conductors. Thus, (IV.59) simplifies to

$$V_{ab} = \frac{1}{2\pi\varepsilon_0} \left(Q_a \ ln \frac{D}{r} + Q_b \ ln \frac{r}{D} \right)$$

Similarly

$$V_{ac} = \frac{1}{2\pi\varepsilon_0} \left(Q_a \ ln \frac{D}{r} + Q_c \ ln \frac{r}{D} \right)$$

Adding these results and using the constraint $Q_a + Q_b + Q_c = 0$ yields

$$V_{ab} + V_{ac} = \frac{1}{2\pi\varepsilon_0} \left[2Q_a \ ln \frac{D}{r} - (Q_b + Q_c) \ ln \frac{D}{r} \right]$$

$$= \frac{3Q_a}{2\pi\varepsilon_0} \ ln \frac{D}{r} \tag{IV.60}$$

Referring to the voltage phasor diagram of Figure IV.9, we observe that

$$V_{ab} + V_{ac} = 3V_{an} = \frac{3Q_a}{2\pi\varepsilon_0} \ ln \frac{D}{r}$$

Hence, the capacitance per-unit length of the line is given by

$$C = \frac{2\pi\varepsilon_0}{ln(D/r)} \quad \text{F/m} \tag{IV.61}$$

Similarly, for the transposed line (Figure IV.8), (IV.58) yields:

for position (1)

$$V_{ab} = \frac{1}{2\pi\varepsilon_0}\left(Q_a \ ln \ \frac{D_{12}}{r} + Q_b \ ln \ \frac{r}{D_{12}} + Q_c \ ln \ \frac{D_{23}}{D_{31}}\right)$$

for position (2)

$$V_{ab} = \frac{1}{2\pi\varepsilon_0}\left(Q_a \ ln \ \frac{D_{23}}{r} + Q_b \ ln \ \frac{r}{D_{23}} + Q_c \ ln \ \frac{D_{31}}{D_{12}}\right)$$

for position (3)

$$V_{ab} = \frac{1}{2\pi\varepsilon_0}\left(Q_a \ ln \ \frac{D_{31}}{r} + Q_b \ ln \ \frac{r}{D_{31}} + Q_c \ ln \ \frac{D_{12}}{D_{23}}\right)$$

Adding the preceding three equation and expressing as an average voltage, we obtain:

$$\left(V_{ab}\right)_{av} = \frac{1}{6\pi\varepsilon_0}\left(Q_a \, ln \ \frac{D_{12}D_{23}D_{31}}{r^3} + Q_b \, ln \ \frac{r^3}{D_{12}D_{23}D_{31}} + Q_c \, ln \ \frac{D_{12}D_{23}D_{31}}{D_{12}D_{23}D_{31}}\right) \tag{IV.62}$$

The last term in (IV.62) is zero. Consequently, we have

$$\left(V_{ab}\right)_{av} = \frac{1}{2\pi\varepsilon_0}\left(Q_a \ ln \ \frac{D_e}{r} + Q_b \ ln \ \frac{r}{D_e}\right) \tag{IV.63}$$

where $D_e = \sqrt[3]{D_{12}D_{23}D_{31}}$. Also,

$$\left(V_{ac}\right)_{av} = \frac{1}{2\pi\varepsilon_0}\left(Q_a \ ln \ \frac{D_e}{r} + Q_c \ ln \ \frac{r}{D_e}\right) \tag{IV.64}$$

Finally, combining (IV.63) and (IV.64), using the constraint that $Q_a + Q_b + Q_c = 0$, and observing that $(V_{ab})_{av} + (V_{ac})_{av} = (V_{an})_{av}$ we obtain

$$3\,V_{an} = \frac{Q_a}{2\pi\varepsilon_0}\left(2 \ ln \ \frac{D}{r} - ln \ \frac{r}{D}\right) = \frac{3Q_a}{2\pi\varepsilon_0} \ ln \ \frac{D}{r} \tag{IV.65}$$

From (IV.65) the per phase capacitance per-unit length of the line is given by

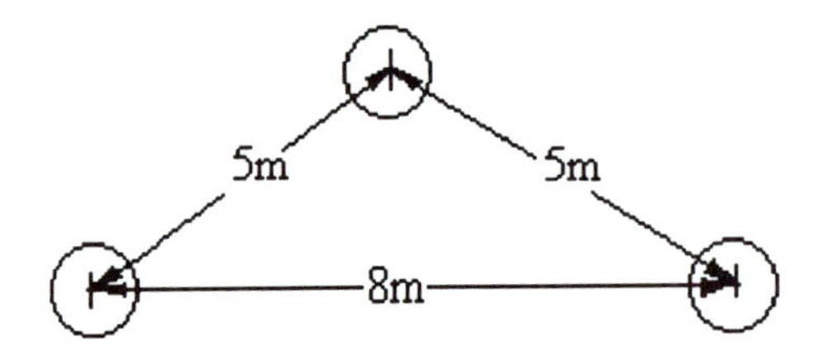

FIGURE IV.12
Example IV.7.

$$C = \frac{2\pi\varepsilon_0}{ln(D_e/r)} \quad \text{F/m} \tag{IV.66}$$

Example IV.7 A single-circuit three-phase 60-Hz transmission line consists of three conductors arranged as shown in Figure IV.12. If the conductors have a diameter of 250 mils, find the inductive and capacitive reactances per kilometer per phase.

Solution

From (IV.45).

$$D_e = (5 \times 5 \times 8)^{1/3} = 5.848 \text{ m}$$

Now, 250 mils = 0.25 in = 0.635 cm; thus

$$r = \frac{1}{2} \times 0.635 \times 10^{-2} \text{ m}$$

And $r' = 0.2473$

$$\frac{D_e}{r'} = \frac{5.848 \times 10^2}{0.2473} = 2365$$

and ln $(D_e/r') = 7.768$. Hence, from (IV.20) (with $\mu_0 = 4\pi \times 10^{-7}$ H/m).

$$L = 2 \times 10^{-7} \times 7.768 \times 10^3 = 1.554 \quad \text{mH/km}$$

Or, inductive reactance per kilometer

$$X_L = \omega L = 377 \times 1.554 \times 10^{-3} = 0.5857 \ \Omega$$

From (IV.66) (with $\varepsilon_0 = 10^{-9}/36\pi$ F/m).

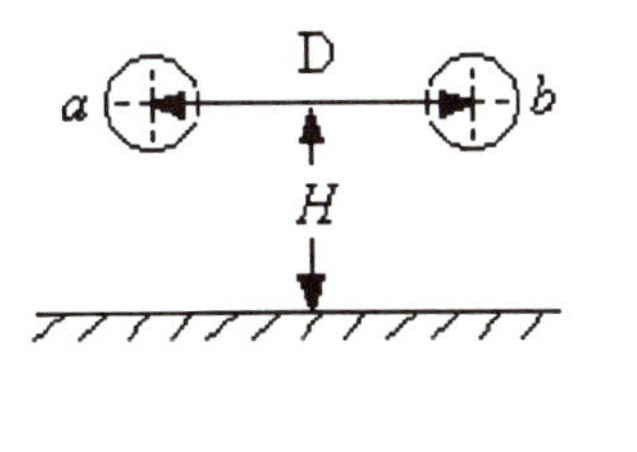

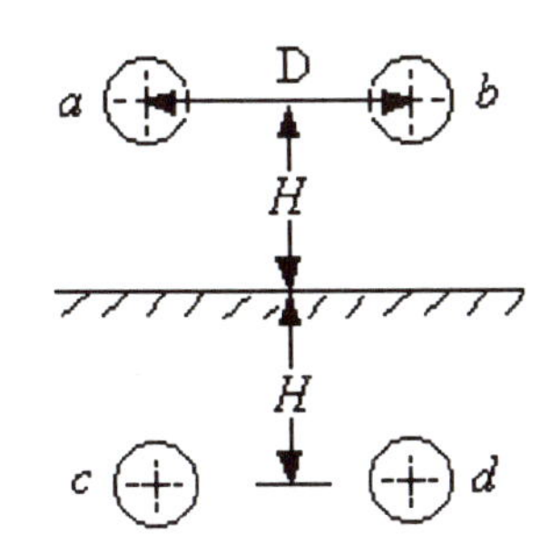

FIGURE IV.13
Example IV-8.

$$C = \frac{2\pi \times 10^{-9}/36\pi}{7.52} \times 10^{3} = 7.387 \times 10^{-9} \text{ F/km}$$

Hence, capacitive reactance per kilometer

$$X_e = \frac{1}{\omega C} = \frac{10^9}{377 \times 7.387} = 0.36 \times 10^6 \ \Omega$$

Example IV.8 A two-conductor single-phase transmission line is H meter above earth as shown in Figure IV.13a. The effect of earth may be simulated by image charges (Figure IV.13b) since earth may be considered to be an equipotential surface.

If the conductors are separated by a distance D and each is of radius r, obtain an expression for the capacitance between the conductors (per-unit length).

From (IV.58) we have

$$V_{ab} = \frac{1}{2\pi\varepsilon_0}\left(Q_a ln \frac{D_{ba}}{D_{aa}} + Q_b ln \frac{D_{bb}}{D_{ab}} + Q_c ln \frac{D_{bc}}{D_{ac}} + Q_d ln \frac{D_{bd}}{D_{ad}}\right)$$

Substituting $Q_a = Q_d = +Q$, $Q_b = Q_c = -Q$, $D_{aa} = D_{bb} = r'$, $D_{ba} = D_{ab} = D$, $D_{ad} = D_{bc}$ and $D_{ac} = D_{bd} = 2H$ yields

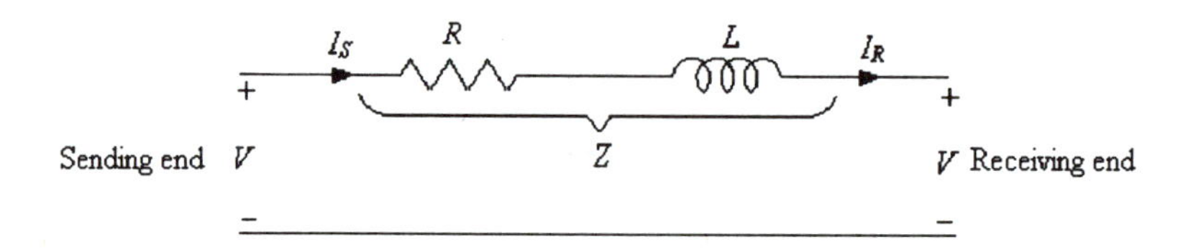

FIGURE IV.14
Representation of a short transmission line (on a per-phase basis).

$$V_{ab} = \frac{Q}{2\pi\varepsilon_0}\left(ln\,\frac{D}{r'} - ln\,\frac{r'}{D} - ln\,\frac{D_{bc}}{2H} + ln\,\frac{2H}{D_{bc}}\right)$$

$$= \frac{2Q}{2\pi\varepsilon_0}\left(ln\,\frac{D}{r'} - ln\,\frac{D_{bc}}{2H}\right)$$

Hence,

$$C = \frac{Q}{V_{ab}} = \frac{\pi\varepsilon_0}{ln(D/r') - ln\,(D_{bc}/2H)} \quad \text{F/m}$$

where $D_{bc} = \sqrt{(4H^2 + D^2)}$.

In this section we have discussed the three basic parameters of transmission lines. A fourth parameter, the leakage resistance or conductance to ground, which represents the combined effect of various current flows from the line to ground, has not been considered because it is usually negligible for most calculations. However, it has been estimated that 132-kV transmission lines, leakage losses vary between 0.3 and 1.0 kW/mile.

IV.2 TRANSMISSION-LINE REPRESENTATION

Returning to the three basic parameters calculated in Example IV.7, we observe that the capacitance is very small for 1-km-long line. In fact, we mentioned in the beginning of this chapter that for a transmission line up to 80 km long, shunt effects, due to capacitance and leakage resistance, are negligible. Such a line is known as a *short transmission line*, and is represented by the lumped parameters R and L, as shown in Figure IV.14. Notice that R is the resistance (per phase) and L is the inductance (per phase) of the entire line, although in Section IV.1 we introduced the transmission-line parameters in terms of per-unit length of the line. The line is shown to have two ends: the sending end at the generator end and the receiving end at the load end. The problems of significance to be solved here are the determination of voltage regulation and efficiency of transmission. These quantities are defined as follows:

$$\text{percent voltage regulation} = \frac{|V_{R(\text{no-load})}| - |V_{R(\text{load})}|}{|V_{R(\text{load})}|} \times 100 \qquad \text{(IV.67)}$$

$$\text{efficiency of transmission} = \frac{\text{power at the receiving end}}{\text{power at the sending end}} \qquad \text{(IV.68)}$$

Often, (IV.67) and (IV.68) are evaluated at full-load values. The calculation procedure is illustrated by the following example.

Example IV.9 Let the transmission line of Example IV.7 be 60 km long. The line supplies a three-phase wye-connected 100-MW 0.9 lagging power factor load at 215 kV line-to-line voltage. If the operating temperature of the line is 60°C, determine the voltage regulation and efficiency of transmission in percent. The per phase line resistance R is 0.62 Ω.

Solution

The inductance L is (from Example IV.7)

$$L = 1.554 \times 10^{-3} \times 60 = 93.24 \text{ mH}$$

and

$$X_L = \omega L = 377 \times 93.24 \times 10^{-3} = 35.15 \ \Omega$$

$$\text{Line current} \quad I \; (= |I_S| = |I_R|) = \frac{100 \times 10^6}{\sqrt{3} \times 215 \times 10^3 \times 0.9} = 298.38 \text{ A}$$

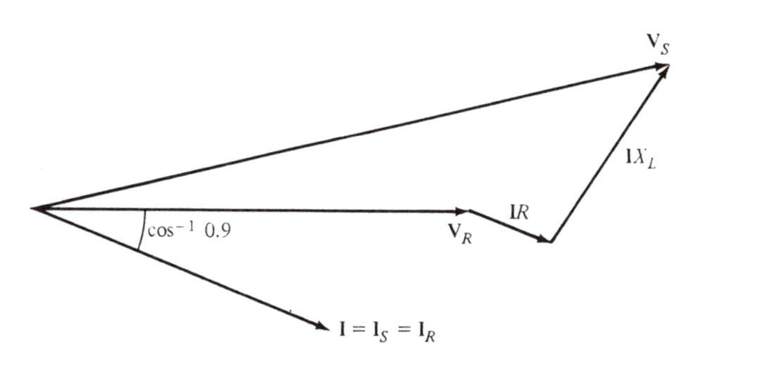

FIGURE IV.15
Example IV.9.

Phase voltage at the receiving end, $|V_R| = \dfrac{215 \times 10^3}{\sqrt{3}} = 124.13 \text{ kV}$

The phasor diagram illustrating the operating conditions is shown in Figure IV.15, from which

$$
\begin{aligned}
V_S &= V_R + I(R + jX_L) \\
&= 124.13 \times 10^3 \angle 0° + 298.38 \angle -25.8° \, (0.62 + j35.15) \\
&\approx 124.13 \times 10^3 \angle 0° + 298.38 \angle -25.8° \times 35.15 \angle 90° \\
&= (128.69 + j9.44) \text{ kV} \\
&= 129.02 \angle 4.19° \text{ kV}
\end{aligned}
$$

Hence

$$
\begin{aligned}
\text{percent voltage regulation} &= \frac{129.02 - 124.13}{124.13} \times 100 \\
&= 3.94 \text{ percent}
\end{aligned}
$$

To calculate the efficiency, we determine the loss in the line as

$$
\begin{aligned}
\text{line loss} &= 3 \times 298.38^2 \times 0.62 = 0.166 \text{ MW} \\
\text{power received} &= 100 \text{ MW (given)} \\
\text{power sent} &= 100 + 0.166 = 100.166 \text{ MW} \\
\text{efficiency} &= \frac{100}{100.166} = 99.83 \text{ percent}
\end{aligned}
$$

The *medium-length transmission line* is considered to be up to 240 km long. In such a line the shunt effect due to the line capacitance is not negligible. Two representations for the medium-length line are shown in Figures IV.16 and IV.17. These are known as the nominal-Π circuit and the nominal-T circuit of the transmission line, respectively. In Figures IV.16 and IV.17, we also show the respective phasor diagrams for lagging power factor conditions. These diagrams aid in understanding the mutual relationships between currents and voltages at certain places on the line. The following examples illustrate the applications of the nominal-Π and nominal-T circuits in calculating the performances of medium-length transmission lines. It is to be noted, however, that the Π-circuit is most commonly used.

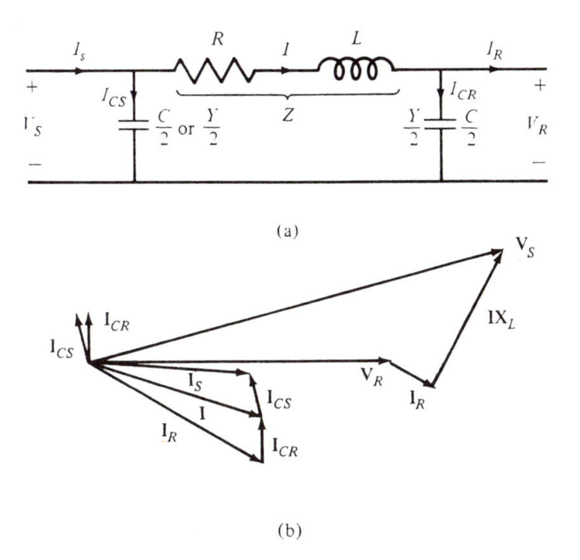

FIGURE IV.16
(a) A nominal-Π circuit; (b) corresponding phasor diagram.

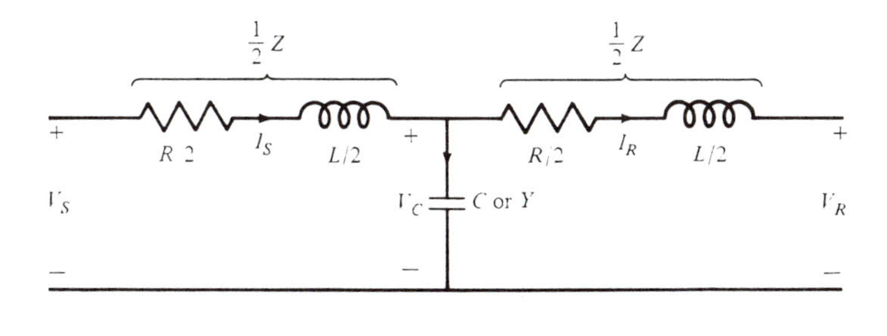

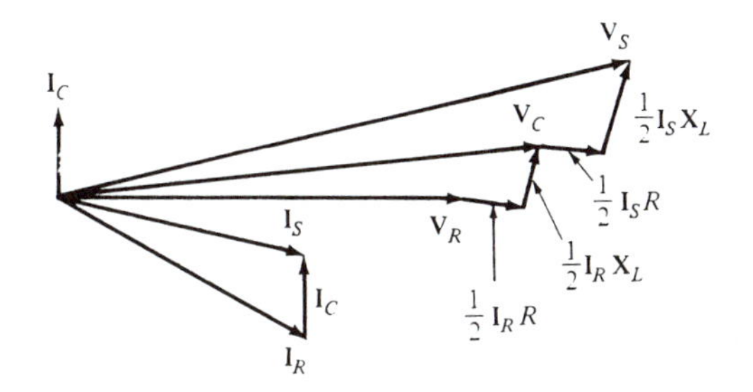

FIGURE IV.17
(a) A nominal-T circuit; (b) corresponding phasor diagram.

Example IV.10 Consider the transmission line of Example IV.9, but let the line be 200 km long. Determine the voltage regulation of the line for the operating conditions given in Example IV.9 using (a) the nominal-Π circuit and (b) the nominal-T circuit.

Solution
From the data of Examples IV.7 and IV.9, we already know that

$$R = 2.07\ \Omega$$

$$L = 310.8\ \text{mH}$$

$$C = 1.4774\ \mu F$$

$$V_R = 124.13\ \text{kV}$$

$$I_R = 298.38\ \angle -25.8°\ \text{A}$$

(a) Using the nomenclature of Figure IV.16, we have

$$I_{CR} = \frac{V_R}{X_{c/2}} = \frac{124.13 \times 10^3\ \angle 0^o}{1/(377 \times 0.5 \times 1.4774 \times 10^{-6})\ \angle 90^o}$$

$$= 34.57\ \angle 90°\ \text{A}$$

$$I = I_R + I_{CR} = 298.38\ \angle -25.8° + 34.57\ \angle 90°$$

$$= 285 \angle -19.5°\ \text{A}$$

$$R + jX_L = 2.07 + j377 \times 0.3108 \simeq 117.19\ \angle 89°\ \Omega$$

$$I\ (R + jX_L) = 285\ \angle -19.5° \times 117.19\ \angle 89°$$

$$= 33.4\ \angle 69.5°\ \text{kV}$$

$$V_s = V_R + I(R + jX_L) = 124.13\ \angle 0° + 33.4\ \angle 69.5°$$

$$= 139.5\ \angle 13.2°\ \text{kV}$$

$$\text{percent regulation} = \frac{139.0 - 124.13}{124.13} \times 100 = 12.38\%$$

(b) Using the nomenclature of Figure IV.17, we have

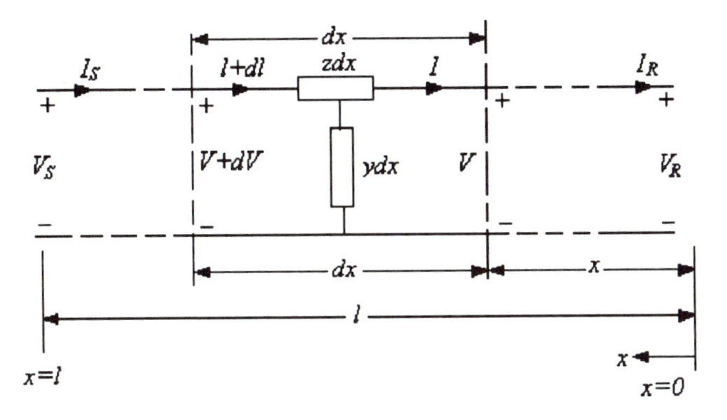

FIGURE IV.18
Distributed parameter long line.

$$V_C = V_R + \frac{1}{2} I_R (R + jX_L)$$

$$= 124.13 \angle 0° + \frac{10^{-3}}{2} \times 298.38 \angle -25.8° \times 117.19 \angle 89°$$

$$= 132.0 + j15.6 = 132.92 \angle 6.74° \text{ kV}$$

$$I_C = \frac{V_c}{X_c} = \frac{132.92 \times 10^3 \angle 6.74°}{1/(377 \times 1.4774 \times 10^{-6}) \angle 90°} = 74.0 \angle 96.74° \text{ A}$$

$$I_S = I_R + I_C = 298.38 \angle -25.8° + 74.0 \angle 96.74° = 266.96 \angle -12.2° \text{ A}$$

$$V_S = V_C + \frac{1}{2} I_S (R + jX_L)$$

$$= 132.92 \angle 6.74° + \frac{10^{-3}}{2} \times 266.96 \angle -12.2° \times 117.19 \angle 89°$$

$$= 139.03 \angle 12.8° \text{ kV}$$

$$\text{percent regulation} = \frac{139.03 - 124.13}{124.13} \times 100 = 12\%$$

Finally, we consider transmission lines over 240 km long. Such a line is known as a *long line*. Parameters of long lines are distributed over the entire length of the line. On a per phase basis, a long line is shown in Figure IV.18. Let z be the series impedance of the line per-unit length, and let y be the shunt admittance per-unit length. With the currents and voltages shown in Figure IV.18, for an elemental length dx, we have

$$dV = Iz\ dx \quad \text{or} \quad \frac{dV}{dx} = zI \tag{IV.69}$$

$$dI = Vy\ dx \quad \text{or} \quad \frac{dI}{dx} = yV \tag{IV.70}$$

Eliminating I from (IV.69) and (IV.70) yields

$$\frac{d^2V}{dx^2} = \gamma^2 V \tag{IV.71}$$

where $\gamma = \sqrt{yz}$, and is known as the *propagation constant.* Solution to (IV.71) may be written as

$$V = C_1 e^{\gamma x} + C_2 e^{-\gamma x} \tag{IV.72}$$

where C_1 and C_2 are arbitrary constants. Differentiating (IV.72) and substituting (IV.69) gives

$$C_1 \gamma e^{\gamma x} - C_2 \gamma e^{-\gamma x} = zI \tag{IV.73}$$

Let $\gamma/z = Z_c = \sqrt{z/y}$. In terms of Z_c, we obtain

$$I = \frac{1}{Z_c}\left(C_1 e^{\gamma x} - C_2 e^{-\gamma x}\right) \tag{IV.74}$$

The quantity Z_c is called the *characteristic impedance* of the line.

To evaluate C_1 and C_2, we refer to Figure IV.18 from which $V = V_R$ and $I = I_R$ at $x = 0$. Substitution of these conditions in (IV.72) and (IV.74) and solving for C_1 and C_2 yields

$$C_1 = \frac{1}{2}(V_R + Z_c I_R)$$

$$C_2 = \frac{1}{2}(V_R - Z_c I_R)$$

In terms of these constants, the voltage and current distributions are given by

$$V = \frac{1}{2} V_R(e^{\gamma x} + e^{-\gamma x}) + \frac{1}{2} I_R Z_c(e^{\gamma x} - e^{-\gamma x}) \qquad \text{(IV.75)}$$

$$I = \frac{1}{2} \frac{V_R}{Z_c} e^{\gamma x} - e^{-\gamma x}) + \frac{1}{2} I_R(e^{\gamma x} + e^{-\gamma x}) \qquad \text{(IV.76)}$$

In terms of hyperbolic functions, (IV.75) and (IV.76) are finally expressed as

$$V = V_R \cosh \gamma x + I_R Z_c \sinh \gamma x \qquad \text{(IV.77)}$$

$$I = I_R \cosh \gamma x + \frac{V_R}{Z_c} \sinh \gamma x \qquad \text{(IV.78)}$$

Since $V = V_s$ and $I = I_s$ at $x = l$, (IV.77) and (IV.78) become

$$V_s = V_R \cosh \gamma l + I_R Z_c \sinh \gamma l \qquad \text{(IV.79)}$$

$$I_s = I_R \cosh \gamma l + \frac{V_R}{Z_c} \sinh \gamma l \qquad \text{(IV.80)}$$

These equations are used in evaluating the performance of long lines. In carrying out numerical computations, the following relationships are often useful:

$$\gamma = \alpha + j\beta$$

$$\cosh \gamma l = \cosh (\alpha l + j\beta l) = \cosh \alpha l \cos \beta l + j \sin \alpha l \sin \beta l$$

$$\sinh \gamma l = \sinh (\alpha l + j\beta l) = \sinh \alpha l \cos \beta l + j \cosh \alpha l \sin \beta l$$

$$\cosh \gamma l = 1 + \frac{(\gamma l)^2}{2!} + \frac{(\gamma l)^4}{4!} + \ldots \simeq 1 + \frac{1}{2} YZ$$

$$\sinh \gamma l = \gamma l + \frac{(\gamma l)}{3!} + \frac{(\gamma l)^5}{5!} + \ldots \simeq \sqrt{YZ}\,(1 + \frac{1}{6} YZ)$$

Example IV.11 The parameters of a 215-kV 400-km 60-Hz three-phase transmission line are as follows:

$$y = j3.2 \times 10^{-6}\ \Omega/\text{km} \qquad z = (0.1 + j0.5)\ \Omega/\text{km}$$

The line supplies a 150-MW load at unity power factor. Determine (a) the voltage regulation, (b) the sending-end power, and (c) the efficiency of transmission.

Solution

$$z = 0.1 + j0.5 = 0.51\ \angle 78.7°$$

$$y = j3.2 \times 10^{-6} = 3.2 \times 10^{-6}\ \angle 90°$$

$$\gamma l = l\sqrt{zy} = 400\sqrt{0.51 \times 3.2 \times 10^{-6}}\ \angle 1/2\ (90 + 78.7)°$$

$$= 0.51\ \angle 84.35° = 0.05 + j0.5 = \alpha + j\beta \text{ rad}$$

$$Z_c = \sqrt{\frac{z}{y}} = \sqrt{\frac{0.51}{3.2 \times 10^{-6}}}\ \angle 1/2\ (78.7 - 90)° = 399.2\ \angle -5.65°\ \Omega$$

$$V_R = \frac{215 \times 10^3}{\sqrt{3}} = 124.13\ \angle 0°\ \text{kV}$$

$$I_R = \frac{150 \times 10^6}{\sqrt{3} \times 215 \times 10^3} = 402.8\ \angle 0°\ \text{A}$$

$$\cosh \gamma l = \cosh 0.05 \cos 0.5 + j\sinh 0.05 \sin 0.5$$

$$= 0.877 + j0.024 = 0.877\ \angle 1.57°$$

$$\sinh \gamma l = \sinh 0.05 \cos 0.5 + j \cosh 0.05 \sin 0.5$$

$$= 0.044 + j0.479 = 0.48\ \angle 84.75°$$

From (IV.79) and (IV.80), we obtain, respectively,

$$V_s = (124.13\ \angle 0° \times 0.877\ \angle 1.57° + 402.8\ \angle 0° \times 10^{-3}$$

$$\times 399.2\ \angle -5.65° \times 0.48\ \angle 84.75°)\ \text{kV}$$

$$= 146.4\ \angle 32.55°\ \text{kV}$$

and

$$I_s = 402.8 \angle 0° \times 0.877 \angle 1.57° + \frac{124.13\angle 0° \times 10^3}{399.2 \angle -5.65°} \times 0.48 \angle 84.75° \text{ A}$$

$$= 386.52 \angle 24.28° \text{ A}$$

Receiving-end voltage on no-load becomes, from (IV.79) (with $I_R = 0$).

$$|(V_R)_{\text{no-load}}| = \frac{|V_s|}{|\cosh \gamma l|} = \frac{146.4}{0.877} = 166.93 \text{ kV}$$

(a) percent voltage regulation $= \dfrac{166.93 - 124.13}{124.13} \times 100 = 34.48\%$

(b) sending-end power $= 3V_s I_s \cos \phi$

$$= 3 \times 146.4 \times 10^3 \times 386.5 \cos(32.55 - 24.28)$$

$$= 167.98 \text{ MW}$$

(c) efficiency of transmission $= \dfrac{\text{power received}}{\text{power sent}} \times 100$

$$= \frac{150}{167.98} = 89.3\%$$

IV.3 TRANSMISSION LINE AS A FOUR-TERMINAL NETWORK

In representing transmission lines by their equivalent circuits, we notice from Section IV.2 that the sending-end voltage and current can be expressed in terms of the receiving-end voltage and current and the line parameters as in (IV.79) and (IV.80). In general, a transmission line may be viewed as a four-terminal network, as shown in Figure IV.19, such that the terminal voltages and currents are mutually related by

$$V_s = AV_r + BI_r \qquad \text{(IV.81)}$$

$$I_s = CV_r + DI_r \qquad \text{(IV.82)}$$

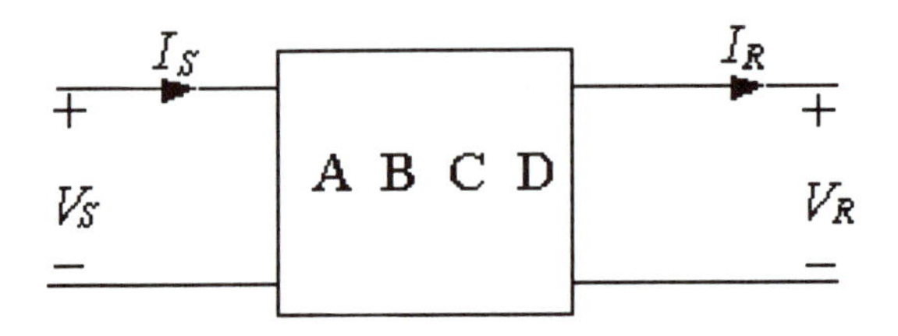

FIGURE IV.19
A transmission line as a four-terminal network.

TABLE IV.2
***ABCD* Constants for Transmission Lines**

Line	Equivalent Circuit	A	B	C	D
Short	Series impedance, Figure IV.14	1	Z	0	1
Medium length	Nominal-Π, Figure IV.16	$1 + \frac{1}{2}YZ$	Z	$Y(1+\frac{1}{4}YZ)$	$1+\frac{1}{2}YZ$
		$1 + \frac{1}{2}YZ$	$Z(1+\frac{1}{4}YZ)$	Y	$1+\frac{1}{2}YZ$
Long	Nominal-T, Figure IV.17	$\cosh \gamma l$	$Z_c \sinh \gamma l$	$\sinh \gamma l / Z_c$	$\cosh \gamma l$
	Distributed parameter, Figure IV.18				

where *ABCD* are constants dependent on line parameters, are called *generalized circuit constants*, and in general are complex. The validity of (IV.81) and (IV.82) is based on the fact that a transmission line can be represented by a linear, passive, and bilateral network. By virtue of reciprocity, the generalized constants are related to each other by the following equation:

$$AD - BC = 1 \tag{IV.83}$$

Transmission lines of various lengths can be represented by equivalent four-terminal networks, as depicted by Figure IV.19. The *ABCD* constants for various lines are summarized in Table IV.2.

IV.4 POWER FLOW ON A TRANSMISSION LINE

Power flow calculations at a point on a transmission line can be conveniently carried out in terms of *ABCD* constants. To show the procedure, let us evaluate the power at the receiving end of a transmission line. Since the *ABCD* constants, in general, are complex, we let

$$A = |A| \angle\alpha \quad \text{and} \quad B = |B| \angle\beta$$

Choosing $\mathbf{V}_R$ as the reference phasor, we assume that

$$V_R = |V_R| \angle 0° \quad \text{and} \quad V_S = |V_s| \angle\delta$$

Consequently, from (IV.81) we obtain

$$I_R = \frac{|V_s|}{|B|} \angle(\delta - \beta) - \frac{|A|\,|V_R|}{|B|} \angle(\alpha - \beta)$$

The complex power $V_R I_R^*$ at the receiving end is given by

$$P_R + jQ_R = \frac{|V_R|\,|V_s|}{|B|} \angle(\beta - \delta) - \frac{|A|\,|V_R|^2}{|B|} \angle(\beta - \alpha) \qquad \text{(IV.84)}$$

which yields

$$P_R = \frac{|V_R|\,|V_s|}{|B|} \cos(\beta - \delta) - \frac{|A|\,|V_R|^2}{|B|} \cos(\beta - \alpha) \qquad \text{(IV.85)}$$

$$Q_R = \frac{|V_R|\,|V_s|}{|B|} \sin(\beta - \delta) - \frac{|A|\,|V_R|^2}{|B|} \sin(\beta - \alpha) \qquad \text{(IV.86)}$$

Figure IV.20(a) depicts (IV.84) phasorially. By shifting the origin from O' to O, Figure IV.20(a) becomes a power diagram, as shown in Figure IV.20(b). Because $O'A = |V_R||V_s|/|B|$ for a given line and a given value of $|V_R|$, the locus of A will be a set of circles (of radii $O'A$) for a set of values of $|V_s|$. Portions of two such loci are given in Figure IV.21. These circles are sometimes called *receiving-end circles*. A number of interesting results may be obtained from the figure. For instance, for the line to transmit maximum power, $O'A$, must become parallel to the real axis; that is, $\beta - \delta = 0$. For this condition, we have, from (IV.85),

$$(P_R)_{\max} = \frac{|V_R|\,|V_s|}{|B|} - \frac{|A|\,|A_R|^2}{|B|} \cos(\beta - \alpha) \qquad \text{(IV.87)}$$

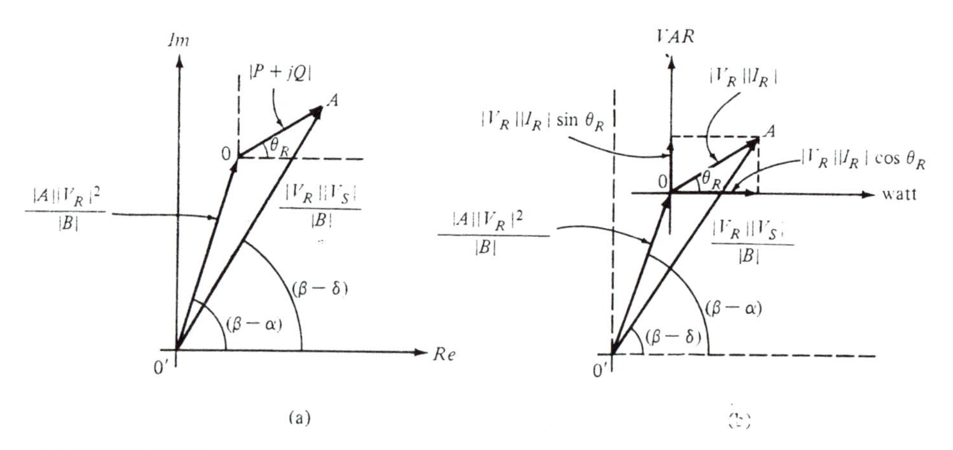

FIGURE IV.20
(a) Phasor diagram representing (IV.84); (b) corresponding power diagram.

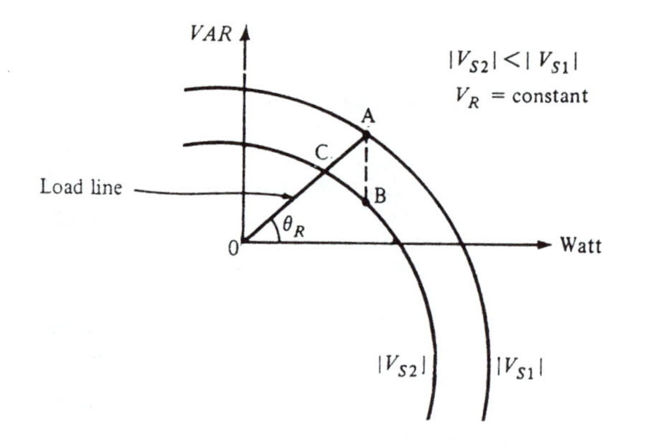

FIGURE IV.21
Receiving end power circles for a constant V_R and two sending-end voltages V_{s1} and V_{s2}.

Line *OA* in Figure IV.21 is the load line the intersection of which with the power circle determines the operating point. Thus, for a load having a lagging power factor angle θ_R, A and C are the respective operating points for sending-end voltages $|V_{s1}|$ and $|V_{s2}|$. These operating points determine the real and reactive powers received at the two sending-end voltages.

From transmission-line models, we know that the receiving-end voltage will tend to go down with the sending-end voltage. Suppose that the sending-end voltage goes down from $|V_{s1}|$ to $|V_{s2}|$. In order to maintain a constant $|V_R|$, the amount of reactive power to be supplied by capacitors at the receiving end (in parallel with the load) is simply given by the length AB, which is parallel to the reactive-power axis. Such information can be obtained very quickly from the circle diagram. Although circle diagrams yield quick estimates of certain results pertinent to transmission lines, digital computers are generally used in practice. Circle diagrams illustrate certain concepts very clearly.

BIBLIOGRAPHY

Bergen, A. R., *Power System Analysis*, Prentice-Hall, 1986.

DelToro, V., *Electric Power Systems*, Prentice-Hall, 1992.

Glover, J. D. and Sarma, M., *Power System Analysis and Design*, (2/e), PWS Publishers, 1994.

Grainger, J. J. and Stevenson, W. D., *Power Systems Analysis*, McGraw-Hill, 1994.

Gross, C. A., *Power System Analysis*, Wiley, 1979.

Horowitz, S. H. and Phadke, A. G., *Power System Relaying*, Res. Studies Press, 1992.

Kundur, P., *Power System Stability and Control*, McGraw-Hill, 1994.

Machowski, J., Bialik, J. W., and Bumby, J. R., *Power System Dynamics and Stability*, Wiley, 1997.

INDEX